Metrology and Instrumentation

Wiley-ASME Press Series

Metrology and Instrumentation

Practical Applications for Engineering and Manufacturing

Samir Mekid
King Fahd University of Petroleum & Minerals
Dhahran, Saudi Arabia

WILEY

The right of Samir Mekid to be identified as the author of this work has been asserted in accordance with law.

Registered Office
John Wiley & Sons, Inc., 111 River Street, Hoboken, NJ 07030, USA

Editorial Office
111 River Street, Hoboken, NJ 07030, USA

For details of our global editorial offices, customer services, and more information about Wiley products visit us at www.wiley.com.

Wiley also publishes its books in a variety of electronic formats and by print-on-demand. Some content that appears in standard print versions of this book may not be available in other formats.

Library of Congress Cataloging-in-Publication Data

Names: Mekid, Samir, author.
Title: Metrology and instrumentation : practical applications for
   engineering and manufacturing / Samir Mekid.
Description: Hoboken, NJ : Wiley, 2021. | Series: Wiley-ASME Press series |
   Includes index.
Identifiers: LCCN 2021034345 (print) | LCCN 2021034346 (ebook) | ISBN
   9781119721734 (hardback) | ISBN 9781119721727 (adobe pdf) | ISBN
   9781119721710 (epub)
Subjects: LCSH: Metrology. | Measuring instruments.
Classification: LCC QC88 .M38 2021 (print) | LCC QC88 (ebook) | DDC
   620.0028/4—dc23
LC record available at https://lccn.loc.gov/2021034345
LC ebook record available at https://lccn.loc.gov/2021034346

Cover image: Courtesy of Abderrahman Mekid
Cover design: Wiley

Set in 9.5/12.5pt STIXTwoText by Straive, Chennai, India   SKY10031532_112221

Contents

Preface

This book is conceived for engineers and technicians operating in various industrial fields. It is also for students of mechanical, production, and other related disciplines in engineering to facilitate understanding of fundamentals of measurements, instruments and governing rules followed by learning various shop-floor required measurement techniques.

The book introduces basic needs from math, statistics, and measurement principles. It discusses errors and their sources in manufacturing while describing the various measurement instruments. Simple physical parameters such as force, torque, strain, temperature, and pressure are explained. The subsequent chapters cover tolerance stack-ups, GD&T, calibration principles in various aspects of manufacturing, and quality standards. ASME and ISO are cited according to needs and to corresponding knowledge throughout the book. Each chapter ends with a set of MCQs with answer tables to help prepare technicians and engineers for various qualification diplomas and certificates. The book adopts an illustrative approach to explain the concepts with solved examples to support understanding.

Chapter 1 of the book introduces the fundamental units and constants needed in metrology supported by the international vocabulary of metrology and international standards.

Chapter 2 emphasizes metrology that covers all scales, starting from nanoscale to large scale. Differences and relationships between scales are introduced to understand the differences and possible complementarity, while Chapter 3 introduces basic math and science background mainly to refresh memories and be a reference in case there is a need to check information. Math and science are of great importance when dealing with measurement since the inception of humanity.

Chapter 4 defines the error and its various possible sources: how error propagates in measurement, errors associated with motion, error classification, and error elimination. An estimation of error, or uncertainty analysis, is a tool for determining the performance capability of machine tools and highlighting potential areas for performance and cost improvement.

In Chapter 5, the measurement and quantification are the fundamental concepts of metrology, including the measurement system characteristics. This considers explicit and internationally accepted definitions, principles, and standards. The purpose of any measurement system is to provide the user with a numerical value corresponding to the variable being measured by the system. What are the international related standards? Examples of length measurement, parts, and machine inspection with reverse engineering are provided.

One of the most significant chapters is Chapter 6 as it introduces the tolerance stacks analysis methods. This chapter establishes uniform practices for stating and interpreting dimensioning, tolerancing, and related requirements for use on engineering drawings and in related documents under ASME Y14.5.1. A brief introduction to geometric dimension and tolerancing is followed by tolerance stacks. This is to assign the right tolerances and to make sure that no unnecessary tight tolerance is selected, leading to costly manufacturing.

Chapter 7 introduces the principles and fundamentals of calibration under the international standards definitions and agreements. It tackles real calibration of machines and instruments in detail for understanding of the process.

Chapter 8 discusses the uncertainty based on the international standards and recent developments followed by the propagation of error with real-world examples. The doubt surrounding this measurement is the uncertainty of measurement. The background and fundamental definition of uncertainty and error will be discussed later based on international standards with all aspects in general practice.

Mechanical measurement for length and others are discussed in two chapters.

Chapter 9 covers some of the instruments used for displacement and length measurements. From length measurement and calibration of instruments such as micrometers, calipers, gages, or tape measures, to high-tech optics-based scales and comparators. The industrial leading dimensional instrument calibration capabilities are available and well designed to reduce risk and inaccuracy in measurements. Chapter 10 covers mechanical instruments measuring other than length measurands. The chapter discusses calibration-related techniques. These are fundamental basis instruments that may be needed by any engineer at any time.

Thermodynamic properties of any material or solution are treated in Chapter 11. They are valuable not only for estimating the usefulness of the material or the feasibility of reactions in solution, but they also provide one of the best methods for investigating theoretical aspects related to the material or solution structure. Thermal properties of materials can be measured directly or indirectly. This includes temperature, developed pressure, calorimetry, and thermal conductivity.

Chapter 12 covers quality management and metrology since they are important components in metrology labs and manufacturing enterprises. This chapter introduces the definition of most components of quality with the related international standards with an overall organization of the lab requirement.

Contemporary digital metrology is discussed in Chapter 13. Digital metrology and its relationship to manufacturing and I4.0 are introduced. This chapter covers the digitalization, automation, and measurements that are becoming extremely important in this era of digital manufacturing and digital twins metrology (DTM). The measurement system is a combination of real-time control system and system for data transmission. Digital computing is the tool for data processing. The technology readiness for most measurement instrumentation exists together with virtual instruments capable of building further the DTM. Since several apps for smart phones have been developed, the last appendix gives a short presentation of 38 apps.

About the Website Materials

This book has an online appendices extension covering smart phone Apps related to various metrology aspects discussed in the book and found in Appendix A, and a technical terms glossary in Appendix B. The link is www.wiley.com\go\mekid\metrologyandinstrumentation.

Acknowledgments

This book is like a vessel of time and knowledge since much of it has been dedicated to compiling information and data, verifying and checking numerous experiments, and ensuring that knowledge is delivered in a simple manner that can be easily captured by readers.

Several specialist companies in the area of metrology and manufacturing have contributed directly or indirectly to this book; hence, I am very thankful to all of them.

I acknowledge the courtesy of Fotofab (Chicago, USA), Leica (UK), and Renishaw (UK) to use some of their materials.

I am thankful to Otila Prian and Nina Fernandez from CREAFORM (AMETEK) for their reverse engineering samples; Ross Snyder, Application Engineer from Sigmetrix (Michigan, USA) for his stack-up analysis contribution. I would also like to thank Mr. Nicolaus Spinner from SPINNER Werkzeugmaschinenfabrik GmbH in Sauerlach (Germany) for providing me with the calibration results of his machines.

My thanks are extended to my Managing Editors: Gabriella Robles and Sarah Lemore from John Wiley & Sons, Inc.

I would also like to acknowledge my MSc student Usman Khan and our departmental secretary Mino Thankachan for the hard work of formatting the chapters. Not to forget Abderrahman Mekid and Khawla Mekid for their editing and revisions.

The author would like to acknowledge the support of King Fahd University of Petroleum and Minerals (KFUPM) through the Deanship of Research Oversight and Coordination (DROC) for their support in research and resources made available to this book under BW#191006.

October 1, 2021   *Professor Samir Mekid*

About the Author

Samir Mekid is professor of mechanical engineering at KFUPM and chartered engineer registered with IMechE (UK) and ASME member. Prior to joining KFUPM, he was assistant professor at UMIST (UK) and The University of Manchester (UK). He has worked with Caterpillar in the design department and has been an expert EU evaluator for various countries to several European Framework Projects Programs, e.g., FP6, FP7. He was member of the Scientific Advisory Board of the Centre of Excellence in Metrology for Micro and Nanotechnology (CEMMNT) in the UK. His area of research includes machine design, manufacturing instrumentation, metrology, mechatronics, smart materials, and sensors design. He has published over 180 publications in professional journals and international conference proceedings and edited three books. He holds more than 30 patents.

During his ongoing career, he taught several courses for undergraduate and graduate students including: metrology, sensors and actuators, together with machine design. He has trained approximately one hundred engineers through short courses on tolerance stack-ups, GD&T, manufacturing systems, and materials selection. He is currently the Founding Director of the Interdisciplinary Research Center for Intelligent Manufacturing and Robotics at KFUPM.

1

Fundamental Units and Constants in Metrology

"When you can measure what you are speaking about and express it in numbers, you know something about it."

—Lord Kelvin (1883). Source: Public Domain.

1.1 Introduction

Metrology is the science of measurement with various applications. It is derived from the Greek words *metro* – measurement and *Logy* – science. The BIMP (Bureau of Weights and Measures in France) defines metrology as "the science of measurement embracing both experimental and theoretical determinations at any level of uncertainty in any field of science and technology."

Five pivots define the functions of metrology:

1) To establish the units of measurements;
2) To replicate these units as standards;
3) To guarantee the measurement uniformity;
4) To develop measurement methods;
5) To investigate the accuracy of methods-related errors.

Based on this, the objectives of metrology are:

1) Selection of proper measuring instrument;
2) Proper measuring standards;
3) Minimizing inspection cost;
4) Defining process capabilities;
5) Standardization;
6) Maintaining accuracy and precision during inspection or as component of an instrument over time of use [1].

Therefore, two types of metrology exist:

1) Deterministic, or industrial, metrology.
2) Legal, or scientific, metrology.

Measurement is the process of revealing a single or multiple values to the characteristics of an object or property by conducting experiments to determine the value of this particular property. These properties may be physical, mechanical, or chemical, such as length, weight, force, strain, volume, angle, and molls.

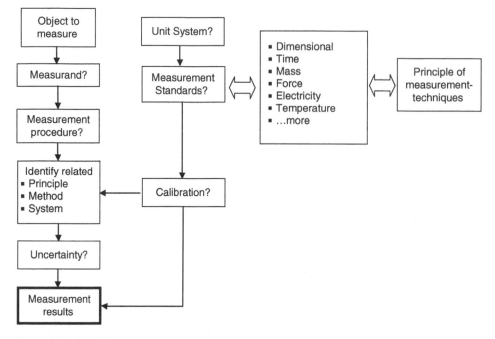

Figure 1.1 Simplified methodology producing measurement results.

Metrology also includes precision, repeatability, and accuracy, which refers to how accurate the measured value is. It establishes a well-known understanding of the measuring process and the related units that are critical in connecting various human activities and ensures that these measurements are linked to reference standards, which is commonly referred to as traceability. For as long as civilization has existed, measurements have been taken. It is necessary for a country's economic and social development. It provides precise measurements that have an impact on the economy, health, safety, and general well-being. It could also be a legal problem. As a result, the topic is always in demand.

This chapter will introduce the fundamental units and constants in metrology through conversions between units and systems. To put measurements into context, a complete methodology of the act of measurement beginning with the object to measure and ending with the result that constitutes the information needed for the object is required. The complete process is summarized in Figure 1.1. The figure depicts a simplified methodology for producing measurement results with minimal conditions such as the units to be known, the calibration of the measurement instruments, and the uncertainty of such measurements. This is to cast the majority of the aspects that engineers conducting measurements must be aware of. A dimension is a non-numerical measure of a physical variable. The unit is used to associate a quantity or measurement with a dimension.

Example 1.1 The mass of an object is a primary dimension, while 15 kg is associated with the quantity 15 of mass with the unit of kg. We need a comparison with some precise unit value to measure the quantity of anything. Body parts (Figure 1.2) and natural surroundings were used by early humans to provide suitable measuring instruments. Elementary measures became essential in the primitive human societies for tasks such as building dwellings, making clothing, bartering for food, and exchanging raw materials.

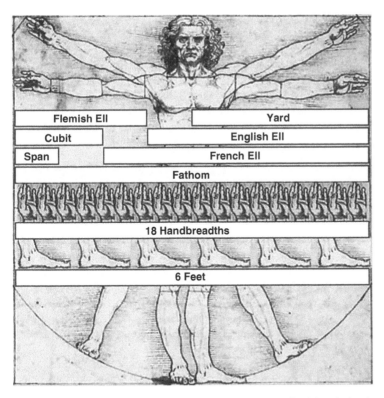

Image labels: Flemish Ell, Yard, Cubit, English Ell, Span, French Ell, Fathom, 18 Handbreadths, 6 Feet

Figure 1.2 Vitruvian man by Leonardo da Vinci showing nine historical units of measurement.
Source: Wikimedia Commons

– According to early Babylonian and Egyptian transcripts, length was first measured with the forearm (cubit), hand (palm and span), and finger (digit).
– The cycles of the celestial bodies such as the sun, moon, and others were used for time measurements.
– Plant seeds were used for the sake of establishing volume measurement, while with the expansion of scales for weighing, seeds and stones became standards. As sample, the carob seed was the base measure for the carat, which is still used as a mass unit in the gemstone industry.

As trade and commerce expanded, it became necessary to standardize measurement systems across many countries. This decreased the possibility of disagreements arising from measurement system misunderstandings.

The international system of units, known as the SI (from French "Système International") unit system, distinguishes physical units into two classes as shown below:

1) Base or primary units; and
2) Derived units.

These two categories cover the most commonly used units, such as time, temperature, length, mass, pressure, and flow rate. The National Institute of Standards and Technology (NIST) [2] introduced the SI units, which can be found at this hyperlink: **SI units** (https://physics.nist.gov/cuu/ Units/). For more information on the SI units, visit the website of the international standards organization known as the Bureau International des Poids et Mesures (BIPM).

Table 1.1 Primary units.

Measurement	Units	Symbol	Description
Unit of length	meter	m	One meter is equal to the length of the path travelled by light in vacuum during a time interval of 1/299792458 of a second.
Unit of mass	kilogram	kg	One kilogram is equal to the unit of mass presented by the international prototype of the kilogram in Figure 1.2. Since 2019, the new definition based on Planck's constant has been used.
Unit of time	second	s	One second is equal to the duration of 9192631770 periods of the radiation corresponding to the transition between the two hyperfine levels of the ground state of the cesium-133 atom.
Unit of electric current	ampere	A	One ampere is defined as follows: the constant current if maintained in two straight parallel conductors of infinite length, of negligible circular cross-section, and placed 1 meter apart in vacuum, will produce between these conductors a force equal to 2×10^{-7} newton per meter of length.
Unit of thermodynamic temperature	kelvin	K	One kelvin is the fraction 1/273.16 of the thermodynamic temperature of the triple point of water.
Unit of amount of substance	mole	mol	One mole is the amount of substance of a system containing as many elementary entities as there are atoms in 0.012 kilogram of carbon-12.
			When the mole is used, the elementary entities must be specified and may be atoms, molecules, ions, electrons, other particles, or specified groups of such particles.
Unit of luminous intensity	candela	cd	One candela is the luminous intensity within one direction, of a source that emits monochromatic radiation of frequency 540×1012 hertz and having a radiant intensity of 1/683 watt per steradian in that direction.

As will be demonstrated later, each measurement unit has a primary quantity that is used by convention. Each primary quantity has only one primary unit. As a result, every primary unit can be decomposed or recomposed further. Table 1.1 shows primary units of different kinds of physical quantities, symbols, and their descriptions. Figure 1.3 depicts the kilogram prototype safely conserved in Paris as a reference unit of kg kept constant in quantity for comparison. The following section discusses derived units, which are shown in Table 1.2.

Figure 1.3 The standard kilogram for mass.

Table 1.2 Derived units.

Derived quantity	Si derived new unit	Symbol	SI units	SI base units
Force	newton	N		$mkgs^{-2}$
Pressure, stress	pascal	Pa	N/m^2	$m^{-1}kgs^{-2}$
Energy, work, quantity of heat	joule	J	Nm	m^2kgs^{-2}
Power	watt	W	J/s	m^2kgs^{-2}
Electric charge	coulomb	C		sA
Electromotive force	volt	V		$m^2kgs^{-3}A^{-1}$
Electric capacitance	farad	F	C/V	$m^{-2}kg^{-1}s^4A^2$
Electric resistance	ohm	Ω	V/A	$m^2kgs^{-3}A^{-2}$
Electric conductance	siemens	S	A/V	$m^{-2}kg^{-1}s^3A^2$
Velocity	meter per second		m/s	
Angular velocity	radian per second		1/s	
Mass flow rate	kilogram per second		kg/s	
Flow rate	liter per second		l/s	

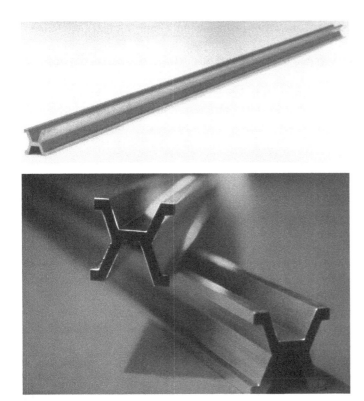

Figure 1.4 The platinum-iridium meter bar reference. *Source:* Wikimedia Commons

The International Prototype Meter bar, shown in Figure 1.4, is made of 90% platinum and 10% iridium alloy and served as the SI (metric system) standard of length from 1889 until 1960, when the SI system switched to a new definition of length based on the wavelength of light emitted by krypton-86. The practical length of the meter was defined by the distance between two fine lines ruled on the central rib of the bar near the ends measured at the freezing temperature of water.

The bar was given an X (Tresca) cross-sectional shape to increase its stiffness-to-weight ratio and improve its thermal accommodation time so the graduation lines could be located on the "neutral" axis of the bar where the change in length with flexure is minimum. The prototype was made in 1889, its length made equal to the previous French standard "Meter of the Archives." At the same time, twenty-nine identical copies were made, which were calibrated against the prototype and distributed to nations to serve as national standards and possibly for comparison after a few years.

1.2 Current Definitions of the Main SI Units

The current definition of the base and primary units are shown in Table 1.1.

1.3 New Definition of Seven Base Units of the SI

Seven base units of the SI are known to be the second, meter, kilogram, ampere, kelvin, mole, and candela. Some have been based on physical constants for a long time. Since 1983, the meter has been defined as the length of the path traveled by light in vacuum over a time interval of $1/299$ $792\,458$ s. However, the four that metrologists have agreed to redefine recently were previously based on something—i.e., an object, experiment, or phenomenon—implying that their value is not universal.

As a result of this decision [3], all seven SI units are currently defined in terms of physical constants.

The meter, symbol m, is the SI unit of length. It is defined by taking the fixed numerical value of the speed of light in vacuum, c, to be 299792458 when expressed in the unit m·s^{-1}, where the second is defined in terms of the cesium frequency $\Delta \nu_{Cs}$. The meter may be expressed directly in terms of the defining constants (Eq.(1.1)):

$$1\,m = \frac{9\,192\,631\,770}{299\,792\,458} \frac{c}{\Delta \nu_{Cs}} \tag{1.1}$$

Previously, one meter was defined as the length traveled by light in 3.335641×10^{-9} s (based on the speed of light in a vacuum). It was also defined as 1,650,763.73 wavelengths in vacuum of the orange red line of the spectrum of krypton-86.

Most affected is the kilogram, which is currently fixed by a 143-year-old platinum alloy cylinder known as the "Le Grand K" and kept at the International Bureau of Weights and Measures (BIPM) in Paris. The kilogram is now defined by Planck's constant, h, recently measured with extraordinary precision. Its agreed value is set at $6.626\,070\,15 \times 10^{-34}$ kg m^2 s^{-1} when expressed in the unit J s, which is equal to kg m^2 s^{-1}, the meter and second being defined in terms of c and $\Delta \nu$. This means that the kilogram is defined in terms of Planck's constant instead of the mass of a cylinder of metal called International Prototype Kilogram.

Meanwhile, the ampere is determined by the elementary electric charge, e, which is given as $1.602\,176\,634 \times 10^{-19}$ when expressed in coulombs. The kelvin is determined by the fixed numerical value of Boltzmann's constant, k, which is $1.380\,649 \times 10^{-23}$ when expressed in units of J K^{-1}, and the mole is determined by Avogadro's constant (N_A), which contains exactly $6.02\,214\,076 \times 10^{23}$

atoms or molecules. This number is the fixed numerical value of the Avogadro constant, N_A, when expressed in units of mol^{-1}.

1.4 Derived International System (SI) Units

A derived SI unit is a measurement unit that is devised for a derived quantity different from primary units shown previously. Derived units combine different base units as described in Table 1.2. These derived units are obtained by simple mathematical transformations.

The Imperial unit system now includes the customary units of the United States in North America. The British Weights and Measures Act of 1824 established the Imperial unit system. Following that, the system was made official throughout the United Kingdom. It should be noted that some units are used in the United States but not in the United Kingdom, and vice versa. The differences are found in the following:

  i) British fluid ounce = 0.961 US fluid ounce;
 ii) US fluid ounce = 1.041 British fluid ounces;
iii) British Imperial gallon = 1.201 US gallons;
 iv) US gallon = 0.833 British Imperial gallon.

1.5 SI Conversion

Converting SI units is very common when considering the SI unit and its related prefix described in Table 1.3.

This system comprises 7 base quantities (common) and 16 prefixes that designate the amount. The base unit and prefixes can be combined to produce the desired result.

Example 1.2 A car's weight can be written as 2000 kg, but it is better expressed in tons. It is no longer appropriate to write the results in grams. The possibilities for combining are limitless. It is

Table 1.3 SI Units and prefixes.

Prefix	Abbreviation	Meaning	Example
tera	T	10^{12}	1 terameter (Tm) = 10^{12} m
giga	G	10^9	1 gigameter (Gm) = 10^9 m
mega	M	10^6	1 megameter (Mm) = 10^6 m
kilo	k	10^3	1 kilometer (km) = 10^3 m
deci	d	10^{-1}	1 decimeter (dm) = 10^{-1} m
centi	c	10^{-2}	1 centimeter (cm) = 10^{-2} m
milli	m	10^{-3}	1 millimeter (mm) = 10^{-3} m
micro	μ	10^{-6}	1 micrometer (μm) = 10^{-6} m
nano	n	10^{-9}	1 nanometer (nm) = 10^{-9} m
angstrom	Å	10^{-10}	1 angstrom (Å) = 10^{-10} m
pico	p	10^{-12}	1 picometer (pm) = 10^{-12} m
femto	f	10^{-15}	1 femtometer (fm) = 10^{-15} m

critical to present the measurement results in a clear and easy-to-understand figure. Based on the previous table:

1 g = 0.001 kg, which can be better presented as $1\,g = 10^{-3}$ kg;
1 nm = 0,000000001 m, which can be better written as $1\,nm = 10^{-9}$ m.

Or 1 000 000 mm = 1 km.

When converting using SI units, the prefix is very important if the user knows the ranking right away.

Example 1.3

567 m = 0.567 km (dividing by 1000 since 1 km = 1000 m)
30 s = 0.5 min (since 1 min = 60 s).

1.6 Fundamental Constants

As a general definition, a fundamental constant refers to a dimensionless physical constant. They are usually assumed to be universal and have constant quantitative values. The numbers are constant and do not involve any physical measurement.

a) **The gravitation constant**
   This is an empirical constant involving gravitational effects and used in Newton's law of universal gravitation, which states that all objects attract each other with a force that is proportional to the product of their masses (m_1 and m_2) and inversely proportional to the square of their distance, as shown in Eq.(1.2).

$$F_1 = F_2 = G\,(m_1 \times m_2)/r^2. \tag{1.2}$$

   Where $G = 6.67430(15) \times 10^{-11}$ with the unit $m^3\,kg^{-1}\,s^{-2}$ in SI units.

b) **The speed of light**
   The speed of light is a constant, denoted by c, and is equal to 299 792 458 m/s (approximately 300,000 km/s, or 186,000 mi/s) in **vacuum**. It is defined as a universal **physical constant** that is important in many areas of **physics**. This constant is exact since the international agreement on the **meter** was defined as the length of the path traveled by light in the vacuum during a time interval of 1/299 792 458 **s**. This constant also features in Einstein's equation of mass-energy equivalence, $E = mc^2$.

c) **Planck's constant**
   Planck's constant, h, can be found in problems classified as quantum physics. It is a **physical constant** representing the **quantum** of electromagnetic action, relating the energy carried by a **photon** to its frequency. The product of Planck's constant by the frequency of a photon gives its energy.
   In quantum mechanics, Planck's constant is of fundamental importance. It serves to define the kilogram in metrology. The value of Planck's constant is exact, with no uncertainty and is given as $h = 6.626\,070 \times 10^{-34}$ J s (or J Hz^{-1}). Planck's constant may be used in the SI unit of frequency, and hence the so-called reduced Planck's constant is used instead, defined as $\hbar = h/2\pi$ ($\hbar$ is pronounced "h-bar").
   The Planck length, denoted ℓ_P, is a unit of **length** describing the distance traveled by light in one unit of **Planck time** in a perfect vacuum. The Planck length ℓ_P is defined as $\ell_P = $ sqrt

($\hbar G/c^3$). This has been considered as the approximate equivalent value of this unit with respect to the meter:

$$\ell_P = 1.616229(38) \times 10^{-35}\,\text{m}, \tag{1.3}$$

where c is the **speed of light** in a vacuum, G is the **gravitational constant**, h is the **reduced Planck's constant**, and the two digits enclosed by parentheses are the standard uncertainty. This length is about 10^{-20} times the diameter of a proton.

The Planck mass, denoted by m_P, is the unit of **mass** in the system of **natural units** of **Planck units**. It is roughly equivalent to 0.021 milligrams (mg). For example, it is roughly the size of a flea egg. It is of the order of 10^{15} (a quadrillion) times larger than the highest energy available to contemporary **particle accelerators**. It is defined as: $m_p = \text{sqrt}\,(\hbar c/G)$, where c is the **speed of light** in a vacuum, G is the **gravitational constant**, and $\hbar$ is the **reduced Planck's constant**.

$$1\,m_p = 2.176435(24) \times 10^{-8}\,\text{kg}. \tag{1.4}$$

The Planck time (t_P) is the **unit of time** in the system of **Planck units** in **quantum mechanics** as expressed in Equation 1.4. A Planck time unit is the **time** needed for **light** to travel a distance of one **Planck length** in a **vacuum**. This time is approximated as 5.39×10^{-44} s.

$$tp = \text{sqrt}\,(\hbar.G/c^5), \tag{1.5}$$

Where $\hbar = h/2\pi$ is the **reduced Planck's constant** (sometimes h is used instead of $\hbar$ in the definition), G = **gravitational constant**, and c = **speed of light** in **vacuum**.

Many other fundamental constants are discussed in their related areas toward the end of this book.

d) **The standard acceleration for gravity**

The standard acceleration for gravity, known as g, varies depending on location and is equal to $9.809\,\text{m s}^{-2}$ in USA.

e) **Avogadro's number**

Avogadro's number refers to the number of units in one mole of any substance, which is also known as the molecular weight in grams. It is defined as L = $6.02214199 \times 10^{23}$. The unit of this depends on the nature of the substance. It can be electrons, atoms, or molecules.

1.7 Common Measurements

The International System of units (SI) is used as a comparison framework for the most commonly used measurements in inspection and testing. It establishes seven fundamental units:

  i) Meter [m] - length;
 ii) Second [s] – time;
iii) Kilogram [kg] - mass;
 iv) Ampere [A] - current;
  v) Candela [cd] - light;
 vi) Kelvin [K] - temperature;
vii) Mole [mole] – amount of substance.

Measurements are carried out in laboratories, outdoor and in situ in plants. Proper equipment is used to measure with a condition that has been previously calibrated.

Indirect measurements can be carried out using equations, with the outcomes being the results of the execution of these equations.

Example 1.4

- voltage [V] = resistance [ohm] × current [A]; hence, current = voltage/resistance.
- area [m^2]= length [m] × length [m]
- pressure [Pa] = force [N] / area [m^2].

Accuracy in measurements is required in many fields, and because all measurements are close approximations, great care must be taken when taking measurements.

Example 1.5 When calibrating, you must generate a known amount of the variable to be measured as well as the SI unit under test.

1.8   Principles and Practices of Traceability

The objective as introduced in this book is to learn and understand measurements and their related calibration and standards, as well as principles and practice of traceability. This is a short introduction.

1.8.1   Definition of Traceability

It is defined as the ability to link the results of the calibration and measurements to the related standard and/or reference through an unbroken chain of comparisons.

The international vocabulary of metrology (VIM) defines traceability as the property of the result of a measurement or the value of a standard that can be related to stated references, usually national or international standards, by an unbroken chain of comparisons, all with stated uncertainties [4]. The unbroken chain of comparisons is called the "traceability chain."

The latter is composed of a number of instruments linked together to supply measurement. The competence and uncertainty are essential elements in the traceability according to ISO 17025 section 5.6.

Because there is always a difference in measurement between the output of the instrument and the true value of the measurand, measurement uncertainty is used to evaluate a quantitative statistical estimate of the limits of that difference. This will be discussed in chapters 3 and 4. VIM defines the measurement uncertainty as a parameter associated with the results of a measurement that characterizes the dispersion of the values that could reasonably be attributed to the measurand.

The calibration is typically performed by measuring a test unit against a known standard or reference. National measurement institutes across countries are typically a source of official approvals and verification for the work performed of various types of measurements, such as NIST (USA), NPL (UK), and BNM (France). The traceability has three essential components described as follows:

- Traceable calibration requiring comparisons with traceable standards or reference materials;
- Traceable calibrations can be performed only by competent laboratories with accreditation to ISO 17025;
- A traceable calibration certificate must contain an estimate of the uncertainty associated with the calibration.

Example of Traceability of Measurement in a Coordinate Measuring Machine (CMM)
Traceability guarantees the proper use of a CMM [5] in a quality management system. It constitutes the most fundamental and important aspects in the operations of a CMM. The aspects of concern are the actual physical chains by which measurement may be related to the SI unit of length. The physical chain of traceability is described as follows:

1) Laser wavelength ref. to atomic clock frequency ($4/10^{14}$);
2) Laser displacement interferometer ($2.5/10^{11}$);
3) Calibrated line scale or step gage (2–$5/10^{8}$);
4) CMM ($1/10^{6}$).

1.8.2 Accreditation and Conformity Assessment

According to the International Organization for Standardization, or ISO, [6]: "Conformity assessment is the term given to different techniques that ensure a product, process, service, management system, person or organization fulfils specified requirements. This section describes the main conformity assessment techniques and explains how to combine them to form conformity assessment schemes. The icons below depict the main conformity assessment techniques and their most common applications. In some instances, one conformity assessment technique may encompass another, e.g. an inspection can include a test technique, or a product evaluation may take into account a test report or an inspection report. How these conformity assessment techniques are used and interrelated is often prescribed in a specific mandatory or voluntary **conformity assessment scheme.**"

Products and equipment must meet specifications, service, and process requirements, especially when sold globally; thus, this business in the global market and world trade necessitates *conformity assessment*. It primarily contains standard measurements based on well-known calibration and references with certifications. As a result, we recognize the significance of international standards in addressing the conformity assessment services provided by ISO.

– The conformity assessment needs are shown in Table 1.4, where their use under ISO standards by first parties that are the suppliers;
– The second parties are defined as the customers, regulators, and trade organizations;
– Third parties are represented by bodies independent from both suppliers and customers.

With the increasing use of these conformity assessment tools, assurance of the competence of conformity assessment bodies (CABs) [7] becomes important, with recognition of this as CABs *accreditation*.

The International Laboratory Accreditation Cooperation (ILAC) [8] is the world's principal international forum for the development of laboratory accreditation practices and procedures. ILAC

Table 1.4 Standards of conformity assessment tools.

Conformity assessment	First parties	Second parties	Third parties	Corresponding ISO standards
Declaration of supplier	✓			ISO/IEC 17050
The calibration and testing	✓	✓	✓	ISO/IEC 17025
The inspection	✓	✓	✓	ISO/IEC 17020
The certification			✓	ISO 17021

promotes laboratory accreditation as a tool for trade facilitation, as well as the recognition of competent calibration and testing facilities worldwide. Table 1.4 shows the conformity assessment tool standards.

Multiple Choice Questions of this Chapter

Multiple Choice Questions are given for each chapter with solutions in an online extension of this book. Please use link: www.wiley.com\go\mekid\metrologyandinstrumentation\

References

1 Holmberg, K., Adgar, A., Arnaiz, A., Jantunen, E., Mascolo, J., and Mekid, S. (Eds), 2010, *E-maintenance*, (ISBN 978-1-84996-204-9), Springer Verlag, New York, USA.

2 NIST, www.nist.gov

3 Banks, M., "Kilogram finally redefined as world's metrologists agree to new formulation for SI units," *Physics World*, Nov. 16, 2018, https://physicsworld.com/a/kilogram-finally-redefined-as-worlds-metrologists-agree-to-new-formulation-for-si-units/.

4 BIPM, *Bureau International des Poids et Mesures* (International Bureau of Weights and Measures, home of the International System of Units, SI), https://www.bipm.org/en/publications/guides/vim.html.

5 Mekid, S., 2008, *Introduction to Precision Machine Design and Error Assessment*, (ISBN13: 9780849378867), CRC Press.

6 ISO, International Organization for Standardization, www.iso.org.

7 Czichos, H., 2011, "Introduction to Metrology and Testing," In: Czichos H., Saito T., Smith L. (Eds.), *Springer Handbook of Metrology and Testing*, Springer Handbooks, Springer, Berlin, Heidelberg.

8 ILAC, (international organization for accreditation bodies operating in accordance with ISO/IEC 17011), http://www.ilac.org.

2

Scales of Metrology

"If you know how to measure big, it does not mean that you know how to measure small!"

—*Prof. SM*

2.1 Introduction to Practical Metrology across All Scales

Although not the topic of this book, commercial exchanges have been one of the first shared activities among humanity. Measurement, whether related to the object itself in terms of size, weight, or its functional characteristic such as performance, is one of the tools to reach agreement and satisfaction to allow transaction. To agree to a given transaction, the counterparty measures the value of the object if this is permissible. Because matter is active and functional at all scales, from less than an atom to very large-scale dimensions such as space, measurement science applies to all scales. This chapter discusses measurement techniques at these different scales and introduces the instruments that are dedicated to these scales. They have been specifically designed to perform these measurements.

Typically, the development of measurement infrastructure, expertise, and standards to assist researchers in properly exploiting technology development at these scales (Figure 2.1) and contributing to the growth and commercialization opportunities in various sectors of our lives such as health, safety, and the environment is of interest in these areas. All scales considered in metrology can be represented as shown in Figure 2.2. Each scale shows the range of operation to its neighborhood.

This chapter will introduce the following metrology scales:

  i) Nanometrology;
  ii) Micrometrology;
 iii) Standard-scale metrology; and
 iv) Large-scale metrology.

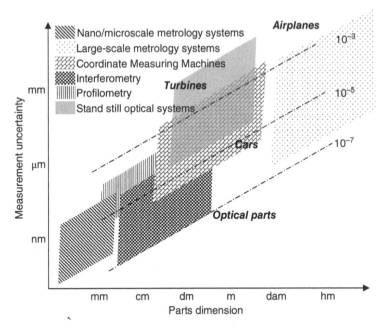

Figure 2.1 Overall scale of metrology.

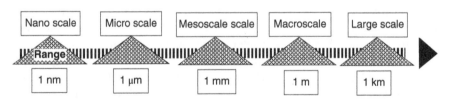

Figure 2.2 Scale range at various levels.

2.2 Nanometrology

2.2.1 Introduction and Need in Industry

Nanotechnology is becoming a large industry sector and is expanding in many ways. Because of their practical utility in everyday life, machines, systems, and devices are shrinking in size. Miniaturization encompasses all materials, processes, and manipulation. The long-term growth of nanotechnology products is dependent on increased investment in infrastructure, including technology, manufacturing, and inspection development. Metrology and measurement standards are required for this type of product conformity evaluation. Quality assurance includes traceability, measurement uncertainty, and standardized definitions to ensure new developments and common understanding.

The current demand is for high-resolution and high-precision measuring machines, which have made excellent progress thus far, while calibration in the nanometric and subnanometric scales has yet to be established. Global efforts to design and integrate nanosystems are only possible if dimensional assessment and characterization procedures are well defined and standardized. Measurement at the nanoscale becomes very important and difficult when rules must be agreed upon and measuring equipment must be developed. The measurement accuracy is expected to be better than 0.1 nm over the defined range of 100 nm for nanoscale. Nanometrology can cover measurements with higher accuracies or uncertainties than previous methods. On the other hand, it contributes to the standard scale by measuring a 1 m mirror segment of a telescope with a precision of 10 nm.

Measurement techniques developed for conventional materials are typically inapplicable to nanostructures. They require specific protocols that take into account the physics and nature of the nanostructured materials, as well as their interactions with any surrounding elements. Serious errors can occur if this is not taken into account. It is important to note that at this scale, many research fields, such as biology, chemistry, and material science, begin to merge, supported by joint physics and computation foundations, to develop new functional materials. It is therefore important to consider a common and agreed perception of a dimensional measurement scale among these research fields. The dimensions affect the properties of nanomaterials, for, (i) a crystal smaller than the electron mean free path reduces the electronic conductivity and the temperature coefficient through grain boundary scattering, (ii) size controls the relaxation of luminescence in oxide nanoparticles and has been used to develop interesting optoelectronic devices, and (iii) very small quantum dots consisting of a few electrons have provided the technology for spintronics and have been used as reductants.

2.2.2 Definition of Nanometrology

Nanometrology is defined as the science of measurement of artifacts at nanoscale. The features of the artifact are between a tenth of a micrometer and a tenth of a nanometer. This can also be extended to the uncertainty of large features. Hence, any engineering dimensional measurement with an uncertainty in the order of a nanometer is part of nanometrology. The nanometrology field explores the measurement of length and characterization of the features at nanoscale. The applications are vast, including micro- and nanotechnologies in manufacturing, biology, and chemical and electrical applications. Nanoscale measurement of dimensions is best achieved with nanometer uncertainty. To best secure this uncertainty, the equipment should contain traceable dimensional calibrations over extended specimen ranges.

2.2.3 Importance of Nanometrology in Science and Technology

The complexity of the nanoworld arises from the very nature of nanostructures [1]. Thus, a nanostructure can be interpreted as an arrangement of atoms or particles in new forms, such as those of fullerenes, core–shell nanoparticles, tangled nanotubes, nanostructured metals, or dendrites, all of which defy the principles of stereology. One important feature of nanomaterials is the special properties they exhibit by virtue of their small size, which have attracted scientists and engineers. However, efficiently exploiting such properties necessitates a thorough understanding of their nature, as well as the ability to characterize the physics or chemistry of very small objects; again, this is impossible with conventional methods and necessitates the development of new, effective methodologies that also enable the production of reproducible nanostructures as a prerequisite for understanding their properties. The new nanostructure preparation and characterization methods should be supplemented by appropriate standards developed in collaboration with the nanometrology concourse. Overall, research-oriented nanometrology is now well established in research institutes and industry. However, there are still a number of constraints that must be overcome before industrial nanometrology can be implemented.

As stated earlier, dimensions play a key role in the properties of nanomaterials. The significance can be illustrated with the following selected examples of properties that arise from the nanometric size of these materials:

a) Atomic diffusion through interfaces becomes an efficient mechanism of mass transfer at the nanoscale. Unlike conventional matter, nanoparticulate matter can be efficiently transferred at low temperatures. Among other applications, this property has been used to enhance the kinetics of hydrogen diffusion in hydrogen storage devices;

b) A crystal size that is smaller than the electron mean free path reduces electronic conductivity and the temperature coefficient through grain boundary scattering;

c) Phonon spectra are modified by effect nanosurfaces and nanoparticles;

d) Band gap changes in nanosized semiconductor particles cause blue-shifts in luminescence signals. Size-induced control of luminescence relaxation in oxide nanoparticles leads to changes in optical properties and has been used to develop interesting optoelectronic devices;

e) Very small quantum dots consisting of a mere few electrons have provided the technology for spintronics and can be used as reductants;

f) Surface effects in magnetic materials govern the magnetic properties of thin layers; in addition, they facilitate the development of more efficient data storage devices and more sensitive magnetic sensors;

g) The mechanical properties of metals less than 100 nm in grain size are dictated by the contribution of grain boundaries. This has facilitated the development of materials with superior strength and ductility and hence improved service performance;

h) Tribological properties are dramatically different when the interacting materials are nanosized. This reduces friction and wear in microelectromechanical systems (MEMS).

Examples of Nanometrology, Parts, and Instruments

There are several manufacturing techniques that result in products or parts with nanometer-level accuracy in length measurement. However, an instrument that can verify that the sizes are within the exact values is also required. This will be supplemented by standards for agreeing on measurement methods using calibrated instruments and a thorough understanding of the measurement procedure. Figure 2.3 depicts a typical experimental setup for a metrological scanning probe microscope measuring slot sizes with ultra-high precision and at the nanoscale with AFM. A sample of SEM images is shown in Figure 2.4.

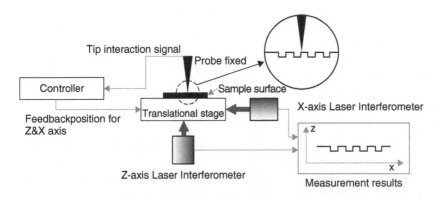

Figure 2.3 Metrological scanning probe microscope.

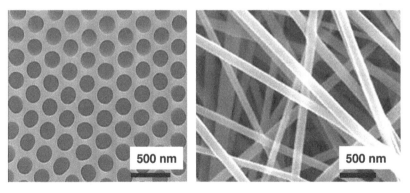

Figure 2.4 SEM image of nanoporous anodic alumina membrane and ZnO fibers.

As instruments, nanomanipulation techniques [2] are extensively in use not only to reveal more characteristics of nano-, micro-, and mesoscopic phenomena but also to build functional nanodevices that can bridge components between two or more scales to diversify the functions. Figure 2.5 depicts an integrated and numerically controlled instrument for nanomanipulation, imaging, in-process inspection with lithography and AFM back-to-back probes, and material characterization at the nanoscale.

A feature of common datum to all operations to avoid accuracy degradation between references is set. Robotic arms secure nanometer accuracy in the manipulation [3, 4]. The variety of operations is supported by multiple end-effector tools, for example, force and temperature measurement. The availability of such an instrument with integrated functions appeared to be extremely beneficial in addressing various fundamental problems in science and engineering, such as understanding, modeling, and testing nanomachining processes, biomimetic design, precise construction of nanostructure arrays, and inspection of devices with complex features [5]. Table 2.1 lists major nanometrological instruments and their related resolution range. These are used for dimensional metrology for nano- and microscales.

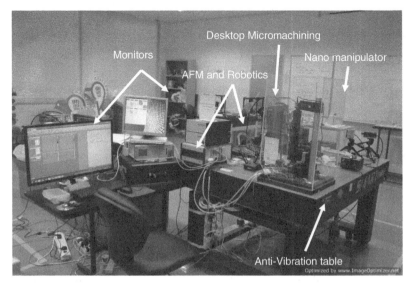

Figure 2.5 Nano-manipulator with monitors and robotics controllers.

Table 2.1 Major nanometrological instruments and their related resolution range.

Measurement	Instrument	Resolution
Dimensional metrology	Transmission electronic microscopy (TEM)	0.5 Å
	Atomic force microscope (AFM)	Depth 0.5–5 nm Lateral 0.2–130 nm
	High-resolution microscopy (HM)	150 nm
	Phase contrast microscope (PCM)	0.1 μm
	Scanning tunneling system (STM)	Depth 0.01 nm Lateral 0.1 nm
	Scanning electron microscope (SEM)	Depth 3 nm Lateral 1–24 nm
	Ellipsometer for thickness of transparent film	Sub nm
	Scanning proximal probe microscope (SPM)	1 nm
	Stylus profilometer for sample surface roughness	Sub nm
	Reflection spectroscopy for thickness of opaque films	nm
	X-ray fluorescence for film thickness (XRF)	100 μm
	Small angle X-ray scattering (SAXS)	Range 2 nm–1 μm

Source: Based on ISO/TR 18196. Nanotechnologies — Measurement technique matrix for the characterization of nano-objects, 2016.

2.3 Standards

The ISO definition of standards states that "ISO standards are internationally agreed upon by experts." They are methods that describe the best way to do something. It could be about manufacturing a product, managing a process, providing a service, or supplying materials—standards cover a wide range of activities.

The ISO adds: "Standards are the distilled wisdom of people with expertise in their subject matter and who know the needs of the organizations they represent—people such as manufacturers, sellers, buyers, customers, trade associations, users or regulators". For instance:

i) Quality management standards to help work more efficiently and reduce product failures;
ii) Environmental management standards to help reduce environmental impacts, reduce waste, and be more sustainable;
iii) Health and safety standards to help reduce accidents in the workplace;
iv) Energy management standards to help cut energy consumption;
v) Food safety standards to help prevent food from being contaminated;
vi) IT security standards to help keep sensitive information secure.

Measurement of materials and products at nanoscale would need these measurements to be standardized to meet specifications. Standardization is a prerequisite for successful development of nanotechnology in all aspects. Nanometrology aims to support the assessment of materials or products irrespective of the method of measurement used. If a statement on a measurement is made as part of a description of the object, it should be final. Nonetheless, the practice requires that measurement results be accompanied by documentation of the method and conditions of measurement.

The standardizations include:

– Scientific instrumentation;
– Measurement methods showing the best way of carrying out measurements;
– Special protocols describing unsuitability of conventional methods translated to nanometrology to avoid gross errors in interpretation.

The following standards are identified by their titles and briefly described in their respective scopes.

1) **ISO/TR 18196:2016(en)**
   Nanotechnologies—Measurement technique matrix for the characterization of nano-objects.
2) **ISO 11952:2019(en)**
   Surface chemical analysis—Scanning-probe microscopy—Determination of geometric quantities using SPM: Calibration of measuring systems.
3) **ISO 25178-601:2010(en)**
   Geometrical product specifications (GPS)—Surface texture: Areal—Part 601: Nominal characteristics of contact (stylus) instruments.
4) **IEC 62622:2012(en)**
   Geometrical product specifications (GPS)—Surface texture: Areal—Part 605: Nominal characteristics of non-contact (point autofocus probe) instrument.
5) **ISO 25178-605:2014(en)**
   Artificial gratings used in nanotechnology—Description and measurement of dimensional quality parameters.

ISO/TR 18196:2016(en)
Nanotechnologies—Measurement technique matrix for the characterization of nano-objects.

SCOPE: This document contains a matrix that directs users to commercially available techniques for measuring common physiochemical parameters of nano-objects. Some techniques can also be used on nanostructured materials. This document connects the nano-object parameters that most commonly need to be measured with corresponding measurement techniques. This document will be a useful tool for nanotechnology interested parties to rapidly identify relevant information for measuring nano-objects. The top row of the Quick-Use-Matrix lists the common nano-object parameters. If a measurement technique listed in the first column of the matrix is applicable, the box in the matrix will be marked. Once a measurement technique of interest is identified, it is recommended that the reader then enters this document's body of text (Clause 5), where the reader will find an alphabetical listing of the measurement techniques and descriptions of the advantages, limitations, relevant standards, measurand(s), and applicable nano-object parameters of each technique. As scientific advances are made and new commercial measurement techniques become available, this document will be reviewed and updated on a regular basis to ensure its relevance.

Many of the techniques listed in this document have not been validated through round-robin testing or any other means for the measurement of nano-objects. This document is intended as a starting point and resource to help identify potentially useful and relevant techniques; it is not an exhaustive or primary source. It is recommended that once a technique has been identified, the reader refers to relevant international standards and conducts a literature search for similar or comparable applications. Other sources of information include application notes and technical literature from instrument manufacturers.

ISO 11952:2019(en)

Surface chemical analysis—Scanning-probe microscopy—Determination of geometric quantities using SPM: Calibration of measuring systems.

SCOPE: This document specifies methods for characterizing and calibrating the scan axes of scanning-probe microscopes (SPMs) for measuring geometric quantities at the highest level. It is applicable to those providing further calibrations and is not intended for general industry use, where a lower level of calibration might be required.

This document has the following objectives:

i) To increase the comparability of measurements of geometrical quantities made using SPMs by traceability to the unit of length;

ii) To define the minimum requirements for the calibration process and the conditions of acceptance;

iii) To ascertain the instrument's ability to be calibrated (assignment of a "calibrate-ability" category to the instrument);

iv) To define the scope of the calibration (conditions of measurement and environments, ranges of measurement, temporal stability, transferability);

v) To provide a model, in accordance with ISO/IEC Guide 98-3, to calculate the uncertainty for simple geometrical quantities in measurements using an SPM;

vi) To define the requirements for reporting results.

IEC 62622:2012(en)

Artificial gratings used in nanotechnology—Description and measurement of dimensional quality parameters.

SCOPE: This technical specification stipulates the generic terminology for the global and local quality parameters of artificial gratings, interpreted in terms of deviations from nominal positions of grating features, and provides guidance on the categorization of measurement and evaluation methods for their determination. This specification is intended to facilitate communication among manufacturers, users and calibration laboratories dealing with the characterization of the dimensional quality parameters of artificial gratings used in nanotechnology.

This specification supports quality assurance in the production and use of artificial gratings in different areas of application in nanotechnology. Whilst the definitions and described methods are universal to a large variety of different gratings, the focus is on one-dimensional (1D) and two-dimensional (2D) gratings.

ISO 25178-605:2014(en)

Geometrical product specifications (GPS)—Surface texture: Areal—Part 605: Nominal characteristics of non-contact (point autofocus probe) instrument.

SCOPE: This part of ISO 25178 describes the metrological characteristics of a non-contact instrument for measuring surface texture using point autofocus probing. The committee responsible for this document is ISO/TC 213 dimensional and geometrical product specifications and verification.

ISO 25178 consists of the following parts, under the general title, Geometrical product specifications (GPS)—Surface texture: Areal:

- Part 1: Indication of surface texture;
- Part 2: Terms, definitions, and surface texture parameters;
- Part 3: Specification operators;
- Part 6: Classification of methods for measuring surface texture;
- Part 70: Material measures;
- Part 71: Software measurement standards;
- Part 601: Nominal characteristics of contact (stylus) instruments;
- Part 602: Nominal characteristics of non-contact (confocal chromatic probe) instruments;
- Part 603: Nominal characteristics of non-contact (phase-shifting interferometric microscopy) instruments;
- Part 604: Nominal characteristics of non-contact (coherence scanning interferometry) instruments;
- Part 605: Nominal characteristics of non-contact (point autofocus probe) instruments;
- Part 606: Nominal characteristics of non-contact (focus variation) instruments;
- Part 701: Calibration and measurement standards for contact (stylus) instruments.

ISO 25178-601:2010(en)

Geometrical product specifications (GPS)—Surface texture: Areal—Part 601: Nominal characteristics of contact (stylus) instruments.

SCOPE: This part of ISO 25178 defines the metrological characteristics of contact (stylus) areal surface texture measuring instruments.

ISO 25178-601 was prepared by Technical Committee ISO/TC 213, dimensional and geometrical product specifications and verification.

ISO 25178 consists of the following parts, under the general title, Geometrical product specifications (GPS) - Surface texture: Areal:

- Part 2: Terms, definitions, and surface texture parameters;
- Part 3: Specification operators;
- Part 6: Classification of methods for measuring surface texture;
- Part 7: Software measurement standards;
- Part 601: Nominal characteristics of contact (stylus) instruments;
- Part 602: Nominal characteristics of non-contact (confocal chromatic probe) instruments;
- Part 603: Nominal characteristics of non-contact (phase-shifting interferometric microscopy) instruments;
- Part 604: Nominal characteristics of non-contact (coherence scanning interferometry) instruments;
- Part 605: Nominal characteristics of non-contact (point autofocusing) instruments;
- Part 701: Calibration and measurement standards for contact (stylus) instruments.

2.4 Micrometrology

2.4.1 Introduction and Need in Industry

Micrometrology deals with metrological measurement at microscale. The miniaturization of products, and thus parts produced in micromanufacturing at the microscale, necessitated the use of an appropriate and related dimensional metrology. As many parts with dimensions of a few micrometers began to be produced, the industry made numerous requests for this. The goal was to assist the manufacturing scale paradigm for quality inspection.

Critical dimensions in semiconductor manufacturing are in the range of submicron and nanometer range. Specialized measurement systems suitable for these types of dimensions have been developed with the necessary accuracy and resolution for process control. In industry, as the dimensions of micromanufacturing extend into the microscale, the technologies and capabilities of semiconductor metrology systems can enable new generations of micromanufacturing metrology systems.

In nondimensional metrology, there was an extreme need to measure low forces from newtons to attonewtons. This is due to the fact that when dealing with nanometer to micrometer dimensions in manufacturing and assembly, related quantities of forces must be measured and known in precise values. In these applications, force control is required, which necessitates measurement. It is possible that effects will appear, such as Van der Waal forces, which are still difficult to evaluate at this scale.

2.4.2 Definition of Micrometrology

Micrometrology is the science of measurement where the range of dimensions of objects are in the sub-millimeter to millimeter range, associated with measurement uncertainty up to 0.1 μm for dimensional measurement. Measurement of dimensions at microscale is well achieved with better than a tenth of a micrometer uncertainty. To reduce uncertainty and improve accuracy, the equipment must have traceable dimensional calibrations over a wide range of specimen ranges.

2.4.3 Examples of Micrometrology of Microparts

Parts with great precision are made through precision chemical etching metals, such as aluminum, brass, copper, tungsten, or titanium. Features are obtained on the parts with micrometer precision (Figures 2.6–2.9).

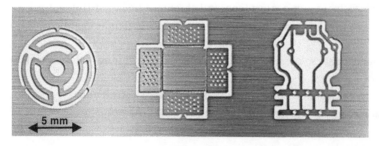

Figure 2.6 Micro-actuator, box and electronic circuit. Courtesy of Fotofab. *Source:* Courtesy of Fotofab

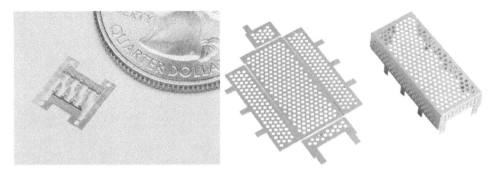

Figure 2.7 Micro-actuator, box and electronic circuit. Courtesy of Fotofab. *Source:* Courtesy of Fotofab

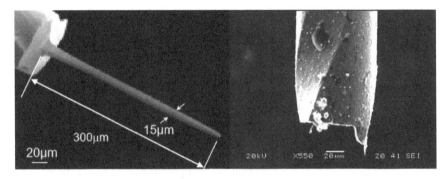

Figure 2.8 Extremely large aspect ratio microtool.

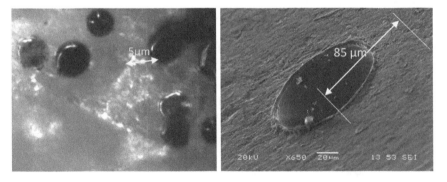

Figure 2.9 Laser printed dots @ 1 μm resolution with CCD imaging. Hole drilled in material.

2.5 Macroscale Metrology

Metrology, as defined and agreed upon by the International Bureau of Measurement, is the science of measurement. It establishes common understanding of units covering most of the human activities in almost all aspects. The inception in dimensional metrology started by the length standard defined from natural source as proposed in France. The decimal-based metric system had been consequently established in 1795. Metrology is not only the ability to measure but to be standardized so that measurements are understood worldwide across civilizations. Furthermore, it encompasses not only experiments but also theoretical determinations at any level of uncertainty in any field of science and technology.

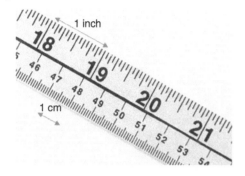

Figure 2.10 Simple ruler in SI and Imperial units to measure length.

The meter [m] is a fundamental unit of length measurement. It used to be defined as one 10-millionth of the distance from the North Pole to the equator. To achieve a new and precise determination, an international prototype of one meter made of platinum (90%) and iridium (10%) with an X shape was created to avoid any distortion in manipulation for comparisons. A ruler sample shown in Figure 2.10.

France has one prototype identical in length to the "metre des archives" as a reference, while USA has received a prototype with a calibrated length of 0.999 9984 m ± 0.2 μm, which is 1.6 μm shorter. The standard scale discussed here is the normal one starting from a few millimeter to around one meter length of objects, after micrometrology and before large-scale metrology.

A coordinate measuring machine (CMM) inspects parts for dimensions and quality (Figure 2.11). A coordinate measuring machine is used to measure the dimensions and characteristics of an object. It can measure a point or a feature formed from several points in a 2D or 3D shape, such as a hole or a whole shape. When tracking map coordinates, the CMM works similarly to your finger. It has three axes that make up the machine coordinates system. A probe is used by the CMM to measure points on a work piece and record their coordinates. Each point on the work piece is unique to the machine's coordinate system.

The ballbar inspects the circular path of a CNC (computer numerical control) machine as well as the positioning performance. This ensures that the programmed circular motion corresponds to the measured one (Figure 2.12). This inspection is performed as part of the CNC machine's diagnostic. It investigates the salient features of the contouring and polar results in relation to machine error sources. ISO 230-4 is the usual standard where the test is described.

Figure 2.11 Coordinate measuring machine inspecting a part. *Source:* ssp48/Envato Elements Pty Ltd

Figure 2.12 Ballbar instrument to inspect motion in CNC machines.

2.5.1 Standards

Most metrology standards are written for metrology in general and are described subsequently.

ISO 16239:2013
Metric series wires for measuring screw threads.

ISO/TS 15530-4:2008
Geometrical Product Specifications (GPS)—Coordinate measuring machines (CMM): Technique for determining the uncertainty of measurement—Part 4: Evaluating task-specific measurement uncertainty using simulation.

ISO 15530-3:2011
Geometrical product specifications (GPS)—Coordinate measuring machines (CMM): Technique for determining the uncertainty of measurement—Part 3: Use of calibrated work pieces or measurement standards.

ISO/TS 15530-1:2013
Geometrical product specifications (GPS)—Coordinate measuring machines (CMM): Technique for determining the uncertainty of measurement—Part 1: Overview and metrological characteristics.

ISO 14978:2018
Geometrical product specifications (GPS)—General concepts and requirements for GPS measuring equipment.

ISO 13385-2:2020
Geometrical product specifications (GPS) - Dimensional measuring equipment—Part 2: Design and metrological characteristics of caliper depth gauges.

ISO 13385-1:2019

Geometrical product specifications (GPS)—Dimensional measuring equipment—Part 1: Design and metrological characteristics of calipers.

ISO 13225:2012

Geometrical product specifications (GPS)—Dimensional measuring equipment; Height gauges—Design and metrological characteristics.

ISO 13102:2012

Geometrical product specifications (GPS)—Dimensional measuring equipment: Electronic digital-indicator gauge—Design and metrological characteristics.

ISO/DIS 12179

Geometrical product specifications (GPS)—Surface texture: Profile method—Calibration of contact (stylus) instruments.

ISO 12179:2000/Cor 1:2003

Geometrical Product Specifications (GPS)—Surface texture: Profile method—Calibration of contact (stylus) instruments—Technical Corrigendum 1.

ISO 12179:2000

Geometrical Product Specifications (GPS)—Surface texture: Profile method—Calibration of contact (stylus) instruments.

ISO/DIS 10360-13

Geometrical product specifications (GPS)—Acceptance and reverification tests for coordinate measuring systems (CMS)—Part 13: Optical 3D CMS.

ISO 10360-12:2016

Geometrical product specifications (GPS)—Acceptance and reverification tests for coordinate measuring systems (CMS)—Part 12: Articulated arm coordinate measurement machines (CMM).

ISO/CD 10360-11.2

Geometrical product specifications (GPS)—Acceptance and reverification tests for coordinate measuring machines (CMM)—Part 11: CMMs using the principle of computed tomography (CT).

ISO/DIS 10360-10

Geometrical product specifications (GPS)—Acceptance and reverification tests for coordinate measuring systems (CMS)—Part 10: Laser trackers.

ISO 10360-10:2016

Geometrical product specifications (GPS)—Acceptance and reverification tests for coordinate measuring systems (CMS)—Part 10: Laser trackers for measuring point-to-point distances.

ISO 10360-9:2013

Geometrical product specifications (GPS)—Acceptance and reverification tests for coordinate measuring systems (CMS)—Part 9: CMMs with multiple probing systems.

ISO 10360-8:2013
Geometrical product specifications (GPS)—Acceptance and reverification tests for coordinate measuring systems (CMS)—Part 8: CMMs with optical distance sensors.

ISO 10360-7:2011
Geometrical product specifications (GPS)—Acceptance and reverification tests for coordinate measuring machines (CMM)—Part 7: CMMs equipped with imaging probing systems.

ISO 10360-6:2001/Cor 1:2007
Geometrical Product Specifications (GPS)—Acceptance and reverification tests for coordinate measuring machines (CMM)—Part 6: Estimation of errors in computing Gaussian associated features—Technical Corrigendum 1.

ISO 10360-6:2001
Geometrical Product Specifications (GPS)—Acceptance and reverification tests for coordinate measuring machines (CMM)—Part 6: Estimation of errors in computing Gaussian associated features.

ISO 10360-5:2020
Geometrical product specifications (GPS)—Acceptance and reverification tests for coordinate measuring systems (CMS)—Part 5: Coordinate measuring machines (CMMs) using single and multiple stylus contacting probing systems using discrete point and/or scanning measuring mode.

ISO 10360-3:2000
Geometrical Product Specifications (GPS)—Acceptance and reverification tests for coordinate measuring machines (CMM)—Part 3: CMMs with the axis of a rotary table as the fourth axis.

ISO 10360-2:2009
Geometrical product specifications (GPS)—Acceptance and reverification tests for coordinate measuring machines (CMM)—Part 2: CMMs used for measuring linear dimensions [6].

ISO 10360-1:2000/Cor 1:2002
Geometrical Product Specifications (GPS)—Acceptance and reverification tests for coordinate measuring machines (CMM)—Part 1: Vocabulary—Technical Corrigendum 1.

ISO 10360-1:2000
Geometrical Product Specifications (GPS)—Acceptance and reverification tests for coordinate measuring machines (CMM)—Part 1: Vocabulary.

ISO 9493:2010
Geometrical product specifications (GPS)—Dimensional measuring equipment: Dial test indicators (lever type)—Design and metrological characteristics.

ISO 8512-2:1990
Surface plates—Part 2: Granite.

ISO 8512-1:1990
Surface plates—Part 1: Cast iron.

ISO 7863:1984
Height setting micrometers and riser blocks.

ISO 5436-2:2012
Geometrical product specifications (GPS)—Surface texture: Profile method; Measurement standards—Part 2: Software measurement standards.

ISO 5436-1:2000
Geometrical Product Specifications (GPS)—Surface texture: Profile method; Measurement standards—Part 1: Material measures.

ISO 3650:1998/Cor 1:2008
Geometrical Product Specifications (GPS)—Length standards—Gauge blocks—Technical Corrigendum 1.

ISO 3650:1998
Geometrical Product Specifications (GPS)—Length standards—Gauge blocks.

ISO/CD 3611
Geometrical product specifications (GPS)—Dimensional measuring equipment: Micrometers for external measurements—Design and metrological characteristics.

ISO 3611:2010
Geometrical product specifications (GPS)—Dimensional measuring equipment: Micrometers for external measurements—Design and metrological characteristics.

ISO 3274:1996/Cor 1:1998
Geometrical Product Specifications (GPS)—Surface texture: Profile method—Nominal characteristics of contact (stylus) instruments—Technical Corrigendum 1.

ISO 3274:1996
Geometrical Product Specifications (GPS)—Surface texture: Profile method—Nominal characteristics of contact (stylus) instruments.

ISO 1502:1996
ISO general-purpose metric screw threads—Gauges and gauging.

ISO 463:2006/Cor 2:2009
Geometrical Product Specifications (GPS)—Dimensional measuring equipment—Design and metrological characteristics of mechanical dial gauges—Technical Corrigendum 2.

ISO 463:2006/Cor 1:2007
Geometrical Product Specifications (GPS)—Dimensional measuring equipment—Design and metrological characteristics of mechanical dial gauges—Technical Corrigendum 1.

ISO 463:2006
Geometrical Product Specifications (GPS)—Dimensional measuring equipment—Design and metrological characteristics of mechanical dial gauges.

2.6 Large-Scale Metrology and Large-Volume Metrology

2.6.1 Introduction and Need in Industry

Large-scale metrology (LSM) was first introduced in 1960 with Berry's report, which stated that precision techniques would be established to support measurement of "Nimrod," the name of the 7 GeV proton synchrotron at the Rutherford High Energy Lab in Harwell (UK).

Most technological advances introduced for conventional-scale metrology and improved over time require additional development to address challenges encountered in large-scale sizes. This can sometimes increase nonlinearly with size.

The best-known challenges in large-scale metrology are:

1) Large measurement sizes in 3D with tight required tolerances;
2) Oversized structures;
3) Complex geometries;
4) Difficult-to-reach features and challenging angles access.

Figure 2.13 shows the structural untwisting that is required to readjust the shell. The main challenge of measurement is the gravitation effects on the large parts. The windmill shows a blade length of 70 m that needs a specific procedure to be measured.

Other challenges related to instruments for large-scale metrology are given below:

(I) He-Ne laser within 3 m can be considered straight but after 3 m it bends;
(II) Accurately measuring the volume of very large tanks for oil, gas, or even cereal grains storage or silos (above or underground) was a challenge in the 70s;
(III) Very large telescopes, known as VLT, need advanced measurement techniques different from standard-scale ones.

The needs are large. This includes the assembly and dimensional inspection of aerospace structures such as space rockets, alignment of jigs, fuselage of airplanes, windmills, gas and steam turbines, and ship construction.

 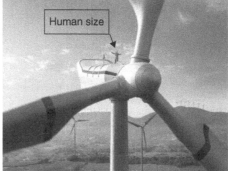

Figure 2.13 Fuselage and wind mill needing LSM.

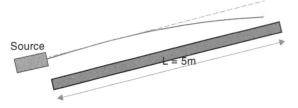

Figure 2.14 Bending of a laser light over a length L.

2.6.2 Definition

The advanced definition of large-scale engineering metrology was introduced since 1978 with reference to the Puttock report [7]. It took its origin from the triangulation methods. Because information between two points in large-scale metrology is transmitted using light rays propagating in the atmosphere, there are limitations due to the medium's variation (Figure 2.14).

Large-scale metrology is mainly used for geometric inspection, and the linear dimensions vary from 1 to 100 m. However, with extension of the scale to GPS and remote sensing, the range extends to kilometers with an accuracy around one meter. Figure 2.13 shows the bending of a laser light over a length L.

Another recent development is the introduction of large-volume metrology (LVM). In principle, this is not dissimilar to LSM. It is defined as metrology for distances greater than one meter to possibly kilometers in 3D coordinate metrology using fixed and portable instruments. This is not a laboratory scale but outdoor measurement and probably not sizes exceeding the size of the earth.

The related issues that may affect measurement under LVM are:

i) Combined algorithms with sensors fusion and higher data rates;
ii) Harsh environment;
iii) Thermal control;
iv) Thermal expansion of some instrument parts [8];
v) Beam bending;
vi) Simultaneous Multiple targets;
vii) Six DOF;
viii) Traceability.

Example 2.1 *Leica Laser Tracker*
A laser tracker is shown in Figure 2.15. The source emitting the laser is in the stand, and a retroreflector is moving with the object or fixed in one position. The objective is to determine the distance d between the laser stand and the object.

$$d = \frac{\lambda_1}{2}N_1 = \frac{\lambda_2}{2}N_2 \qquad (2.1)$$

where:

d: distance being measured;
λ_1 and λ_2: two path lengths when signals are in phase;
N_1 and N_2: corresponding integer numbers of wavelength over the length d.

$$N_2 = N_1 + 1 \qquad (2.2)$$

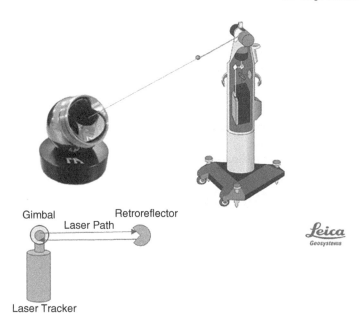

Gimbal

Laser Path

Retroreflector

Laser Tracker

Figure 2.15 Spherical retro reflector for laser trackers (courtesy of API). *Source:* Courtesy of Leica

Substituting Eq.(2.2) in Eq.(2.1):

$$N_1 = 2 \cdot d \cdot f_1/c \tag{2.3}$$

$$N_2 = 2 \cdot d \cdot f_2/c \tag{2.4}$$

Where c is the speed of light and f_1 and f_2 are the respective frequencies. Substituting Eqs. (2.3) and (2.4) into (2.2) leads to:

$$d = c/2\left(f_2 - f_1\right) \tag{2.5}$$

From the practice, the uncertainty for a distance of 10 m is around 15 µm to 25 µm according to Faro laser trackers.

Example 2.2   *Long length measurement*

An instrument measures the distance between two points to be 5000 mm with a tolerance of ±3 mm, which makes the range of acceptance [4997–5003] mm. The measurement shows 5008 mm; hence, it can be concluded that the part is out of tolerance (Figure 2.16). A closer observation shows that the slag of the tape is 6 mm; if taken away, it makes the real measurement 5002 mm. Hence, it is within tolerance.

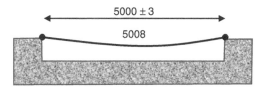

Figure 2.16   Bending measuring tape over the whole length to be measured.

2.6.3 Verification Standards

Because the tests should be performed in accordance with the manufacturer's recommendations for normal instrument operation, the standards use simple decision rules to determine conformance. Tests should be carried out in accordance with normal operation of the instrument as recommended by the manufacturer.

1) **ASME B89.4.19 Performance Evaluation Tests and Geometric Misalignments in Laser Trackers.**
2) **ISO 10360-10:2016 Geometrical product specifications (GPS)**—Acceptance and reverification tests for coordinate measuring systems (CMS)—Part 10: Laser trackers for measuring point-to-point distances.
3) **ISO 17123-1:2014 Optics and optical instruments—Field procedures for testing geodetic and surveying instruments**—Part 1: Theory
4) **ISO 17123-3:2001 Optics and optical instruments—Field procedures for testing geodetic and surveying instruments** - Part 3: Theodolites.

1- ASME B89.4.19 Performance Evaluation Tests and Geometric Misalignments in Laser Trackers
This standard details the verification procedures specific to spherical coordinate systems under the configuration of laser trackers and laser radar. It describes three kinds of tests to be performed on laser trackers: ranging tests, length measurement system tests, and two-face system tests.

The ranging tests evaluate the tracker's ability to measure distance in a purely radial direction. Length measurement system tests are performed to assess the tracker's ability to measure different lengths within the work volume. Because these tests exercise the kinematic links in the tracker, they are sensitive to most of the tracker's geometric misalignments. The standard requires length measurement system tests to be performed in 33 predetermined and two user-defined positions. There are a number of geometric misalignments that produce angle errors that reverse in sign between a front face and back face measurement of the tracker. Two-face tests are therefore excellent diagnostics of these geometric misalignments. The standard requires that two-face errors be measured at 36 predetermined positions [9]. The current time-consuming B89.4.19 verification process is a pass/fail test. Other standards exist, such as VDI/VDE 2617 part 10 and ISO 10360—Part 10.

2-ISO 10360-10:2016Geometrical product specifications (GPS)—Acceptance and reverification tests for coordinate measuring systems (CMS)—Part 10: Laser trackers for measuring point-to-point distances

SCOPE: This part of ISO 10360 specifies the acceptance tests for verifying the performance of a laser tracker by measuring calibrated test lengths, test spheres and flats according to the specifications of the manufacturer. It also specifies the reverification tests that enable the user to periodically reverify the performance of the laser tracker. The acceptance and reverification tests given in this part of ISO 10360 are applicable only to laser trackers utilizing a retroreflector as a probing system. Laser trackers that use interferometry (IFM), absolute distance meter (ADM) measurement, or both can be verified using this part of ISO 10360. This part of ISO 10360 can also be used to specify and verify the relevant performance tests of other spherical coordinate measurement systems that use cooperative targets, such as "laser radar" systems.

NOTE: Systems such as laser radar systems, which do not track the target, will not be tested for probing performance.

This part of ISO 10360 does not explicitly apply to measuring systems that do not use a spherical coordinate system (i.e., two orthogonal rotary axes having a common intersection point with a third linear axis in the radial direction). However, the parties can apply this part of ISO 10360 to such systems by mutual agreement.

This part of ISO 10360 specifies:

- Performance requirements that can be assigned by the manufacturer or the user of the laser tracker;
- The manner of execution of the acceptance and reverification tests to demonstrate the stated requirements;
- Rules for proving conformance, and;
- Applications for which the acceptance and reverification tests can be used.

3-ISO 17123
This standard refers to theodolite-type instruments.

ISO 17123-1:2014 Optics and optical instruments—Field procedures for testing geodetic and surveying instruments—Part 1: Theory

This part of ISO 17123 gives guidance to provide general rules for evaluating and expressing uncertainty in measurement for use in the specifications of the test procedures of ISO 17123-2, ISO 17123-3, ISO 17123-4, ISO 17123-5, ISO 17123-6, ISO 17123-7, and ISO 17123-8.

ISO 17123-2, ISO 17123-3, ISO 17123-4, ISO 17123-5, ISO 17123-6, ISO 17123-7, and ISO 17123-8 specify only field test procedures for geodetic instruments without ensuring traceability in accordance with ISO/IEC Guide 99. For the purpose of ensuring traceability, it is intended that the instrument be calibrated in the testing laboratory in advance.

This part of ISO 17123 is a simplified version based on ISO/IEC Guide 98-3 and deals with the problems related to the specific field of geodetic test measurements.

ISO 17123 consists of the following parts, under the general title Optics and optical instruments—Field procedures for testing geodetic and surveying instruments:

- Part 1: Theory;
- Part 2: Levels;
- Part 3: Theodolites;
- Part 4: Electro-optical distance meters (EDM instruments);
- Part 5: Electronic tacheometers;
- Part 6: Rotating lasers;
- Part 7: Optical plumbing instruments.

4-ISO 17123-3:2001 Optics and optical instruments—Field procedures for testing geodetic and surveying instruments—Part 3: Theodolites

SCOPE: This part of ISO 17123 specifies field procedures to be adopted when determining and evaluating the precision (repeatability) of theodolites and their ancillary equipment when used in building and surveying measurements. Primarily, these tests are intended to be field verifications of the suitability of a particular instrument for the immediate task at hand and to satisfy the requirements of other standards. They are not proposed as tests for acceptance or performance evaluations that are more comprehensive in nature.

This part of ISO 17123 can be thought of as one of the first steps in the process of evaluating the uncertainty of a measurement (more specifically a measurand). The uncertainty of a measurement result is dependent on a number of factors. These include among others: repeatability (precision), reproducibility (between-day repeatability), traceability (an unbroken chain to national standards) and a thorough assessment of all possible error sources, as prescribed by the ISO Guide to the expression of uncertainty in measurement (GUM).

These field procedures have been developed specifically for in situ applications without the need for special ancillary equipment and are purposefully designed to minimize atmospheric influences [2, 9].

2.7 Instruments Techniques

LSM has moved from triangulation methods using multiple direction measurements to tracking laser interferometers, allowing 3D position by direction and range. An accuracy of 10s of micrometers over 10s of meters in range can be achieved, but this depends on the environment and configuration of the parts to be measured.

The instruments are classified based on the measurement technique as shown in Figure 2.17. They can be centralized or distributed. The centralized systems are stand-alone units independently measuring the coordinates of a point or object needing one or more ancillary devices, for instance, laptop and reflectors. The second group of distributed instruments consists of several independent instruments that contribute measurements to be processed in order to determine the coordinate of a point or object [10].

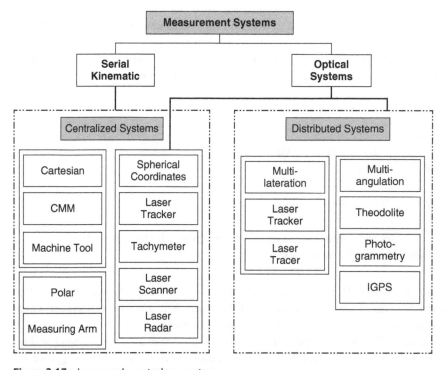

Figure 2.17 Large-scale metrology systems.

2.7.1 Large Coordinate Measuring Machines

Coordinate measuring machines (CMM) have the advantage of better precision when measuring in large ranges. For large and heavy parts, bridge and gantry types are used. CMMs are programmable, flexible instruments used to collect and report on dimensional data for virtually any type of manufactured component, for example, engine blocks, door panels, camera bodies, and turbine blades. CMMs typically collect their data by touching a component with a calibrated probe as directed by an operator. The CMM records the location of these touch points.

CMMs can also use machine vision to collect data without contacting the component (video probes and scanners). CMMs remove the requirement for dedicated gauging in manufacturing. Instead of buying one gage to measure 20 mm holes, another to measure 35 mm holes, and yet another to measure a surface's flatness (all of which may require replacement if your component changes), the CMM can be programmed to measure any combination of features. The CMM is reprogrammed whenever the requirements change. Figure 2.18 shows a large CMM inspection of the shape of a Jaguar car.

2.7.2 Laser Trackers

Lau and Hoken [11] pioneered the use of laser trackers in 1980. They are constructed with a laser interferometer housed inside a tube reflected off a mirror to determine the position of optical targets known as retroreflective targets positioned within a large object. The accuracy of laser trackers is in the order of 25 μm over a distance of several meters.

2.7.3 Theodolite

A theodolite is a precision instrument that can measure both vertical and horizontal planes using a triangulation method. A straight-line beam of light is emitted to a peg to determine distance and angles (Figure 2.19). Surveyors use them in open areas, and they have also been used in metrology and rocket launch technology. Figure ISO17025 and recently ISO 17123 give formal recognition of competence of a theodolite calibration laboratory.

Figure 2.18 CMM for car geometry inspection (LK).

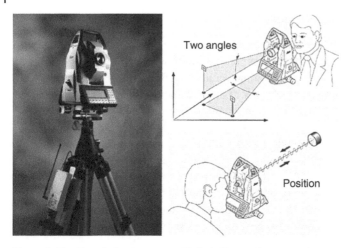

Figure 2.19 Theodolite (courtesy of Leica). *Source:* Courtesy of Leica

Global Positioning Systems (GPS)

GPS is an example of using highly redundant length data from a constellation of instruments distributed around a target point. It is possible to locate a receiver in the open air with an accuracy of 30 cm. GPS was originally intended to be a military system, but it was changed to a civilian system after the KAL007 disaster in 1983, with limited availability beginning in 2000. It is still improving as new satellites are launched. The upgrade includes a wide-area augmentation system as well as the introduction of time in global navigation satellite systems.

Indoor GPS (iGPS)

In addition to the large-scale metrology used to measure the dimensions and positions of their components, the assembly of large structures is a challenge. Blocks are used in the assembly of wind turbines and airplanes. The latter are assembled iteratively using theodolites, measuring tapes, and laser levels as measurement systems. The development of measurement-assisted assembly(MAA) has resulted in a paradigm shift. As a result, the complementarity to large scale [12] using MAA can also be done indoors using iGPS, or indoor GPS.

Figure 2.20 shows the transmitter allowing two measurement values: the azimuth angle (ϕ) and elevation angle (θ). When the coordinate system of the transmitter is fixed, the sensor position

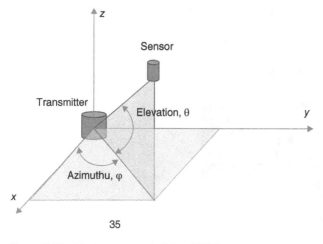

35

Figure 2.20 Measurement principle of iGPS.

can be located in a line related to the transmitter with respect to these two angles. If two or more transmitters are placed in the measuring volume, the accurate position of the sensor is at the line of intersection.

Multiple Choice Questions of this Chapter

Multiple Choice Questions are given for each chapter with solutions in an online extension of this book. Please use link: www.wiley.com\go\mekid\metrologyandinstrumentation\

References

1 Herrera-Basurto, R., and Simonet, B.M., 2013, "Nanometrology," doi:10.1002/9780470027318 .a9177.
2 S. Mekid, 2020, "Integrated nanomanipulator with in-process lithography inspection," *IEEE Access*, vol. 8, pp. 95378–95389, doi: 10.1109/ACCESS.2020.2996138.
3 Mekid, S., and Lim, B., 2004, "Characteristics comparison of piezoelectric actuators at low electric field: Analysis of strain and blocking force," *Smart Materials and Structures*, 13(5), pp. N93–N98.
4 Mekid, S., and Shang, M., 2015, "Concept of dependent joints in functional reconfigurable robots," *Journal of Engineering, Design and Technology*, 13(3), pp. 400–418.
5 https://www.iso.org/obp/ui/#iso:std:iso:tr:18196:ed-1:v1:en:sec:5.
6 https://www.iso.org/standard/40954.html, ISO 10360-2:2009.
7 Puttock, M. J., 1978, "Large-scale metrology," Ann. CIRP, 2711: 351-6.
8 Mekid, S., 2010, "Spatial thermal mapping using thermal infrared camera and wireless sensors for error compensation via open architecture controllers," *Proceedings of the Institution of Mechanical Engineers. Part I: Journal of Systems and Control Engineering*, 224(7), pp. 789–798.
9 "ASME B89.4.19, performance evaluation tests and geometric misalignments in laser trackers," *Journal of Research of the National Institute of Standards and Technology*, 2009, 114, pp. 21–35.
10 S. Mekid, 2005, "Design Strategy for Precision Engineering: Second Order Phenomena." *J. Engineering Design*, 16(1).
11 Lau, K.C., and Hocken, R.J., 1987, "Three and five axis laser tracking systems," December 22, 1987, US Patent No. 4714339.
12 Peggs, G.N., Maropoulos, P., Hughes, B., Forbes, A., Robson, S., Ziebart, M., and Muralikrishnan, B., 2009, "Recent developments in large-scale dimensional metrology," Proceedings of the Institution of Mechanical Engineers - Part B." *Journal of Engineering Manufacture* 223, pp. 571–595. doi: 10.1243/09544054JEM1284.

3

Applied Math and Statistics

"Do not say you can't do maths!"

—Eligible Professor

3.1 Introduction

Math and science are of great importance when dealing with measurement since the inception of humanity. Exchanging or swapping goods with equivalent value was the initial business since money did not exist. Hence, people were trading gold for salt and alternatively, ivory, skins, or pepper for equivalent salt or gold. This is to say that value of things matters, and this initiated quantitative measure of one product against the other. Math has then played a great role.

This chapter aims to introduce basic math and science background mainly to refresh memories and be a reference in case there is a need to check information.

3.2 Scientific and Engineering Notation

Numbers are written in various ways whether they are too large or too small. However, they can be written in a reduced way depending on the quantity. Hence, agreed notations are used to ease writing.

Example 3.1 The following two numbers although very small or very high, can be written easily as:

$$0.000\ 000\ 023 = 23 \times 10^{-9}$$
$$25\ 400\ 000\ 000 = 2.54 \times 10^{10}$$

Scientific notation shows a number as a value between 1 and 10, but not including 10, multiplied by a power of 10.

Example 3.2

$$269870000 = 2.6987 \times 10^8.$$

An engineering notation converts a very large or small number into a value between 1 and 1000 using a power of 10 usually in increments of 3 as follows:

$$269870000 = 269.87 \times 10^6.$$

Other examples may be written with the following scientific notation:

A converted decimal number: $0.0005 = 5 \times 10^{-4}$.

Or a converted fraction number: $4/10 = 0.4 = 4 \times 10^{-1}$

3.3 Imperial/Metric Conversions

The Imperial system is an Anglo-Saxon system of units. A conversion is sometimes needed either way and as a quick reference, Tables 3.1 and 3.2 are given below for this purpose.

Example 3.3 Conversion from metric system to Imperial system:

$$1 \text{ cm} = 0.39370 \text{ in}$$
$$1 \text{ in} = 2.54 \text{ cm}$$

The method is to find the right conversion number that you multiply by your number.

Table 3.1 Table conversion of distance imperial/metric.

METRIC CONVERSIONS					
1 centimeter	=	10 millimeters	1 cm	=	10 mm
1 decimeter	=	10 centimeters	1 dm	=	10 cm
1 meter	=	100 centimeters	1 m	=	100 cm
1 kilometer	=	1000 meters	1 km	=	1000 m

IMPERIAL CONVERSIONS					
1 foot	=	12 inches	1 ft	=	12 in
1 yard	=	3 feet	1 yd	=	3 ft
1 chain	=	22 yards	1 ch	=	22 yd
1 furlong	=	220 yards (or 10 chains)	1 fur	=	220 yd (or 10 ch)
1 mile	=	1760 yards (or 8 furlongs)	1 mi	=	1760 yd (or 8 fur)

METRIC -> IMPERIAL CONVERSIONS					
1 millimeter	=	0.03937 inches	1 mm	=	0.03937 in
1 centimeter	=	0.39370 inches	1 cm	=	0.39370 in
1 meter	=	39.37008 inches	1 m	=	39.37008 in
1 meter	=	3.28084 feet	1 m	=	3.28084 ft
1 meter	=	1.09361 yards	1 m	=	1.09361 yd
1 kilometer	=	1093.6133 yards	1 km	=	1093.6133 yd
1 kilometer	=	0.62137 miles	1 km	=	0.62137 mi

IMPERIAL -> METRIC CONVERSIONS					
1 inch	=	2.54 centimeters	1 in	=	2.54 cm
1 foot	=	30.48 centimeters	1 ft	=	30.48 cm
1 yard	=	91.44 centimeters	1 yd	=	91.44 cm
1 yard	=	0.9144 meters	1 yd	=	0.9144 m
1 mile	=	1609.344 meters	1 mi	=	1609.344 m
1 mile	=	1.609344 kilometers	1 mi	=	1.609344 km

Table 3.2 Table conversion of mass imperial/metric.

Please note, long tons are used in the UK, short tons are used in the US.			
METRIC CONVERSIONS			
1 gram	= 1000 milligrams	1g	= 1000 mg
1 kilogram	= 1000 grams	1 kg	= 1000 g
1 ton (1 megagram)	= 1000 kilograms	1 ton (1 Mg)	= 1000 kg
IMPERIAL CONVERSIONS			
1 ounce	= 16 drams	1 oz.	= 16 dr.
1 pound	= 16 ounces	1 lb.	= 16 oz.
1 stone	= 14 pounds	1 st.	= 14 lb.
1 quarter	= 2 stone	1 qr.	= 2 st.
1 hundredweight	= 4 quarters (or 8 stone)	1 cwt	= 4 qr. (or 8 st.)
1 ton, long	= 20 hundredweight (160 stone)	1 ton	= 20 cwt (or 160 st.)
METRIC -> IMPERIAL CONVERSIONS			
1 gram	= 0.035274 ounces	1 g	= 0.035274 oz.
1 kilogram	= 2.20462 pounds	1 kg	= 2.20462 lb.
1 kilogram	= 35.27396 ounces	1 kg	= 35.27396 oz.
1 ton	= 0.9842 ton, long	1 ton	= 0.9842 ton, long
1 ton	= 157.47304 stone	1 ton	= 157.47304 st.
IMPERIAL -> METRIC CONVERSIONS			
1 ounce	= 28.34952 grams	1 oz	= 28.34952 g
1 pound	= 453.59237 grams	1 lb	= 453.59237 g
1 pound	= 0.45359 kilograms	1 lb	= 0.45359 kg
1 stone	= 6.35029 kilograms	1 st	= 6.35029 kg
1 hundredweight	= 50.8023 kilograms	1 cwt	= 50.8023 kg
1 ton, long	= 1.01605 tonnes	1 ton, long	= 1.01605 ton
1 ton, short	= 0.90718 tonnes	1 ton, short	= 0.90718 ton

Example 3.4

$$25.5 \text{ cm} = \mathbf{0.39370} \times 25.5 = 10.03935 \text{ in}$$

3.4 Ratio

A ratio shows a comparison between two numbers to indicate how many times the first number is larger or smaller than the second number. For example, a box of 10 balls where 2 are red and 8 are black can show a ratio of red balls to black balls as 2:8 or 1:4 while the other ratio of black balls to the red ones is 8:2 or 4:1.

In other representation as percentage with respect to overall, 2/10 or 20% are red and the rest, 8/10 or 80%, are black.

Sometimes, three numbers A, B, and C need to be compared, for example, thickness: width: length = 1:4:8.

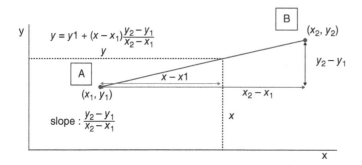

Figure 3.1 Calibration figure.

3.5 Linear Interpolation

Linear interpolation is a method of curve fitting that uses linear polynomials to construct needed data points as an interpolation within the range of a discrete set of known data points. Assume two known points $A(x_1,y_1)$ and $B(x_2,y_2)$. There is a need to estimate the y value we would get for some x value that is between x_1 and x_2. This y value estimate is the *interpolated* value.

The simple method is to draw a straight line between $A(x_1,y_1)$ and $B(x_2,y_2)$. The y value is expected to be on the line for the chosen x, hence the name *linear interpolation*. It is trivial to show that the formula of the line between A and B is (Figure 3.1):

$$y = y_1 + (x - x_1)(y_2 - y_1)/(x_2 - x_1) \tag{3.1}$$

Example:

x	y
2	4
6	7

– What is the corresponding value y if $x = 4$?

The interpolation formula in Eq. (3.1) can be used to find y:

$$y = 4 + (4 - 2)(7 - 4)/(6 - 2) = 11/2.$$

3.6 Number Bases

A base is the number of different digits or combination of digits and letters that a system of counting uses to represent numbers. For example, the most common base used today is the decimal system, as we have ten fingers. Because "**dec**" means 10, it uses the 10 digits from 0 to 9.

Other different bases are also used in computers.

- Binary is base 2 used for example in computers, which deal only with 0s and 1s.
  A scale can be drawn showing the correspondence between decimal and binary numbers.

Decimal:	0	1	2	3	4	5	6	7	8	9	10	11	12	13	14	15
Binary:	0	1	10	11	100	101	110	111	1000	1001	1010	1011	1100	1101	1110	1111

- Hexadecimal (base 16) is used because of how computers group binary digits together. The following table gives the correspondence between decimal and hexadecimal numbers.

Decimal:	0	1	2	3	4	5	6	7	8	9	10	11	12	13	14	15
Hexadecimal:	0	1	2	3	4	5	6	7	8	9	A	B	C	D	E	F

Examples:

1) **Base 10** e.g. $10^0 = 1$, $10^1 = 10$, $10^2 = 100$
2) **Base 2** e.g. $2^0 = 1$, $2^1 = 2$, $2^4 = 16$.

3.7   Significant Figures, Rounding, and Truncation

When precise measurements and rounded or estimated figures are involved, it is important to know and understand the technical aspects of these numbers. There are three related aspects of numbers: significant digits, rounding, and truncating.

i) Significant digits

A number is expressed in scientific notation where the number of significant digits (or figures) is the number of digits needed to express the number to within the uncertainty of calculation. All non-zero digits and zero digits between two non-zero digits are significant.

Example:
a) 1.225 ± 0.005; this number has four significant digits.
b) 358 has three significant digits
c) 47,569 has five significant digits
d) 120,548,120 has nine significant digits

ii) Rounding

Rounding a number is replacing it with a different number that is approximately equal to the original. If the number to round is followed by 5, 6, 7, 8, or 9, then it is rounded up. Hence, 27 can be rounded up to 30. If the number is followed by 0, 1, 2, 3, or 4, then it is rounded down, for example 22 becomes 20, and 23.13 becomes 23.1.

iii) Truncating.

This is shortening or reducing a number to a number of digits right of the decimal point. A method of approximating a decimal number by dropping all decimal places past a certain point without rounding. As an example; 3.14159265 becomes 3.1415, and if rounded then 3.1416.

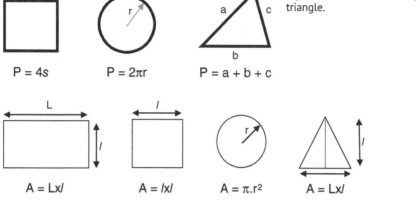

Figure 3.2 Perimeter of square, circle, and triangle.

P = 4s   P = 2πr   P = a + b + c

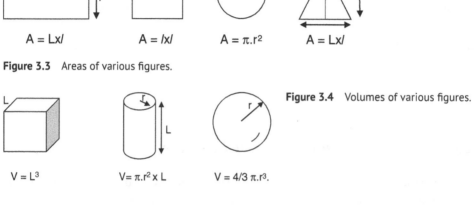

A = Lx*l*   A = *l*x*l*   A = π.r²   A = Lx*l*

Figure 3.3 Areas of various figures.

V = L³   V= π.r² x L   V = 4/3 π.r³.

Figure 3.4 Volumes of various figures.

3.8 Geometry and Volumes

3.8.1 Perimeter

The perimeter is the length surrounding the area in 2D space (Figure 3.2). It is measured in linear units, such as meters.

Example:

- A square with s as length of one side has a perimeter of P = 4s.
- A circle with radius r has a perimeter (or circumference) of P = 2πr.
- A triangle with sides of lengths a, b, and c has a perimeter P = a + b + c.

3.8.2 Volume and Area

The area is a size of the surface of a shape. It is measured in square units. This is presented in 2D. The volume is the amount of space measured in cubic units and it is presented in 3D space. Examples are given in Figures 3.3 and 3.4 but many can be found in other literature.

3.9 Angular Conversions

Angles can be converted from degrees to radians and vice versa (Table 3.3). In addition, an angle can be converted into length if the radius is known (called "arc").

A full circle covers 360 degrees, or 2π radians. Hence, 90 degrees is π/2 radian, for example.

Table 3.3 Angular conversion to degrees.

from	unit	To (°)	Multiply by (degrees)	Multiply by (degrees)
degree	□°	degree	1	1
minutes	□'	degree	1/60	0.016667
seconds	□"	degree	1/3600	$2.777778 \cdot 10^{-4}$
quadrant	quadrant	degree	360/4	90
revolution	r	degree	360/1	360
mil	mil	degree	360/6400	0.05625
radian	rad	degree	180/π	57.29578

3.10 Graphs and Plots

With large data to analyze, graphs and plots become important to visualize the trend of the collected data. One of the important tasks is to extract information to help understanding how the measurements describe the system behavior. Hence, either the measurand or instrument need to be considered. Venn diagrams, bar graphs, pie charts, stack plots, and axis plots are examples to extract information and draw conclusions. The developed data literacy allows the user to ask and answer meaningful questions by collecting, analyzing, and making sense of the recorded measurements in real life.

Excel spreadsheet is the most common software needing hands-on experience in collecting and organizing data and most importantly using various graphical representations.

Several ways to represent data exist. The next table shows the most usual ways to visualize data. Some of them as defined hereafter.

a) **Line graph**: A line graph is used to visualize the value of something over time or other parameter. The line graph consists of a horizontal x-axis and a vertical y-axis.

b) **Bar chart:** A bar chart, or bar graph, presents categorical data with rectangular bars with heights or lengths proportional to the values that they represent. The bars can be shown vertical or horizontal.

c) **An OHLC chart** is another type of bar chart showing open, high, low, and closing (OHLC) values for each period, for example, or other range. OHLC charts are useful since they show the four major data points over a period.

d) **A Band chart** is a line chart where we add shaded areas to display the upper and lower boundaries of the defined data ranges. The shaded areas are the bands.

e) **Spider chart** (radar chart) is a graphical method exhibiting multivariate data in the form of a two-dimensional chart of three or more quantitative variables represented on axes starting from the same point.

f) **A multi-pane chart** shows a distribution of a collection of series between several panes, thus presenting data in a clear and uncluttered way. It is very much appreciated in detailed analyses.

Table 3.4 shows all different charts needed in engineering.

Table 3.4 Table of various chart needed in Engineering, courtesy of tmssoftware.[1]

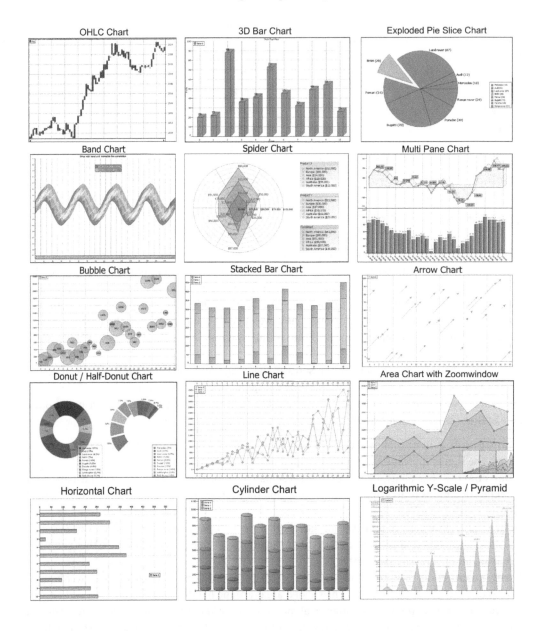

Source: Courtesy of Tmssoftware

1 https://www.tmssoftware.com/site/newsletters/20090323.htm.

3.11 Statistical Analysis and Common Distributions

3.11.1 Definition of Measurement Data

Measurement data is quantitatively conveying an information about the measurement being carried out for any physical measurand. The measurand is the quantity intended to be measured. Measurement data are widely used in statistics since measurements are done several times to calculate the average, variance, and so forth. The quantity can be represented in four scales:

- Nominal: scale with different categories grouped, e.g., black, red, yellow, or just one category: hair color.
- Ordinal: similar to nominal but there is a clear order in the data, e.g., very large, large, medium, small.
- Interval: also called continuous, is used to show variation, and
- Ratio, called percentage, has a defined zero point.

Single measurements are not reliable in general since they are surrounded by many sources or errors, as we will discuss later. Hence, as a first approach assuming experience, calibrated instruments, and minimum standards in measurement, it is required to do more than one measurement for the same measurand under similar conditions and deliver an average of these measurements.

This means that a large number of measurements may require statistical analysis. Data needs to be collected and organized. In this case, we use descriptive statistics. Hence, by definition, it is used to organize a particular set of measurements, for example, calculating central tendency (mean, median, and mode) and variability (standard deviation, kurtosis, and skewness). We can make inferences, based on data gathered from a sample, about a larger population from which the sample was drawn; this is called inferential statistics.

3.11.2 Statistical Measurements

In the context of metrology within this book, measurements have to be conducted by the efficient application of scientific principles. High accuracy is achieved either by refining an existing product, that is, a machine or system, or designing a new product that meets initial specifications. The engineering method to formulate and to converge to targets encompasses the following steps:

a) Develop a clear and concise description of what to measure.
b) Identify key factors that may affect the measurements.
c) Identify key instrument(s) required for the measurements.
d) Conduct appropriate experiments and acquire data properly.
e) Identify the more appropriate model for error assessment.
f) Draw conclusions and report quality of the results.
g) Set up actions if is necessary to adjust the system to comply with initial specifications.

3.11.3 Statistical Analysis of Measurements

Statistical analysis is used to understand variability that exists in the system that is under investigation, for example, a machine tool. Successive inspection measurements on the latter may not produce exactly the same result. Hence, the existing techniques will be able to construct a clear picture of the existing defect if it exists within the machine. The variations are usually observed in engineering measurements repeatedly taken under identical conditions. The source of the variation can be identified within:

a) The measurement system in which resolution and repeatability are key parameters.
b) The measurement procedure and technique mainly represented by repeatability.
c) The measured variables including temporal variation (t) and spatial variation (x).

The statistical analysis provides estimates of the following:

a) A single representative value that best characterizes the data set;
b) Some representative value that provides the variation of the data; and
c) an interval (or range) about the representative value in which the true value is expected to be.

The statistical analysis helps understanding experimental and interpreting data mainly in:

a) Extracting the best value of a quantity from a set of measurements by estimating a parameter from a data sample. The value of the parameter using all of the possible data, not just the sample data, is called the population parameter, or true value of the parameter. An estimate of the true parameter value is made using the sample data.
b) Deciding whether the experiment (i.e., measurements) is consistent (with theory, other experiments, etc.).

3.11.4 Probability

We introduce a couple of aspects of probability. Assume trials are being carried out. The probability of the specified event, E, to occur e times having N trials is defined as follows:

$$P(E) = \frac{e}{N} \quad N \to \infty \tag{3.2}$$

The probability, P, is a number between 0 and 1, i.e., $0 \leq P \leq 1$, with $P = 0$ for an event that never occurs and $P = 1$ for an event that always occurs.

The events A and B are independent if $P(A \cap B) = P(A) \cdot P(B)$. They are mutually exclusive if

$$P(A \cap B) = 0, \text{ or } P(A \cup B) = P(A) + P(B).$$

This probability could be either discrete or continuous:

a) Discrete probability: P has certain values only

Example:

a) Discrete: a fair coin, for which $P(x_i) = \frac{1}{2}$ (x_i heads or tail).
b) Continuous: P could have any number between 0 and 1: $0 \leq P \leq 1$
   - Probability density function (pdf); $f(x)$

$$f(x)dx = dP(x \leq a \leq x + dx) \tag{3.3}$$

   - Probability for x to be in the range $a \leq x \leq b$ is defined as:

$$P(a \leq x \leq b) = \int_a^b f(x) \cdot dx \tag{3.4}$$

where $f(x)$ is normalized to one, $\int_{-\infty}^{+\infty} f(x) \cdot dx = 1$ and $\int_a^a f(x) \cdot dx = 0$
$P(x)$: Discrete: Binomial or Poisson
$f(x)$: Continuous: Uniform, Gaussian, exponential, or chi-square.

A probability distribution is described by its mean, mode, median, and variance. Mean and variance will be introduced later in this chapter.

3.11.5   Sample and Population

Before going into further explanation, it is necessary to introduce both terms "*sample*" and "*population,*" which are generally used in statistical analysis of error. In a manufacture, a batch of ball bearings had been made, and an inspector takes a small number of bearings for quality check. So it can be said a sample of size n of bearings had been drawn from a finite population of size p ($n \ll p$) of ball bearings (Fig. 3.5).

Note that the following conditions are applied:

1) The sample is used to estimate properties of the population, and additional data cannot be added to the population. For example, no more bearings can be added to the bag.
2) The sample size is assumed to be very small compared to the population size: $n \ll p$.
3) A finite number of items, n, is randomly drawn from a population, which is assumed to be with indefinite size.

The most commonly used measure of a measurand is the mean. The true mean is the sum of all the members of the given population divided by the number of members in the population. As it is typically impractical to measure every member of the population, a random sample is drawn from the population. The sample mean is calculated by summing the values in the sample and dividing by the number of values in the sample. This sample mean is then used as the point estimate of the population mean.

3.11.6   Formulation of Mean and Variance for Direct Measurements

The population mean is the expected outcome, such that if an infinite number of measurements are made, the average of the infinite measurements is the mean. In a real-world situation, it is impractical and impossible to extract statistical information from the entire population. Take the bearing manufacturing, for example, it is virtually impossible to make measurement on every single bearing if there is more than a million in a batch. Nevertheless, this will provide good background information on how to establish the size of error. This represents the *true* value of a measurement. If a large number of measurements are taken then the arithmetic average of these n measurements is:

$$\mu \approx \sum_{i=1}^{n} \frac{x_i}{n} \tag{3.5}$$

where

μ is the population mean
x_i is the ith value of the quantity being measured
n is the total number of measurements.

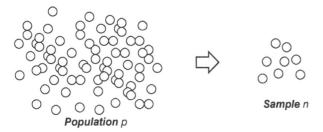

Figure 3.5   Sample taken from a population.

The amount of error in a ***single*** measurement is given by:

$$d = x - \mu \tag{3.6}$$

where:

d is deviation
x is a particular value of the quantity being measured
μ is the population mean

The standard deviation of population σ is an estimate of the average error of measurements of a very large sample and can be found from square root of the variance (σ^2).

The variance is the summing of the deviations divided by number of measurements.

$$\sigma \approx \sqrt{\frac{d_1^2 + d_2^2 + d_3^2 + \cdots + d_n^2}{n}} \tag{3.7}$$

The standard deviation is a very important parameter in establishing the likely size of error.

3.11.7 Mean and Variance Based on Samples

In real life, it is usual to deal with samples from a population and not the population itself. So sample mean $\bar{x}$ and sample standard deviation S_x would be used as an approximation to population mean μ and population standard deviation σ, respectively. The population mean is usually denoted μ and is the expected value $E(x)$ for a measurement. The sample standard deviation S_x is defined in Eq. 2.11.

$$S_x = \sqrt{\frac{1}{n-1} \sum_{i=1}^{n} (x_i - \bar{x})^2} \tag{3.8}$$

where:

n is the number of measurements
x_i is the ith value of the quantity being measured
$\bar{x}$ is the sample mean

The denominator of standard deviation $(n - 1)$ is called the number of degrees of freedom (d.o.f), which is a measure of how much precision an estimate of variation has. The d.o.f. can be viewed as the number of independent parameters available to fit a model to data. Usually, the more parameters you have, the more accurate your fit will be. However, for each estimate made in a calculation, you remove one degree of freedom. This is because each assumption or approximation you make puts one more restriction on how many parameters are used to generate the model. Put another way, for each estimate you make, your model becomes less accurate. The difference between the two definitions of standard deviation for a sample or for a population is almost numerically insignificant. The comparison is between $\sqrt{n}$ and $\sqrt{n-1}$.

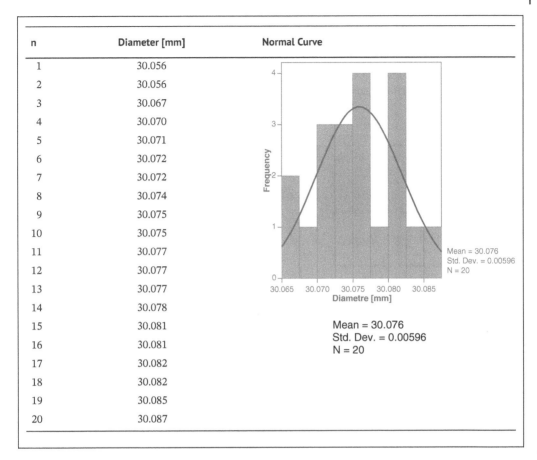

n	Diameter [mm]	Normal Curve
1	30.056	
2	30.056	
3	30.067	
4	30.070	
5	30.071	
6	30.072	
7	30.072	
8	30.074	
9	30.075	
10	30.075	
11	30.077	Mean = 30.076
12	30.077	Std. Dev. = 0.00596
13	30.077	N = 20
14	30.078	
15	30.081	Mean = 30.076
16	30.081	Std. Dev. = 0.00596
17	30.082	N = 20
18	30.082	
19	30.085	
20	30.087	

Figure 3.6 Statistical data for diameter measurements.

3.11.8 The Standard Deviation of the Mean

The standard deviation σ_x characterizes the average uncertainty of the separate measurements x_1, $x_2, x_3, \ldots, x_i$, whose average is the best estimate average.

Now, for the whole population, the best estimation is still the population average but with the uncertainty on the average $\sigma_{\bar{x}} = \sigma_x / \sqrt{n}$ also called standard deviation of the mean (SDOM) or standard error of the mean. Hence,

$$x_{value} = \bar{x} \pm \sigma_{\bar{x}} \tag{3.9}$$

Example: The inspection of a component diameter has been made automatically 20 times. The values are recorded on Fig. 3.6. Hence, the mean diameter d is 30.076 mm and the uncertainty in any one measurement is therefore 0.006 mm with 68% confidence that d lies within the range 30.076 ± 0.006 mm.

3.12   Formulation of the Standard Uncertainty and Average of Indirect Measurements

Assume that the measurand $g(X)$ is not measured directly but is determined indirectly from the quantity X measured. The Taylor series expansion gives:

$$g(X) = \underbrace{\frac{(X-a)^0}{0!} \cdot g(a)}_{zero-order} + \underbrace{\frac{(X-a)^1}{1!} \cdot \frac{\partial g}{\partial X}(a)}_{first-order} + \underbrace{\frac{(X-a)^2}{2!} \cdot \frac{\partial^2 g}{\partial X^2}(a)}_{second-order} + \cdots + \underbrace{\frac{(X-a)^n}{n!} \cdot \frac{\partial^n g}{\partial X^n}(a)}_{n^{th}-order} + R_n$$

(3.10)

where a is a constant about which the expansion is carried out. R_n is a remainder term known as the Lagrange remainder, defined by [2] as

$$R_n = \underbrace{\int \cdots \int_{x_0}^x g^{(n+1)}(X)(dX)^{n+1}}_{n+1}.$$

With respect to the current investigation, it seems obvious that the expansion should be carried out about the mean (i.e., central value). The average of measurement is obtained through the expectation value defined by the first statistical moment as:

$$x = E(X) = \int_{-\infty}^{\infty} x \cdot f_X(x) \cdot dx$$

(3.11)

where $f_X(x)$ is the probability density function (PDF).

The standard deviation is known to be the square root of the variance, which in turn is written as the second statistical moment:

$$Var(X) = E(X - E(X))^2 = \int_{-\infty}^{\infty} (x - \mu_x)^2 \cdot f_x(x)dx,$$

(3.12)

where $\mu_x \equiv E(X)$. It also expresses the uncertainty of measurement $u(x)$ defined in GUM [3], and the standard deviation σ_x, hence:

$$Var(X) = \sigma_x^2 = u^2(x)$$

(3.13)

3.12.1   How to Determine the Measured Value and Random Error?

Random errors, E_e, (also referred to as precision errors) are different for each successive measurement but have an average value of zero. If enough readings are taken, the distribution of precision errors will become apparent. A reliable estimate of random errors can be obtained with appropriate statistical methods being applied.

There are different approaches in making measurements, such as:

a) *Repeated measurements of one single quantity.*
b) *Measurement of two different variables or more that are related.*

Both approaches will be discussed in detail in the following sections.

3.12.2   Repeated Measurements of One Single Quantity

Although this is not the most general situation that we will encounter in practice, it is the easiest one, and it is necessary to fully understand this before moving to a more general and complex one.

a) How to Determine Measured Values $x_{measured}$

As introduced previously, the value of a measurand could be represented as:

$$x_{true} = x_{measured} \pm E_s \pm E_r \tag{3.14}$$

For repeated measurements, the previous equation becomes:

$$x_{true} = \bar{x} \pm E_s \pm E_r \tag{3.15}$$

where:

x_{true} is the true value
$\bar{x}$ is the mean value
E_s is the systematic error, or component of laboratory bias
E_r is the random error under repeatability conditions.

The total uncertainty could be expressed as the quadratic sum of random and systematic errors. However, there is an alternative method to determine the "best estimate" if the following conditions are met:

 i) Total number of measurements n is very large.
 ii) Value in the sample occurs more than once.

$$\bar{x} = \frac{\sum\limits_{i=1}^{n} (f_i x_i)}{\sum\limits_{i=1}^{n} f_i} \tag{3.16}$$

where:

f_i is the occurrence of the ith value
$f_i x_i$ is the product of ith value and its corresponding occurrences

b) How to determine random errors?

The uncertainty calculated previously for the standard deviation of the mean is also considered as the random uncertainty component in a measurement. The random uncertainty could also be computed using distribution functions. The latter are used in statistical analysis of error, such as binomial, exponential, hyper-geometric, chi-square, Poisson, and so forth. Not all distributions have a symmetric bell shape; the Poisson distribution is usually not symmetric, for example.

If a set of measurement results is affected by many small sources of random error and negligible systematic error, the measured values are presented in a bell-shaped curve centered on the true value of x. Two particular distributions will be discussed in this book, but others could be found in many other specialized books on statistics and probability for engineers. The distributions considered here are either the normal distribution and Student's t-distribution.

3.12.3 Normal Distribution

When the measurements sets contain only random errors, their distribution will be a symmetric bell-shaped curve and centered on the mean value with zero error. Such distribution is called *Normal* (also known as *Gaussian*) distribution (Fig. 3.7).

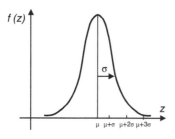

Figure 3.7 Normal Distribution.

The probability distribution of x values is described by a *probability density function* (PDF). For an infinite population, the mathematical expression for the Normal PDF is:

$$f(x) = \frac{1}{\sigma\sqrt{2\pi}} \exp\left[-\frac{(x-\mu)^2}{2\sigma^2}\right] \tag{3.17}$$

or

$$f(z) = \frac{1}{\sqrt{2\pi}} \cdot \exp\cdot\left[-\frac{z^2}{2}\right] \tag{3.18}$$

where:

x is a particular value of the quantity being measured

$$z = \frac{x-\mu}{\sigma} \tag{3.19}$$

σ is the standard deviation of population
μ is the population mean

Confidence Intervals (Size of Random Error) for Population As introduced previously, a confidence interval gives an estimated range of values, where the population mean μ is located. The confidence level defines how sure you can be. It is a percentage of certainty that a measurement will lie within the confidence interval. The confidence level is defined in terms of probability to be within the confidence interval as $(1 - \alpha)$, where α is the probability to be outside the interval.

The probability of $(1 - \alpha)$ to select a sample that will produce a range containing the value of x is defined as (Fig. 3.8).

$$\begin{cases} P(L \leq x \leq U) = 1 - \alpha \\ \quad 0 < \alpha < 1 \end{cases} \tag{3.20}$$

This range is mentioned as $100(1 - \alpha)\%$ confidence interval (CI) for the parameter x. If a sampling distribution, x, is normal with a mean μ and variance σ^2, then if $z = (x - \mu)/\sigma$, the probability could be written as:

$$P\left(-z_{\frac{\alpha}{2}} \leq z \leq z_{\frac{\alpha}{2}}\right) \tag{3.21}$$

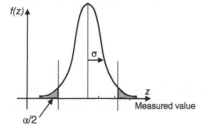

Figure 3.8 Probability density function.

Table 3.5 Variable *z* vs. confidence level.

Variable, z	Confidence level (%)
±0.6754	50
±1	68.3
±1.6449	90
±1.96	95
±3	99.7
±4	99.994

and *x* could be located as:

$$P\left(\mu - z_{\frac{\alpha}{2}} \cdot \sigma \leq x \leq \mu + z_{\frac{\alpha}{2}} \cdot \sigma\right) \tag{3.22}$$

The mathematical expression for the confidence interval (within which the true population *μ* is located) is given by the expression:

$$\mu - z_{\frac{\alpha}{2}}\sigma < x < \mu + z_{\frac{\alpha}{2}}\sigma \tag{3.23}$$

where

σ standard deviation of population
μ population mean
x a particular value of the quantity being measured
z a variable
c =1 − *α* the confidence level

The variable *z* is defined as:

$$z_{\frac{c}{2}} = \frac{x - \mu}{\sigma} \tag{3.24}$$

The width of the confidence interval depends on the confidence level required. As the width of the confidence interval increases, the probability that the population mean *μ* will fall within the interval increases. The following table 3.5 is a summary of various confidence levels for the normal distribution:

For example, if a population has a normal distribution, then the probability that a single measurement has an error greater than the standard deviation ±σ is 31.7%. On the other hand, it can be said that the probability that the error will be within ±σ is 68%: confidence level, c.

In most precision engineering applications, we would be satisfied with 99.7% confidence level (3σ). (Figure 3.9 together with Table 3.6.) With such a confidence level, the odds that measurement *x* is greater than 3σ is 1 in 370.

3.12.4 Student's *t*-distribution

When the sample is very small, as is usual in engineering applications, the standard deviation S_x does not provide a reliable estimate of the standard deviation σ of the population; therefore, the previous equations (normal distribution) should not be used. Hence, when it is required to calculate the size of sample needed to obtain a certain level of confidence in measurements, it requires prior knowledge of the population standard deviation σ, but the latter is unknown.

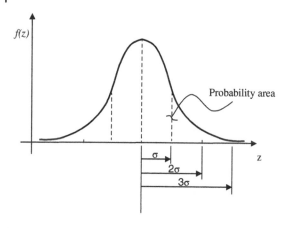

Figure 3.9 Areas under the curve between 0 and various values.

Often a preliminary sample will be conducted so that a reasonable estimate of this critical population parameter can be made. If confidence intervals for the population mean are required using an unknown σ, then the distribution known as the **Student's *t*-distribution**[2] can be used.

Some important properties of the Student's *t*-distribution are given below.

a) Student's *t*-distribution is different for different sample sizes.
b) The Student's *t*-distribution is usually shown as a bell-shaped but with smaller sample sizes. The distribution is less peaked than a normal distribution and with thicker tails. As the sample size increases, the distribution approaches a normal distribution. For *n* > 30, the differences are almost negligible.
c) The mean is in this case zero.
d) The population standard deviation is unknown.

The distribution of the quantity *t* is defined as:

$$t = \frac{\bar{x} - \mu}{S_{\bar{x}}}$$

(3.25)

where

$S_{\bar{x}}$ is standard deviation of the sample mean.
μ is the population mean (unknown)
$\bar{x}$ is the sample mean

Such distribution (Figure 3.10) depends on the number of samples taken through the degrees of freedom *v*, which is (*n* – 1). As the number of sample increases toward 30, the *t*-distribution will be exactly as a normal distribution.

Confidence Intervals for Large Number of Samples (*n* > 30)
It is known that *mean* is the sum of the values of quantity being measured divided by total number of measurements *n*. However, if additional measurements were made, the new *mean* value would differ from the first one. Furthermore, if the test (with *n* number of measurements) were repeated many times, a set of samples for the mean values would be obtained (sampling errors with large population). Such samples of mean would also show dispersion about a central value. A profound

---

2 By W.S. Gosset, an amateur statistician, writing under the pseudonym "Student" in 1908.

Table for Normal distribution

Table 3.6 Table of Normal z distribution.

z	0.00	0.01	0.02	0.03	0.04	0.05	0.06	0.07	0.08	0.09
0.0	0.0000	0.0040	0.0080	0.0120	0.0160	0.0199	0.0239	0.0279	0.0319	0.0359
0.1	0.0398	0.0438	0.0478	0.0517	0.0557	0.0596	0.0636	0.0675	0.0714	0.0753
0.2	0.0793	0.0832	0.0871	0.0910	0.0948	0.0987	0.1026	0.1064	0.1103	0.1141
0.3	0.1179	0.1217	0.1255	0.1293	0.1331	0.1368	0.1406	0.1443	0.1480	0.1517
0.4	0.1554	0.1591	0.1628	0.1664	0.1700	0.1736	0.1772	0.1808	0.1844	0.1879
0.5	0.1915	0.1950	0.1985	0.2019	0.2054	0.2088	0.2123	0.2157	0.2190	0.2224
0.6	0.2257	0.2291	0.2324	0.2357	0.2389	0.2422	0.2454	0.2486	0.2517	0.2549
0.7	0.2580	0.2611	0.2642	0.2673	0.2704	0.2734	0.2764	0.2794	0.2823	0.2852
0.8	0.2881	0.2910	0.2939	0.2967	0.2995	0.3023	0.3051	0.3078	0.3106	0.3133
0.9	0.3159	0.3186	0.3212	0.3238	0.3264	0.3289	0.3315	0.3340	0.3365	0.3389
1.0	0.3413	0.3438	0.3461	0.3485	0.3508	0.3531	0.3554	0.3577	0.3599	0.3621
1.1	0.3643	0.3665	0.3686	0.3708	0.3729	0.3749	0.3770	0.3790	0.3810	0.3830
1.2	0.3849	0.3869	0.3888	0.3907	0.3925	0.3944	0.3962	0.3980	0.3997	0.4015
1.3	0.4032	0.4049	0.4066	0.4082	0.4099	0.4115	0.4131	0.4147	0.4162	0.4177
1.4	0.4192	0.4207	0.4222	0.4236	0.4251	0.4265	0.4279	0.4292	0.4306	0.4319
1.5	0.4332	0.4345	0.4357	0.4370	0.4382	0.4394	0.4406	0.4418	0.4429	0.4441
1.6	0.4452	0.4463	0.4474	0.4484	0.4495	0.4505	0.4515	0.4525	0.4535	0.4545
1.7	0.4554	0.4564	0.4573	0.4582	0.4591	0.4599	0.4608	0.4616	0.4625	0.4633
1.8	0.4641	0.4649	0.4656	0.4664	0.4671	0.4678	0.4686	0.4693	0.4699	0.4706
1.9	0.4713	0.4719	0.4726	0.4732	0.4738	0.4744	0.4750	0.4756	0.4761	0.4767
2.0	0.4772	0.4778	0.4783	0.4788	0.4793	0.4798	0.4803	0.4808	0.4812	0.4817
2.1	0.4821	0.4826	0.4830	0.4834	0.4838	0.4842	0.4846	0.4850	0.4854	0.4857
2.2	0.4861	0.4864	0.4868	0.4871	0.4875	0.4878	0.4881	0.4884	0.4887	0.4890
2.3	0.4893	0.4896	0.4898	0.4901	0.4904	0.4906	0.4909	0.4911	0.4913	0.4916
2.4	0.4918	0.4920	0.4922	0.4925	0.4927	0.4929	0.4931	0.4932	0.4934	0.4936
2.5	0.4938	0.4940	0.4941	0.4943	0.4945	0.4946	0.4948	0.4949	0.4951	0.4952
2.6	0.4953	0.4955	0.4956	0.4957	0.4959	0.4960	0.4961	0.4962	0.4963	0.4964
2.7	0.4965	0.4966	0.4967	0.4968	0.4969	0.4970	0.4971	0.4972	0.4973	0.4974
2.8	0.4974	0.4975	0.4976	0.4977	0.4977	0.4978	0.4979	0.4979	0.4980	0.4981
2.9	0.4981	0.4982	0.4982	0.4983	0.4984	0.4984	0.4985	0.4985	0.4986	0.4986
3.0	0.4987	0.4987	0.4987	0.4988	0.4988	0.4989	0.4989	0.4989	0.4990	0.4990
3.1	0.4990	0.4991	0.4991	0.4991	0.4992	0.4992	0.4992	0.4992	0.4993	0.4993
3.2	0.4993	0.4993	0.4994	0.4994	0.4994	0.4994	0.4994	0.4995	0.4995	0.4995
3.3	0.4995	0.4995	0.4995	0.4996	0.4996	0.4996	0.4996	0.4996	0.4996	0.4997
3.4	0.4997	0.4997	0.4997	0.4997	0.4997	0.4997	0.4997	0.4997	0.4997	0.4998

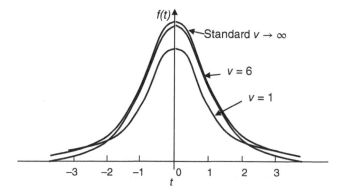

Figure 3.10 Probability distribution for the t-statistic.

theorem of statistics shows that if number of measurement n for each sample is very large, then the distribution of the mean values is normal and it has a standard deviation.

$$\sigma_\mu = \frac{\sigma}{\sqrt{n}} \tag{3.26}$$

where

σ_μ is standard deviation of the population mean
σ is the sample deviation of the population
n is the total number of measurements

So the confidence interval now becomes:

$$\bar{x} - z_{\frac{c}{2}}\sigma_\mu < \mu < \bar{x} + z_{\frac{c}{2}}\sigma_\mu \tag{3.27}$$

$$\bar{x} \pm \frac{z\sigma}{\sqrt{n}} \tag{3.28}$$

where

σ_μ is standard deviation of population mean
μ is the population mean
c is the confidence level
$\bar{x}$ is the sample mean

Unfortunately, σ is usually unknown, but a reasonable approximation to σ is S_x when n is large. This gives:

$$\bar{x} - z_{\frac{c}{2}}S_{\bar{x}} < \mu < \bar{x} + z_{\frac{c}{2}}S_{\bar{x}} \tag{3.29}$$

with standard deviation of sample mean $S_{\bar{x}}$ equal to:

$$S_{\bar{x}} = \frac{S_x}{\sqrt{n}} \tag{3.30}$$

Equations (3.32) and (3.33) provide confidence interval for the sample mean $\bar{x}$ when the standard deviation of the population, σ, is known or can be approximated by standard deviation of sample mean, $S_{\bar{x}}$. However, these equations are only valid when the sample is "large," i.e., $n > 30$. In most engineering experiments, n is usually less than that, and it would be more suitable to employ Student's t-distribution (Table 3.7).

Table for *t*-distribution

Table 3.7 Table of Student's *t*-distribution.

ν	0.10	0.05	0.025	0.01	0.005	0.001
1	3.078	6.314	12.706	31.821	63.657	318.313
2	1.886	2.920	4.303	6.965	9.925	22.327
3	1.638	2.353	3.182	4.541	5.841	10.215
4	1.533	2.132	2.776	3.747	4.604	7.173
5	1.476	2.015	2.571	3.365	4.032	5.893
6	1.440	1.943	2.447	3.143	3.707	5.208
7	1.415	1.895	2.365	2.998	3.499	4.782
8	1.397	1.860	2.306	2.896	3.355	4.499
9	1.383	1.833	2.262	2.821	3.250	4.296
10	1.372	1.812	2.228	2.764	3.169	4.143
11	1.363	1.796	2.201	2.718	3.106	4.024
12	1.356	1.782	2.179	2.681	3.055	3.929
13	1.350	1.771	2.160	2.650	3.012	3.852
14	1.345	1.761	2.145	2.624	2.977	3.787
15	1.341	1.753	2.131	2.602	2.947	3.733
16	1.337	1.746	2.120	2.583	2.921	3.686
17	1.333	1.740	2.110	2.567	2.898	3.646
18	1.330	1.734	2.101	2.552	2.878	3.610
19	1.328	1.729	2.093	2.539	2.861	3.579
20	1.325	1.725	2.086	2.528	2.845	3.552
21	1.323	1.721	2.080	2.518	2.831	3.527
22	1.321	1.717	2.074	2.508	2.819	3.505
23	1.319	1.714	2.069	2.500	2.807	3.485
24	1.318	1.711	2.064	2.492	2.797	3.467
25	1.316	1.708	2.060	2.485	2.787	3.450
26	1.315	1.706	2.056	2.479	2.779	3.435
27	1.314	1.703	2.052	2.473	2.771	3.421
28	1.313	1.701	2.048	2.467	2.763	3.408
29	1.311	1.699	2.045	2.462	2.756	3.396
30	1.310	1.697	2.042	2.457	2.750	3.385

Confidence Intervals for Small Number of Samples on the Unknown Mean ($n < 30$)

Also the confidence interval is given by:

$$\bar{x} - t_{\frac{\alpha}{2},\nu} S_{\bar{x}} < \mu < \bar{x} + t_{\frac{\alpha}{2},\nu} S_{\bar{x}} \tag{3.31}$$

$$\bar{x} \pm \frac{ts}{\sqrt{n}} \tag{3.32}$$

where:

$S_{\bar{x}}$ is the standard deviation of the sample mean
c is the confidence level
$\alpha = 1 - c$ is the level of significance
n is the sample size
$v = n - 1$ are the degrees of freedom

Example: If twelve values in a sample have a sample mean $\bar{x} = 1.009$ and a standard deviation $S_x = 0.04178$, what is the confidence interval for the true mean value μ with confidence level= 95%?

- Degree of freedom $v = 12 - 1 = 11$
- Level of significance $\alpha = 1 - 0.95 = 0.05$.

Following table 3.8 is an extracted part of the table 3.7:

Table 3.8 Excerpt from Student's t distribution table.

v	$t_{0.10}$	$t_{0.05}$	$t_{0.025}$	$t_{0.01}$	$t_{0.005}$	v
10	1.372	1.812	2.228	2.764	3.169	10
11	1.363	1.796	2.201	2.718	3.106	11
12	1.356	1.782	2.179	2.681	3.055	12

From the table, value for $t_{0.025,11} = 2.201$, therefore the confidence interval is:

$$\bar{x} - 2.201\frac{S_x}{\sqrt{12}} < \mu < \bar{x} + 2.201\frac{S_x}{\sqrt{12}}$$

$$0.98245 < \mu < 1.03555$$

Multiple Choice Questions of this Chapter

Multiple Choice Questions are given for each chapter with solutions in an online extension of this book. Please use link: www.wiley.com\go\mekid\metrologyandinstrumentation\

4

Errors and their Sources

"If you wanna master the error, look for its source!"

INTRODUCTION

Metrology's measurement error is an important subset. Manufacturing of parts or artifacts will always result in dimensional lengths that are within or outside of the specified tolerance, so it is necessary to measure the current length and determine its true measurement. The measurement process, on the other hand, is always subject to some degree of uncertainty. Measurements are not exact, and they should be reported as a range of values rather than a single value. Several examples can be found in our daily lives. If a manufactured part is given to two technicians to be dimensioned, the results will be slightly different. Furthermore, if you measure the speed of your car using the in-car speed meter, the result may differ slightly from the police speed gun reading outside. In this case, we changed the instrument, introducing a new type of error.

This chapter will define error and its various possible sources, how error propagates in measurement, errors associated with motion, error classification, and error elimination. An estimation of error or uncertainty analysis is a tool for determining the performance capability of machine tools and highlighting potential areas for performance and cost improvement. Machine designers are keenly interested in the implementation of this process. Aspects of measurement characterization used in this book will be defined later.

4.1 Definition of the Error and Their Types

The error is defined as the amount by which a measured value deviates from its true value. The error is inextricably linked to the accuracy. It is attributed to three major causes: human measurement errors, the measuring method, and the instrument. The errors are divided into three categories: gross, random, and systematic.

- Gross errors are a consequence of human mistakes: wrong measurement readings and recording incorrect data. Readings should be taken with care, and several readings of the same quantity should be taken and recorded.

– Random errors are caused by unknown changes in the experiment and hence, are unpredictable. They are related to the precision of results. Statistical analysis and averaging may be used to improve the precision of the best estimate value. The magnitude of such errors may be assessed from the results of repeated measurements and reduced with a high number of observations.

– Systematic errors occur consistently at every measurement. As a result, they are predictable and consistent, and they cannot be discovered solely by examining the results. Systematic errors are also the main reason why providing good accuracy is much more expensive than providing good precision or repeatability. Unlike random errors, systematic errors cannot be reduced by increasing the number of observations.

4.1.1 Systematic Errors

A systematic error occurs consistently in only one direction each time the measurement is performed, that is, the value of the measurement is either greater or lower than the true value. This is due to the defects in the instrumentation or the method of measurement.

Examples of conditions that allow systematic errors are:

1) Measuring a distance using the worn end of a meter stick;
2) Using a non-calibrated or bad calibrated instrument;
3) Using a measuring tape that is stretched resulting in the measuring results always be lower than the true value;
4) Neglecting the effects of air index variation in laser interferometry;
5) Any measuring instrument or measurand having variable effects, such as friction and air resistance, will result in a systematic shift of measurement outcomes.

In practice, predicting the sources and the magnitude of systematic errors is a difficult task, but whenever possible, attempts should be made to identify them and quantify their effect. It is always recommended for the sake of true measurement and minimum disturbance of experiments to be careful to eliminate as many of the systematic and random errors as possible. Proper calibration and adjustment of the equipment will help reduce the systematic errors leaving only the accidental and human errors to cause any spread in the data. Statistical methods typically allow for the reduction of random errors but not always below the limit of the measuring instruments' precision.

The following are some parameters to check:

i) **Instrument drift:** Some electronic instruments may have readings that drift over time. This can be quantified, and it must be considered whether it is a significant source of error.

ii) **Hysteresis and lag time:** Hysteresis can appear either in the system under inspection or within some add-ons to the instrument. It is within the instrument as an embedded hysteresis through initial measurement tests and needs to be differentiated from measuring devices that may have low response time when measuring and storing data.

iii) **Missing factor:** The measuring operation should account for all possible factors involved in the experiment except the independent variable that is being analyzed. In laser interferometry, for example, considering the effects of temperature, pressure, or humidity on the laser beam becomes important. Hence, corrections are sometimes applied to account for an error that was not detected.

iv) **Failure to calibrate or to check the zero of instrument:** The calibration of instruments used for measurements should be carried out prior to measurements. Accuracy could be checked with another calibrated instrument if there is no calibration standard for the instrument in use. The instrument's zero must be checked to ensure that it is the true zero and used for initial readings.

4.1.2 Random Errors

Random errors may result from human and inadvertent errors. Changing experimental conditions are the cause of the inadvertent errors and are beyond the control of the user. Examples are:

1) High pressure applied on the caliper jaw by the user;
2) Temperature or humidity effect of the measurement instrument.

Human errors may involve miscalculations in analyzing data, incorrect reading of an instrument, and possibly selecting some readings out of others without clear rules. It is impossible to quantify random errors precisely because the magnitude of the random errors and their effect on the experimental values varies throughout the measurement's repetition. Statistical methods are used in this case to estimate the random errors in the measurement. As a result, the following definitions are required:

  i) **Resolution of the instrument:** All instruments have finite precision that limits the ability to resolve small measurement differences.
 ii) **Physical variations:** Statistical measurements over the entire range are important to detect any variation that may be important to investigate locally.
iii) **Environmental factors**: Any inspection requires the experiment to be protected from environmental changes that could affect the measurement. Temperature, pressure, humidity, vibrations, drafts, electronic noise, and other effects from nearby devices are examples of these parameters.
 iv) **Parallax factor**: It can occur whenever there is some distance between the measuring scale and the indicator used to obtain a measurement. The reading may introduce some errors.
  v) **User errors:** These are errors incorrectly introduced for many reasons but mainly related to the user skills and aptitude to properly carry out measurement.

4.1.3 Components of Motion Error Assessment

Any mechanical system is designed to satisfy specifications including a number of functions characterized by defined accuracy. The error of the mechanical system is usually transferred to the product manufactured if the machine is a CNC machine-tool or to a key motion performed by the machine, for example, a telescope, radar, and robot. The following parameters will be normally assessed for any mechanical system:

a) **Manufactured product:** This includes:

   i) Dimensional precision;
  ii) Angular precision;
 iii) Form precision;
 iv) Surface roughness.

b) **Machine motion**: This includes:

   i) Kinematic precision, motion per axis, such as positioning, straightness, angular position.
  ii) Surface layer alteration in 3D space, for example.

4.2 Measurement Characteristics

4.2.1 Characterization of the Measurement

A number of criteria defined below characterize a measurement. Hence, it becomes important to understand the definition and related quantification of each aspect. These terms are frequently used interchangeably, but they have distinct meanings according to international standards. The definitions are based on the International Vocabulary of Basic and General Metrology Terms (VIM) or BS 5725. For example, it is critical to distinguish between high resolution and high accuracy. Accuracy is much more difficult and costly to achieve than resolution, and having one does not always guarantee the other.

Accuracy
Accuracy is defined as the closeness of agreement between a test and the accepted reference value.

a) Accuracy of Measurement The accuracy of measurement is the closeness of agreement between a quantity value obtained by measurement and the true value of the measurand (quantity intended to be measured). It is important to note that:

i) Accuracy cannot be expressed as a numerical value;
ii) Accuracy is inversely related to both systematic error and random error;
iii) The term "accuracy of measurement" should not be used for trueness of measurement and the term "measurement precision" should not be used for "accuracy of measurement."

b) Accuracy of a Measuring System It is the ability of a measuring system to provide a quantity value measured close to the true value of a measurand. It is important to note that:

i) Accuracy is greater when the quantity value is closer to the true value;
ii) The term "precision" should not be used for "accuracy";
iii) This concept is related to accuracy of measurement.

BS 5725 [1] uses the two terms "trueness" and "precision" to describe the accuracy of a measurement method. "Trueness" refers to the closeness of agreement between the arithmetic mean of a large number of test results and the true or accepted reference value. "Precision" refers to the closeness of agreement between test results.

Precision of Measurement
The precision is the closeness of agreement between quantity values obtained by replicate measurements of a quantity, under specified conditions. The measurement precision is usually expressed numerically by measures of imprecision, such as standard deviation, variance, or coefficient of variation under the specified conditions of measurement. Usually, when the term "accuracy" is applied to a set of test results, it will involve a combination of random components and a common systematic error or bias component. Figure 4.1 shows the various situations of the combined trueness and precision.

4.2.2 Resolution, Error Uncertainty, and Repeatability

a) Resolution
Resolution is defined as the error of mobility or the smallest generable movement. In servo-control, it is defined as the smallest value, either digital or analog, that the sensor can indicate, which is commonly referred to as the noise level (Fig.4.2).

i) Resolution of a Measuring System It is the smallest change in the value of a quantity being measured by a measuring system that causes a perceptible change in the corresponding indication. It is noteworthy that the resolution of a measuring system may depend on, for example, noise (internal or external) or friction. It may also depend on the value of the quantity being measured.

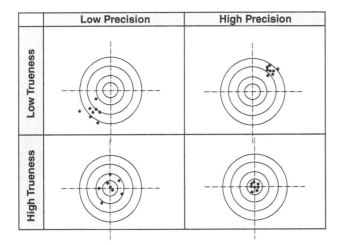

Figure 4.1 Example of measurements.

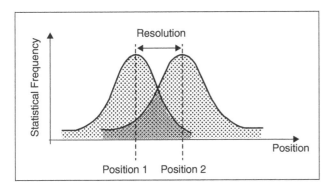

Figure 4.2 Graphical representation of a resolution.

ii) Resolution of a Displaying Device In the case of a displaying device, the resolution is the smallest difference between indications or dials that can be meaningfully distinguished. Examples are:

- A ruler has a resolution of one division, the smallest unit, usually 1 mm.
- Numerically, the resolution of a controller having N bits can be determined as the smallest recognized step by the controller over the intended stroke distance defined by Eq. 4.1. For a 16 bits controller used to cover a stroke of 100 mm, the minimum displacement resolved by this controller will be: $100/2^{16}$ mm $= 1.5$ μm.

$$step = \frac{distance}{2^N} \tag{4.1}$$

b) Measurement Errors

i) Error of Measurement The error of measurement is the deviation of quantity value obtained by the measurement and the true value of the measurand (feature being measured). It is necessary to distinguish "error of measurement" from "relative error of measurement".

c) Error Sources
The following are the error sources, which may be the cause of the errors:

1) Calibration Errors Several sources of error could easily affect the measurement system during the process of calibration:

– Experience in the calibration method;
– The bias and precision errors of the reference used in the calibration;
– The way the reference is used.

ii) Data Acquisition Errors Data acquisition errors are errors caused by the act of recording measurement data, such as power settings, environmental conditions, and sensor locations.

iii) Data Reduction Errors These are the errors resulting from the curve fits and correlations with their associated unknowns.

c) Uncertainty
The uncertainty is the parameter that characterizes the dispersion of the quantity values that are being attributed to a measurand, based on the information used. The uncertainty can be propagated through a relationship equation of parameters to a result. The following are examples:

– Surface area uncertainty derived from measured diameter.
– Normal stress uncertainty derived from force and cross-sectional area measurements, for instance, $\sigma = f(F, A)$.

d) Repeatability
The repeatability is the dispersion of readings or the bandwidth of uncertainty (Figure 4.3) obtained by repeatedly positioning a worktable to a single position. It is also the difference between two attempts to move the carriage to the same position.

i) Repeatability of Measurement The repeatability of measurement is a condition of measurement in a set of measurements under the same measurement procedure, same operator, same measuring system, same operating conditions, same location, and replicated measurements over a short period. If not repeatable, then there is an issue to be discussed.

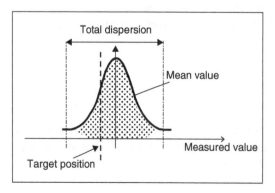

Figure 4.3   Repeatability.

ii) Measurement Repeatability (Repeatability)   The measurement repeatability is the measurement precision under repeatability conditions of measurement, and it is expressed as standard deviation. The conditions of performed measurement are as follows:

– Measured with the same method,
– On identical test items,
– In the same laboratory,
– By the same operator,
– Using the same equipment,
– Within short intervals of time (in particular the equipment should not be recalibrated between the measurements).

The repeatability depends on:

– Resolution;
– Drifts of displacement transducer, thermal, mechanical and electronics;
– Deformation of the machine structure;
– Combination of limiting friction of guide ways and elastic compliance of actuators.

The repeatability does not include hysteresis or drifts and can be estimated in either a uni-directional or bi-directional way depending on the application.

4.2.3   Model of Measurement

The measured value $x_{measured}$ from an instrument must be identical to the true value x_{true} theoretically:

$$x_{true} = x_{measured} \tag{4.2}$$

However, these two values are not identical in general. The deviation from the true value is the accuracy figure for an instrument, or more generally, described as "*error*".

$$x_{true} = x_{measured} + Error. \tag{4.3}$$

Note that in most textbooks, both terms "*error*" and "*uncertainty*" are used interchangeably, but the term "*error*" will be chosen here.

The term "error" can refer to a variety of errors, but the majority of them can be classified as *systematic errors*, E_s, or *random errors*, E_r. If estimates for both random and systematic uncertainties are known and the method for combining random and systematic uncertainties are not completely clear, it is stated as (Fig. 4.4, Fig. 4.5):

$$x_{true} = x_{measured} \pm E_s \pm E_r \tag{4.4}$$

where:

x_{true}: The true value of measurements
E_s: The systematic error, or component of laboratory bias
E_r: The random error under repeatability conditions.

For example, the error of measurement could be defined as: $E_s + E_r$. (Fig. 4.6)
The error depends on the following characteristics:

1) Resolution of the sensor;
2) Positioning repeatability;
3) Static and dynamic stiffness of machine element;
4) Abbé offset error.

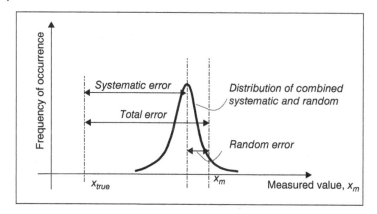

Figure 4.4 Systematic error is larger than the typical random error.

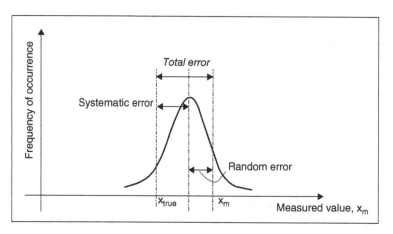

Figure 4.5 Typical random error is larger than the systematic error.

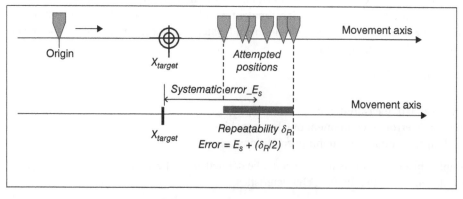

Figure 4.6 Example of error on linear movement.

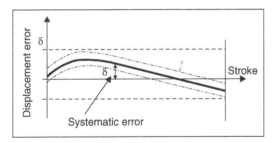

Figure 4.7 Total error of a linear slide. One-dimensional uncertainty.

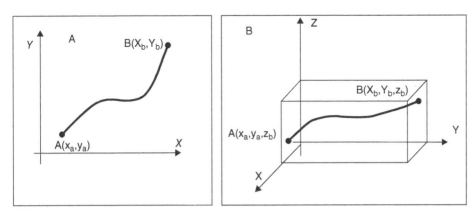

Figure 4.8 a) Planar error, b) Volumetric error.

In the case of a displacement of a linear slide along a distance D, Figure 4.7 shows the displacement error and systematic error. Since all objects could have six degrees of freedom in 3D space, the error could be defined in a planar or volumetric movement as shown in Figure 4.8.

4.3 Propagation of Errors

When a quantity Q depends on a couple of other quantities a, b, c, each one with its uncertainty δa, δb, δc; the uncertainty of Q will depend on the uncertainty of these components, meaning that these uncertainties will propagate to the uncertainty of Q. The related rules are summarized hereafter.

a) Algebraic Addition
If Q has an algebraic addition of components, such as a, b, and c as in Eq. (4.5), then the uncertainty of Q is expressed in Eq. (4.6).

$$Q = a + b - c \tag{4.5}$$

$$\delta Q = \sqrt{(\delta a)^2 + (\delta b)^2 + (\delta c)^2} \tag{4.6}$$

b) Product or Division
If Q is composed in product and division of components, such as a, b, c, and d as in Eq. (4.7), then the uncertainty of Q is expressed in Eq. (4.8).

$$Q = \frac{ab}{cd} \tag{4.7}$$

$$\frac{\delta Q}{|Q|} = \sqrt{\left(\frac{\delta a}{a}\right)^2 + \left(\frac{\delta b}{b}\right)^2 + \left(\frac{\delta c}{c}\right)^2 + \left(\frac{\delta d}{d}\right)^2} \tag{4.8}$$

Example 4.1 The volume of a box having the following dimensions: L = 2 ± 0.01 m, h = 3 ± 0.01 m, and t = 5 ± 0.01 m is

$$V = K \times H \times t = 30 \text{ m,}$$

$$\delta V = \sqrt{(0.01)^2 + (0.01)^2 + (0.01)^2} = 0.017.$$

As a result, the volume is expressed as: V = 30 ± 0.017 m or, rounded up, V = 30 ± 0.02 m.

c) Power

If Q is expressed as in Eq. (4.9) with an exact number n, then the uncertainty is given in Eq. (4.10).

$$Q = x^n, \tag{4.9}$$

$$\frac{\delta Q}{|Q|} = |n| \frac{\delta x}{|x|}. \tag{4.10}$$

Example 4.2 The period of oscillation is measured to be T = 0.5 ± 0.02 s; what is the uncertainty in the frequency?
The frequency is defined by: $f = 1/T = (T)^{-1}$

$$n = -1,$$

$$f = 1/T = 2.0 \text{ Hz.}$$

Applying Eq. (4.10), $\delta f = (1)(0.02/0.5)(2) = 0.08$ Hz
Hence, the frequency can be written as: $f = 2.0 \pm 0.08$ Hz.

d) Generalized Technique: First-Order Average and Uncertainty of Measurement

As an indirect method, the quantity to be measured and the uncertainty assessed through its components can be carried out using the equation of the measurand. The expression of the measurand is treated as follows, taking into account the first-order truncation of the Taylor's expansion (Eq. 4.11) and the application of the average definition (Eq. 4.12).

The measurand $g(X)$ is measured indirectly and determined from the quantity X. The Taylor's series expansion gives:

$$g(X) = \underbrace{\frac{(X-a)^0}{0!} \cdot g(a)}_{Zero-order} + \underbrace{\frac{(X-a)^1}{1!} \cdot \frac{\partial g}{\partial X}(a)}_{first-order} + \underbrace{\frac{(X-a)^2}{2!} \cdot \frac{\partial^2 g}{\partial X^2}(a)}_{second-order} + \dots + \underbrace{\frac{(X-a)^n}{n!} \cdot \frac{\partial^n g}{\partial X^n}(a)}_{n^{th}-order} + R_n \tag{4.11}$$

where a is a constant about which the expansion is carried out. R_n is a reminder term known as the Lagrange remainder, defined by [2] as:

$$R_n = \int \dots \int_{xo}^{x} g^{n+1}(X)(dX)^{n+1}.$$

With respect to the current investigation, it seems obvious that the expansion should be carried out about the mean (i.e., central value). The average of measurement is obtained through the expectation value defined by the first statistical moment as:

$$x = E(X) = \int_{-\infty}^{\infty} x \cdot f_X(x) \cdot dx, \tag{4.12}$$

where $f_X(x)$ is the probability density function (PDF).

The standard deviation is defined as the square root of the variance, which is also known as the second statistical moment:

$$Var(X) = E(X - E(X))^2 = \int_{\infty}^{-\infty} (x - \mu_x)^2 \cdot f_X(x) \cdot dx, \tag{4.13}$$

where $\mu_x \equiv E(X)$. It also expresses the uncertainty of measurement $u(x)$ defined in GUM [3], and the standard deviation σ_x; hence:

$$Var(X) = \sigma_x^2 = u^2(x), \tag{4.14}$$

Evaluating the expectation on either side of (Eq. 4.11) truncated to the first order will give:

$$y = g(x), \tag{4.15}$$

where x is the best estimate of measurement (i.e., average of measurement).

The combined variance and standard uncertainty of $g(X)$ are calculated by subtracting $E(g(X))$ from either side of equation (4.11) truncated to the first order, squaring both sides, and writing the expectation from either side to finally obtain in terms of uncertainty:

$$u_c^2(y) = \left(\frac{\partial g}{\partial x}\right)^2 \cdot u^2(x). \tag{4.16}$$

Hence,

$$u_c(y) = \left(\frac{\partial g}{\partial x}\right) \cdot u(x) \tag{4.17}$$

or

$$\delta Q = \left|\frac{dQ}{dx}\right| \cdot \delta x. \tag{4.18}$$

Example 4.3 If $Q = A \cdot x$, the uncertainty $\delta Q = |A| \cdot \delta x$ similar to Eq. (4.5).

Example 4.4 Find the uncertainty of $\sin(\alpha)$ in radians if the angle is measured in degrees with uncertainty of $1°$, $\alpha = 15 \pm 1°$
$\sin(\alpha) = 0.258819$
The uncertainty of $\sin(\alpha)$ is:

$$\delta(\sin(\alpha)) = \left|\frac{d}{d\alpha} \sin(\alpha)\right| \cdot \delta\alpha$$
$$= |\cos(\alpha)| \cdot \delta\alpha$$

$$\delta\alpha = 1° = 0.0174532\,rad$$

$$\delta(\sin(\alpha)) = |\cos(15)| \times 0.0174532 = 0.9659258 \times 0.0174532 = 0.0168584.$$

Hence, $\sin(\alpha) = 0.258819 \pm 0.0168584$

Example 4.5 The volume of a cube is determined by $V_c = x^3$, and the length x has an uncertainty of δx. Establish the equation that will be used to calculate the uncertainty of the volume.

Based on Eq. 4.17:

$$\delta V_c = 3x^2 \delta x$$

e) Function with two variables or more

Given a functional relationship between several measured variables (x, y), $Q = f(x, y)$, the uncertainty in Q if the uncertainties in x, y are as follows:

The variance in Q:

$$\sigma_Q^2 = \sigma_x^2 \left(\frac{\partial Q}{\partial x}\right)^2 + \sigma_y^2 \left(\frac{\partial Q}{\partial y}\right)^2 + 2\sigma_{xy} \left(\frac{\partial Q}{\partial x}\right) \left(\frac{\partial Q}{\partial y}\right) \tag{4.19}$$

If the variables x and y are uncorrelated, $\sigma_{xy} = 0$; therefore:

$$\sigma_Q^2 = \sigma_x^2 \left(\frac{\partial Q}{\partial x}\right)^2 + \sigma_y^2 \left(\frac{\partial Q}{\partial y}\right)^2$$

where:

$$\sigma_Q = \sqrt{\sigma_x^2 \left(\frac{\partial Q}{\partial x}\right)^2 + \sigma_y^2 \left(\frac{\partial Q}{\partial y}\right)^2}. \tag{4.20}$$

For more variables $(x, y, ..., z)$ if independent (uncorrelated):

$$\delta Q = \sqrt{\left(\frac{\partial Q}{\partial x}\cdot\right)^2 \delta x^2 + \left(\frac{\partial Q}{\partial y}\right)^2 \delta y^2 + ... \left(\frac{\partial Q}{\partial z}\right)^2 \delta z^2}. \tag{4.21}$$

Q at the average values

If x and y have several measurements: $x(x_1, x_2, x_3, ..., x_n)$ and y $(y_1, y_2, y_3, ..., y_n)$, the average of x and y will be:

$$\mu_x = \frac{1}{n}\sum_1^n x_i \quad \text{and} \quad \mu_y = \frac{1}{n}\sum_1^n y_i. \tag{4.22}$$

The evaluation of Q at the average values: $Q \equiv f(\mu_x, \mu_y)$.

The expansion of Q_i about the average using a Taylor expansion (Eq. 4.11) is given as:

$$Q_i = f(\mu_x, \mu_y) + (x_i - \mu_x)\left(\frac{\partial Q}{\partial x}\right)\Big|_{\mu_x} + (y_i - \mu_y)\left(\frac{\partial Q}{\partial x}\right)\Big|_{\mu_y} + \textit{higher order terms} \tag{4.23}$$

If the measurements are close to the average values: $Q_i - Q = (x_i - \mu_x)\left(\frac{\partial Q}{\partial x}\right)\Big|_{\mu_x} + (y_i - \mu_y)\left(\frac{\partial Q}{\partial y}\right)\Big|_{\mu_y}$ and its variance is:

$$\sigma_Q^2 = \frac{1}{n}\sum_{i=1}^n (Q_i - Q)^2 \tag{4.24}$$

or:

$$\sigma_Q^2 = \sigma_x^2 \left(\frac{\partial Q}{\partial x}\right)^2_{\mu_x} + \sigma_y^2 \left(\frac{\partial Q}{\partial y}\right)^2_{\mu_y} \tag{4.25}$$

for uncorrelated variables.

Higher-order expressions for average and uncertainty are developed for two variables in [4].

Example 4.6 Calculate the variance in the power using error propagation: $P = I^2R$ with $I = 1.0 \pm 0.1$ A and $R = 10.0 \pm 1.0$ Ω.

Ans. The power is calculated; $P = 10.0$ W

$$
\begin{aligned}
\sigma_P^2 &= \sigma_I^2 \left(\frac{\partial P}{\partial I}\right)_{I=1}^2 + \sigma_R^2 \left(\frac{\partial P}{\partial R}\right)_{R=10}^2 \\
&= \sigma_I^2 (2IR)^2 + \sigma_R^2 (I^2)^2 \\
&= (0.1)^2 (2 \cdot 1 \cdot 10)^2 + (1)^2 (1^2)^2 = 5W^2.
\end{aligned}
$$

Hence, $P = 10.0 \pm 2.2$ W.

If the true value of P is 10.0 W, measured several times with an uncertainty of ± 2.2 W, from the previous statistics (e.g., normal), 68% of the measurements will lie in the range [7.8, 12.2] W.

Example 4.7 The variables x and y were measured for the quantity $Q = x^2y - xy^2$ as follows: $x = 3.0 \pm 0.1$ and $y = 2.0 \pm 0.1$. What is the value of Q and its uncertainty?

Ans. Q = 6.0 ± 0.9

Example 4.8 An object is placed at a distance p from a lens, and an image is formed at a distance q from the lens, the lens's focal length is expressed as follows:

$$
f = \frac{pq}{p + q}
$$

Using the general rule for error propagation, derive a formula for the uncertainty δf in terms of p, q, and their uncertainties. Calculate the quantity f and its uncertainty if $p = 10.1 \pm 0.3$ and $q = 4.0 \pm 0.5$.

4.4 Sources of Errors

Errors always exist to some degree of uncertainty resulting from measurement. It is possible to state that all measurements are inaccurate to some extent. Measurement error is the difference between the indicated and actual values of the measurand. The error could be expressed as an absolute error or on a relative scale.

It is critical to thoroughly investigate the measurement system errors that cause these uncertainties, as well as the meaning and interpretations of these errors and methods for reducing or avoiding them. Each component of the measuring system has sources of errors that can contribute to measurement error.

Error sources can be classified into several categories. They must, however, be identified in any application and budgeted to compensate for or correct errors.

4.4.1 Static Errors and Dynamic Errors

The errors of movement are classified into two categories: errors linked to static sources and errors linked to the dynamic behavior of the machine. It is brought to the attention of the designer that often these errors are linked to each other. Actually, the static errors sometimes generate the dynamic errors. These two types of errors affect the precision and the repeatability of the movement.

a) Quasi-Static Errors

The quasi-static errors appear at very low speeds and are found in:

i) The conceptual design that may induce the Abbé error, assembling of parts inducing errors from the tolerances and excess in tightened torques;
ii) Machining errors (planarity, parallelism, straightness, surface waviness, etc.);
iii) Load errors (own weight, external charges, bearing preloads, cutting forces, etc.);
iv) Errors due to reversible influences (thermal dilatations) and nonreversible influences (material aging);
v) Errors due to dimensional instability of materials;
vi) Metrology errors (sensor precision and Abbé errors).

b) Dynamic Errors

The dynamic errors are generated by the power supplied from external forces, which often are not controllable:

i) Rolling on surface waviness;
ii) Machining processes (e.g., cutting force);
iii) Noise source induced by control system components (motors, actuators) pressure pulsations of hydraulic or pneumatic supply circuits, fluid turbulence, and so forth.

Example: *Accuracy in Work Pieces*

The dimensional and form accuracies of a machined part are affected by various sources of errors: geometric and kinematic errors, thermal errors, load induced errors, and environmental errors, as shown in Figure 4.9.

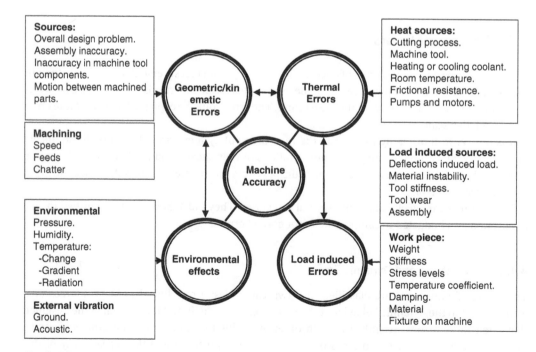

Figure 4.9 Various sources of errors affecting the accuracy of a work piece.

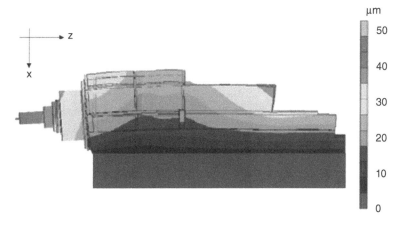

Figure 4.10 Typical thermal deformation of horizontal motor spindle.

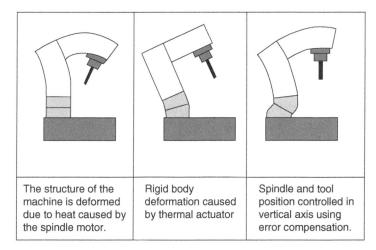

The structure of the machine is deformed due to heat caused by the spindle motor.	Rigid body deformation caused by thermal actuator	Spindle and tool position controlled in vertical axis using error compensation.

Figure 4.11 Thermal Errors affecting structure of a machine tool.

Thermal errors are thought to be the most difficult to compensate for, due to the variety of heating sources found in a machine tool, for example, and the selection of the appropriate temperature variables for a given model (Fig. 4.10). Structural deformation with error compensation is shown in Fig. 4.11.

c) Abbé Error

Dr. Ernst Abbé noted: "If errors in parallax are to be avoided, the measuring system must be placed coaxially with the axis along which displacement is to be measured on the work-piece." Abbé error is present in various mechanical systems that generate linear movement to varying degrees. The distance traveled from the source magnifies the angular errors. As a result, there is a significant impact on positioning.

In Figure 4.12, the following definitions are needed:

- ***Reference point (R):*** This point is defined as the effective point if no Abbé error occurred.
- ***Sensing point (S):*** This point is defined as the point where the measurement is done (where the sensor/probe is measuring.

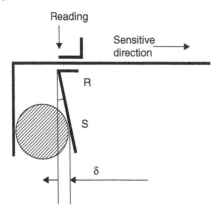

Reading

Sensitive direction

R

S

δ

Figure 4.12 Examples of Abbé errors in measurement using a caliper.

– *Abbé offset:* The distance from the sensing point to the reference point. The following are the actions that the designer must consider:
 1) Design with bearings and actuators near the activity of the machine;
 2) Measure near the source.

In Figure 4.19 a measurement error δ due to Abbe is obtained. More information is given in [5].

d) Noise

Noise is typically produced by the servo control when parts of the machine begin to move, as well as by the electronics installed on or near the machine. When measurements are taken over a wide range of frequencies, the influence of this noise can be seen. The spectrum in Figure 4.13 shows the various sources of noise and external excitations affecting the system.

The analysis of this measurement spectrum shows the following:

- Sources of noise are shown in this noise power spectrum.
- Peak at 50 Hz could be:

– main pick-up from the main supply on the sensor circuitry, or
– due to motion measurement.

- Noise at high frequencies could be due to electromagnetic interference.
- Single high frequencies could be due to processor clocks or external interference.
- Real analysis is limited to the functional bandwidth rather than the whole spectrum.

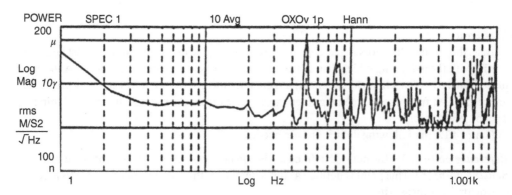

Figure 4.13 Measurement spectrum.

4.5   Error Budget

To predict the performance of the proposed design, the conventional error budget combines errors for each axis of motion or a related function in a machine. As a result, it is critical to comprehend the errors introduced during manufacturing and by the environment. Donaldson [6] has defined the error budget as "a system analysis tool, used for prediction and control of the total error of a system at the design stage for systems where accuracy is an important measure of performance. Given a system-error goal, an error budget can be used in a control mode to set individual subsystem error limits, while also making trade-offs that balance the level of difficulty among the subsystems. In a predictive mode, proposed subsystem designs can be assessed for error contributions, leading to a predicted overall system error." The "Guide to the Expression of Uncertainty in Measurement" [3] has methods to test the error budget. A root mean square (rms) value is determined for the set of individual error limits. The rms value is an estimate of the total machine error caused by the individual errors. This method is conservative and does not necessarily give a clear understanding of the effects of individual errors in systems that are more complex.

The method of error budget can be applied when designing virtual machines and evaluating the distribution of errors associated with a particular design before manufacturing the parts and assembling the machine. Manufacturing tolerances and assembly-based errors are taken into consideration. The level of contribution of each component on the total budget error in various configurations of the machine can be known and hence possibly adjusted to meet its error allocation and so that the budget error limit will not be exceeded.

A body in 3D space has six degrees of freedom (DOFs). Every axis of motion also has six DOFs. In the design of a very accurate motion in one axis, the objective would be to precisely control one or two of these DOF and to minimize the effect of the remaining DOFs. If only positioning is controlled in the x-axis, the other five, two straightness and three angles, will be the parasitic motion and be part of the x-axis performance.

In a planar motion XY or multi-axis motion, the unwanted parasitic movements generated by the DOFs of each axis combine statistically to create the system's error budget. These parasitic movements result from several sources, such as imperfections in the bearings, deflections due to loads and thermal distortions.

In the error budgeting, the objective is to allocate allowable values for each error source and select components such that the ability of each component to meet its error allocation is not exceeded.

4.5.1   Components of the Error Budget

All of the types of errors defined thus far can be included in the error budget and must be adequately budgeted. Budgeting necessitates a thorough understanding of the error sources and how they interact with one another. Any mistake will engender an unknown statistical error that cannot be compensated for.

For example, the contributions of the static and dynamic errors are estimated to be 65–70% and 30–35%, respectively. The possible sources of errors that affect the accuracy are as follows:

– Geometric errors of machine components and structure, e.g., cutting forces;
– Machine assembly, e.g., tool wear;
– Fixtures, e.g., controllers;
– Kinematic errors, e.g., actuators and sensors;
– Thermal distribution, e.g., instrumentation;
– Material instability, e.g., errors due to second-order phenomena;
– Structural stiffness.

4.5.2 Example of Error-Budget Table

Having all the errors identified in any mechanical system assessed can be processed in two steps; the first one is to fill a preliminary table with all existing errors, while the second is to properly combine them in the final error-budget table after an evaluation. Tables 4.1, 4.2, and 4.3 are examples of these tables.

a) Preliminary Table

Table 4.1 is an example of a preliminary table.

b) Budget Table

Tables 4.1, 4.2, and 4.3 are examples of budget tables.

Table 4.1 Preliminary error budget table.

Subsystem	Error name	Value	Unit	Range of motion [mm]	Method of measurement	Environment	Standard used
A_i	x-Positioning	5	µm	200		21°C	BS230
	y-Straightness	3		200	Laser interferometry		BS230
	z-Straightness	4		200			BS230

Table 4.2 Error-budget table.

Manipulators R1, R2 & R3	Range	Abbé error (after compensation)	Positioning	Resolution	Repeatability	24H drift [nm]
x-axis	20 mm	<1 µm	< 0.1 µm	1 nm	5 nm	
y-axis	20 mm	0.5 µm	< 0.1 µm	1 nm	5 nm	
z-axis	20 mm	0.5 µm	< 0.1 µm	1 nm	5 nm	
Rotary table, R0						
θ-axis	360°	n/a	5 arcsec		±2 arcsec	
x-axis	40 mm	n/a	1 µm	0.5 µm	±0.5 µm	
y-axis	40 mm	n/a	1 µm	0.5 µm	±0.5 µm	
z-axis	-	n/a	0.1 µm	-	-	
AFM, R4 & R5						
x-axis	110 µm	± 1 nm	<0.6 %	1 nm	-	1.5 nm
y-axis	110 µm	± 1 nm	<0.6 %	1 nm	-	1.5 nm
z-axis	20 µm	± 1 nm	90pm	1 nm	-	-
Manipulator/Rotary table						
X, Y, Z position			Locked within <1 µm	Manipulator position error <0.6 µm		
AFM/Rotary table						
X, Y, Z position			Locked within <1 µm	AFM position error <2 nm		

Table 4.3 Error-budget table.

Axis	Component	Geometric effect	Non-rigid effect
X axis	Pitch	−8 arcsec	10 arcsec
	Yaw	5 arcsec	neglect
	Roll	−17 arcsec	10 arcsec
	Linear positioning	55 μm	−40 μm
	Horizontal Straightness	3 μm	neglect
	Vertical Straightness	−9 μm	neglect
Y axis	Pitch	−8 arcsec	10 arcsec
	Yaw	5 arcsec	10 arcsec
	Roll	−17 arcsec	10 arcsec
	Linear positioning	55 μm	neglect
	Horizontal Straightness	3 μm	neglect
	Vertical Straightness	−9 μm	neglect
Z axis	Pitch	−8 arcsecs	neglect
	Yaw	–	neglect
	Roll	23 um	
	Linear positioning	55 μm	neglect
	Horizontal Straightness	3 μm	neglect
	Vertical Straightness	−9 μm	
	XY Squareness	−8 arcsecs	10 secarc
	XY Squareness	5 arcsecs	
	XZ Squareness	−17 arcsecs	
Total	21		

4.6 Error Elimination Techniques

Because errors in a machine tool or any other mechanical system vary continuously, there is a need to devise a method to either reduce to a bare minimum, avoid the errors, or better yet, compensate for them all to achieve zero error in the system. In the literature [7–9], the methods are referred to as error avoidance and error compensation.

4.6.1 Methods

Error Avoidance
The goal of the machine design process, part manufacturing, and assembly techniques is to ensure the fewest errors possible, which is referred to as error avoidance. Because of the high accuracy [10], the cost of this type of investigation may rise.

Error Compensation
This method is the ultimate after error avoidance and is much needed by various industries to secure machines that are more accurate, thereby ensuring high performance production. The prin-

ciple is that the errors involved in a machine's functional task are measured, possibly in-process, and appropriately compensated globally or as the task progresses.

Precalibrated Error Compensation

The precalibrated error is measured either before or after an assigned motion and is used to modify or calibrate that motion during subsequent operations. The assumption of repeatability is made here.

Active Error Compensation

The active error is observed during the operation and is used to correct the process as it runs. This technique implies that a high precision machine is not required but rather, a high precision controller capable of achieving precision motions with active compensation as the motion progresses in a low-grade machine. This is critical in today's industry to achieve high output at a low cost. As the active compensation operates instantly or better in real time, the ideal case would be to capture not only static errors but also dynamic errors varying over time, as well as the interaction between sources of errors. It is a continuous process of eliminating all dynamic errors including vibration, thermal, and cutting force–based errors.

Two approaches are being used to carry out this active compensation:

1) Parametric Error Measurement Approach

i) Identification or modeling of the error in question;
ii) Measurement and mapping of the various components of error that have been identified;
iii) Development of suitable control systems or networks to compensate for these errors.

2) Master Part Tracking Approach
The used methods for the identification of errors based on the type of error to be monitored are shown below:

i) Grid calibration method: This method is commonly used for geometric error modeling. It calibrates the error at discrete grid points of the working volume and interpolates the estimated error for the actual needed position.
ii) Error synthesis method: This method acquires the total error in terms of individual error components. It is commonly used for geometric and thermal error modeling.
iii) Designed artifact method: This method measures dimensions of especially designed objects instead of direct measurement of the errors. It is used to model local geometric and thermal errors.
iv) Metrology frame method: It is used to measure partial geometric and thermal errors online using optical systems mounted on the machine, thereby eliminating some off-line measurements.
v) Finite element method: This method is employed to estimate thermal errors through thermo-elastic deformation and heat transfer analysis of the machine structure. No experiments are involved in this method of error estimation.

To summarize the whole process of error compensation, here are the steps to be considered (also shown in Figure 4.14):

i) Machine calibration;
ii) Zeroing the machine axes;
iii) Identification of the linkage structure and derivation of an error synthesis model using homogeneous transformation matrices (HTMs);

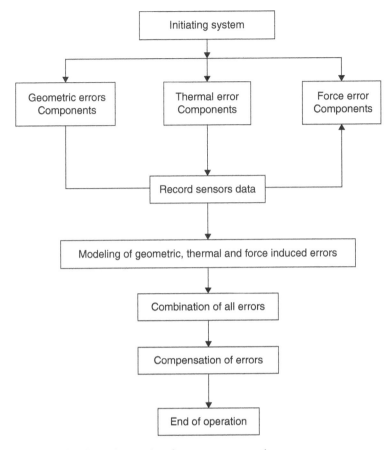

Figure 4.14 General procedure for error compensation.

iv) Sensors optimal location for data acquisition;
v) Measurement and record of each error component of the key moving stage and environmental conditions for each machine configuration and/or position;
vi) Definition of the error component model and combination with all existing models for all axes of the machine into one overall error model, and;
vii) Loading of the error compensation system in the machine controller.

4.7 Model of Errors in CNC Using HTM

Prior to modeling the error in 3D space in a machine, it is important to understand how the motion is combined in the three or more axes that exist in the machine, for example, CNC machine tool or telescope.

Kinematic Displacement of a Rigid Body

In a reference system (Fig. 4.15), a rigid body has its own coordinates with respect to that original coordinate system. R_0 is a reference system defined as R_0 $(0, i, j, k)$ with the unity vectors: $i = (1,0,0)_{R0}$, $j = (0,1,0)_{R0}$, and $k = (0,0,1)_{R0}$.

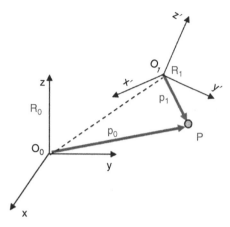

Figure 4.15 Transformation from original to current coordinates.

Any motion transformation of the coordinate system from R_0 to R_1 will define the position of P with respect to the frame R_0 as:

$$O_0P = O_0O_1 + [R_{R0 \to R1}] \cdot O_1P \tag{4.26}$$

Or

$$P_o = O_0O_1 + [R^0{}_1] \cdot P_1 \tag{4.27}$$

The inverse transformation can be obtained by premultiplying both sides by $[R^0{}_1]^T$:

$$P_1 = -[R^0{}_1]^T O_0O_1 + [R^0{}_1]^T \cdot P_o \tag{4.28}$$

$$P_1 = -[R^1{}_0] O_0O_1 + [R^1{}_0] \cdot P_o \tag{4.29}$$

We define $[R^0{}_1]$ as the rotation matrix of frame R_1 with respect to frame R_0, and the rotation from R_0 to R_1 is as follows:

The components of each vector unit are the direction cosines of the axes of frame (O_1, i_1, j_1, k_1) with respect to (O, i, j, k) in an orthonormal frame and are defined as:

$$i_1 = a_{11}i + a_{12}j + a_{13}k$$
$$j_1 = a_{21}i + a_{22}j + a_{23}k$$
$$k_1 = a_{31}i + a_{32}j + a_{33}k \tag{4.30}$$

The rotation matrix is extracted from the previous coordinate system and therefore expressed as:

$$R = \begin{bmatrix} a_{11} & a_{21} & a_{31} \\ a_{12} & a_{22} & a_{32} \\ a_{13} & a_{23} & a_{33} \end{bmatrix} \tag{4.31}$$

Where the vector units components are vertical. The matrix is orthogonal: $i_1 . j_1 = 0$, and so forth, and $R^T . R = I$. For a compact representation of the relationship between the coordinates of the same point in two different frames, we use a homogenous representation of a generic point P:

$$\mathbf{O_0P} = \{x_P, y_P, z_P, 1\}^T$$

$$O_0P = \begin{bmatrix} xp \\ yp \\ zp \\ 1 \end{bmatrix} . \tag{4.32}$$

The 4th component is the scale factor, i.e., 1 in this case.

The HTM of any transformed point with respect to the original coordinate system moved with combined translation and rotation can be written as:

$$
\begin{Bmatrix} x \\ y \\ z \\ 1 \end{Bmatrix} = \begin{bmatrix} a_{11} & a_{21} & a_{31} & r_{xo1} \\ a_{12} & a_{22} & a_{32} & r_{yo1} \\ a_{13} & a_{23} & a_{33} & r_{zo1} \\ 0 & 0 & 0 & 1 \end{bmatrix} \begin{Bmatrix} x_1 \\ y_1 \\ z_1 \\ 1 \end{Bmatrix}.
$$

(4.33)

Where r_x, r_y, r_z are coordinates of the origin O_1 in (O, x, y, z), and a_{ij} are the direction cosines of the (O_1, x_1, y_1, z_1) system axis with respect to (O, x, y, z).

The translation vector from the original coordinate to the current one is defined as:

$$
OO' = \sqrt{(r_{xo1})^2 + (r_{yo1})^2 + (r_{zo1})^2}.
$$

(4.34)

Basic Translation Matrices

For better understanding of the motion transformation, let us take one motion per axis and adapt the matrix shown before. Moving along single axes x, y, or z is shown below (Fig. 4.16):

$$
M_{Tx} = \begin{bmatrix} 1 & 0 & 0 & x \\ 0 & 1 & 0 & 0 \\ 0 & 0 & 1 & 0 \\ 0 & 0 & 0 & 1 \end{bmatrix} \quad M_{Ty} = \begin{bmatrix} 1 & 0 & 0 & 0 \\ 0 & 1 & 0 & y \\ 0 & 0 & 1 & 0 \\ 0 & 0 & 0 & 1 \end{bmatrix} \quad M_{Tz} = \begin{bmatrix} 1 & 0 & 0 & 0 \\ 0 & 1 & 0 & 0 \\ 0 & 0 & 1 & z \\ 0 & 0 & 0 & 1 \end{bmatrix}.
$$

(4.35)

Therefore, their combination into translations only gives only gives Eq.(n):

$$
M_T = \begin{bmatrix} 1 & 0 & 0 & x \\ 0 & 1 & 0 & y \\ 0 & 0 & 1 & z \\ 0 & 0 & 0 & 1 \end{bmatrix}
$$

Basic Rotation Matrices

Rotations matrices M around x, y, or z are given below (also shown in Fig. 4.17):

$$
M_{R/x} = \begin{bmatrix} 1 & 0 & 0 & 0 \\ 0 & \cos(\theta x) & -\sin(\theta x) & 0 \\ 0 & \sin(\theta x) & \cos(\theta x) & 0 \\ 0 & 0 & 0 & 1 \end{bmatrix} \quad M_{R/y} = \begin{bmatrix} \cos(\theta y) & 0 & \sin(\theta y) & 0 \\ 0 & 1 & 0 & 0 \\ -\sin(\theta y) & 0 & \cos(\theta y) & 0 \\ 0 & 0 & 0 & 1 \end{bmatrix}
$$

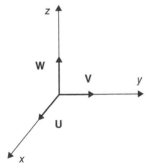

Figure 4.16 3D axis system with unit vectors.

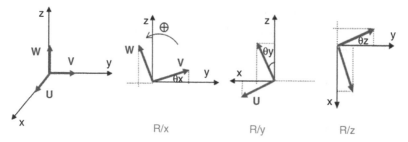

Figure 4.17 3D axis system with unit vectors.

$$M_{R/z} = \begin{bmatrix} \cos(\theta z) & -\sin(\theta z) & 0 & 0 \\ \sin(\theta z) & \cos(\theta z) & 0 & 0 \\ 0 & 0 & 1 & 0 \\ 0 & 0 & 0 & 1 \end{bmatrix} \tag{4.36}$$

$$R = R_z(\alpha)R_y(\beta)R_z(\gamma) = \overbrace{\begin{bmatrix} \cos\alpha & -\sin\alpha & 0 \\ \sin\alpha & \cos\alpha & 0 \\ 0 & 0 & 1 \end{bmatrix}}^{\text{yaw}} \overbrace{\begin{bmatrix} \cos\beta & 0 & \sin\beta \\ 0 & 1 & 0 \\ -\sin\beta & 0 & \cos\beta \end{bmatrix}}^{\text{pitch}} \overbrace{\begin{bmatrix} 1 & 0 & 0 \\ 0 & \cos\gamma & -\sin\gamma \\ 0 & \sin\gamma & \cos\gamma \end{bmatrix}}^{\text{roll}} \tag{4.37}$$

$\alpha = \theta x \cdot \beta = \theta y, \ and \ \gamma = \theta z.$

Note that the matrix multiplication is inverted and the product will result in:

$$R = \begin{bmatrix} \cos\alpha\cos\beta & \cos\alpha\sin\beta\sin\gamma - \sin\alpha\cos\gamma & \cos\alpha\sin\beta\cos\gamma + \sin\alpha\sin\gamma \\ \sin\alpha\cos\beta & \sin\alpha\sin\beta\sin\gamma + \cos\alpha\cos\gamma & \sin\alpha\sin\beta\cos\gamma - \cos\alpha\sin\gamma \\ -\sin\beta & \cos\beta\sin\gamma & \cos\beta\cos\gamma \end{bmatrix}. \tag{4.38}$$

The coordinates of the point P in (x, y, z) written in (O_1, x_1, y_1, z_1) can then be explicitly written as Eq. (4.33):

$$x_1 = a_{11}x + a_{12}y + a_{13}z + r_x$$
$$y_1 = a_{21}x + a_{22}y + a_{23}z + r_y$$
$$z_1 = a_{31}x + a_{32}y + a_{33}z + r_z.$$

Case of 3D Axis Machine

A machine tool has geometric errors that are generated by the structural elements of the machine tools. These errors affect the machine repeatability and kinematic accuracy. In a 3-axis CNC machine, 21 error components exist comprising 3 linear positioning errors, 6 straightness errors, 9 angular errors, and 3 squareness errors.

Over one axis, the angular errors, that is, pitch, roll, and yaw, are common sources of errors affecting positioning. Similarly, these angles on the spindle will affect its tool tip and hence error in machining. The tool path or contouring accuracy of the work piece is degraded by the large straightness and hence the machine performance. Usually, the wear and sudden impacts of the guide ways can be the main cause of damage triggering these errors. These can happen sometimes from the structure of the machine or the foundation that allows a bowing effect when the slide moves. Also, out-of-squareness between axes can seriously degrade machine tool performance

Error component	Linear displacement error			Angular error roll, pitch, yaw		
x-axis	$\delta x(x)$	$\delta y(x)$	$\delta z(x)$	$\varepsilon x(x)$	$\varepsilon y(x)$	$\varepsilon z(x)$
y-axis	$\delta x(y)$	$\delta y(y)$	$\delta z(y)$	$\varepsilon x(y)$	$\varepsilon y(y)$	$\varepsilon z(y)$
z-axis	$\delta x(z)$	$\delta y(z)$	$\delta z(z)$	$\varepsilon x(z)$	$\varepsilon y(z)$	$\varepsilon z(z)$
Squareness error	$\Phi x(y); \Phi x(z); \Phi y(z)$					

Figure 4.18 Geometric error components for a 3-axis machine.

by producing dimensional errors in produced parts. Figure 4.18 summarizes all errors described here:

i) Nominal positions of every point are shown by x, y, z;
ii) Positional errors of each point along x, y, and z directions are $\delta x(x); \delta y(y)$;
iii) $\delta z(z)$, where the first subscript refers to error direction and the second refers to moving direction.

In the same way:

i) $\delta y(x), \delta z(x), \delta x(y), \delta z(y), \delta x(z), \delta y(z)$ are straightness errors;
ii) $\varepsilon x(x), \varepsilon y(x), \varepsilon z(x), \varepsilon x(y), \varepsilon y(y), \varepsilon z(y), \varepsilon x(z), \varepsilon y(z), \varepsilon z(z)$ are angular errors; and
iii) $\phi x(y), \phi x(z), \phi y(z)$ are squareness errors.

The squareness error $\phi x(y)$ between X-axis and Y-axis is shown in Figure 4.26. The squareness errors can be calculated by applying the least square curve fit to obtain the mean straight line as shown in Figure 4.19. The squareness error can be expressed as Eq. (4.40).

$$S_{xy} = (\pi/2 - \phi_{target}) + \theta_x + \theta_y, \tag{4.39}$$

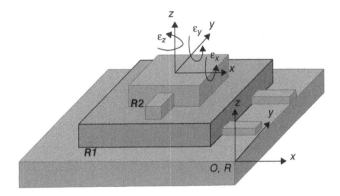

Figure 4.19 Six degrees of freedom error motion of 3D stage.

where θx and θy are produced angles between reference axis and the mean straight line, and φ_{target} is the edge angle of target.

The transformation from stage 1 to stage 2 is then written as $[^1T_2]_{actual}$:

$$[^1T_2]_{actual} = \begin{bmatrix} 1 & -\varepsilon_{z2}(x) & \varepsilon_{y2}(x) & x + \delta_{x2}(x) + a_2 \\ \varepsilon_{z2}(x) & 1 & -\varepsilon_{x2}(x) & \delta_{y2}(x) + b_2 \\ -\varepsilon_{y2}(x) & \varepsilon_{x2}(x) & 1 & \delta_{z2}(x) + c_2 \\ 0 & 0 & 0 & 1 \end{bmatrix} \tag{4.40}$$

where:

$\varepsilon_{x2}(x)$: Roll error of X-axis.
$\varepsilon_{y2}(x)$: Pitch error of X-axis.
$\varepsilon_{z2}(x)$: Yaw error of X-axis.
a_2: Constant offset in X direction between $R1$ and $R2$.
b_2: Constant offset in Y direction between $R1$ and $R2$.
c_2: Constant offset in Z direction between $R1$ and $R2$.
$\delta_{x2}(x)$:Linear displacement error of X axis.

$$\delta_{y2}(x) = \delta'_{y2}(x) + \alpha_{xy}X \tag{4.41}$$

$$\delta_{z2}(x) = \delta'_{z2}(x) + \alpha_{xz}X \tag{4.42}$$

Where:

$\delta'_{y2}(x)Y$: straightness of X-axis as it moves in X direction.
$\delta'_{z2}(x)Z$: straightness of X-axis as it moves in X direction.
α_{xy}: Squareness error between X- and Y-axis.
α_{xz}: Squareness error between X- and Z-axis.

^{x}Nominal X axis position which amplifies α_{xy} to yield and Abbé error in Y direction [8] in Eq.(4.41).

^{x}Nominal X axis position which amplifies α_{xz} to yield and Abbé error in Z direction in Eq. (4.42).

$$[^RT_1]_{actual} = \begin{bmatrix} 1 & -\varepsilon_{z1}(y) & -\varepsilon_{y1}(y) & \delta_{x1}(y) + a_1 \\ \varepsilon_{z1}(y) & 1 & -\varepsilon_{x1}(y) & y + \delta_{y1}(y) + b_1 \\ -\varepsilon_{y1}(y) & \varepsilon_{x1}(y) & 1 & \delta_{z1}(y) + c_1 \\ 0 & 0 & 0 & 1 \end{bmatrix} \tag{4.43}$$

$$[^RT_3]_{actual} = \begin{bmatrix} 1 & -\varepsilon_{z3}(z) & \varepsilon_{y3}(z) & \delta_{x3}(z) + a_3 \\ \varepsilon_{z3}(z) & 1 & -\varepsilon_{x3}(z) & \delta_{y3}(z) + b_3 \\ -\varepsilon_{y3}(z) & \varepsilon_{x3}(z) & 1 & z + \delta_{z3}(z) + c_3 \\ 0 & 0 & 0 & 1 \end{bmatrix} \tag{4.44}$$

Where:

$\varepsilon_{y1}(y)$: Roll error of Y-axis;
$\varepsilon_{x1}(y)$: Pitch error of Y-axis;
$\varepsilon_{z1}(y)$: Yaw error of Y-axis;
a_1: Constant offset in X direction between R and $R1$;

b_1: Constant offset in Y direction between R and $R1$;

c_1: Constant offset in Z direction between R and $R1$;

$\delta_{y1}(y)$: Linear displacement error of Y axis;

$$\delta_{x1}(y) = \delta'x_1(y) + \alpha_{xy}y \tag{4.45}$$

$$\delta_{z1}(y) = \delta'z_1(y) + \alpha_{zy}y \tag{4.46}$$

$$^{y}T_z = \begin{bmatrix} 1 & -\varepsilon_{zz}(z) & \varepsilon_{yz}(z) & \delta_{xz}(z) + a_z \\ \varepsilon_{zz}(z) & 1 & -\varepsilon_{xz}(z) & \delta_{yz}(z) + b_z \\ -\varepsilon_{yz}(z) & \varepsilon_{xz}(z) & 1 & z + \delta_{zz}(z) + c_z \\ 0 & 0 & 0 & 1 \end{bmatrix} \tag{4.47}$$

Where:

$$^{2}T_{Work} = \begin{bmatrix} W_x \\ W_y \\ W_z \\ 1 \end{bmatrix} \tag{4.48}$$

$$^{3}T_{Work} = \begin{bmatrix} T_x \\ T_y \\ T_z \\ 1 \end{bmatrix} \tag{4.49}$$

where W_x, W_y, W_z are the elements of work coordinates on the table and T_x, T_y, and T_z are the elements of tool coordinators for X, Y, and Z directions, respectively, as shown in Figures 4.58 and 4.59. The remaining effect is only squareness error between moving axes according to Eq. (4.40), which should be added to the summation as Eq. (4.50).

$$\begin{bmatrix} e_x \\ e_y \\ e_z \end{bmatrix} = ^{R}\begin{bmatrix} e_x \\ e_y \\ e_z \end{bmatrix}_{Tool} + ^{R}\begin{bmatrix} e_x \\ e_y \\ e_z \end{bmatrix}_{Work} + \begin{bmatrix} e_x \\ e_y \\ e_z \end{bmatrix}_{Squareness} \tag{4.50}$$

If the tool error and the work error are also measured from the R base, then the same size errors mean precise product because the tool and the work part are translated with the same error distance in the same direction. The errors have to be defined exactly using the defined transformation matrices. Compensation of errors can be added [7–10].

4.8 Case Study of Errors Budget

Two objectives are to be learned in this case study: errors modeling for a combined motion system and its corresponding budget error table as a result of the errors analysis.

4.8.1 Description of the Designed System

As shown in Figure 4.20, a nanomanipulator is composed of three sets of interactive systems: a manipulation area with three robotic arms RA1, RA2, or RA3 to operate under a CCD camera, an AFM with two probes PR4 and PR5, and a visual evaluation area. To ensure precision in positioning,

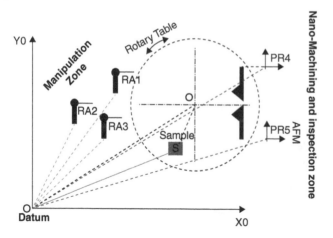

Figure 4.20 Schematic of the integrated manipulator and overall coordinate system.

all three zones are located on a moving rotary table that is synchronized with all manipulators under one controller. The reference datum localized on the granite platform holding all subsystems is the common reference to all local coordinate system of subsystems.

As an example, the sample S shown in Figure 4.20 is held on the rotary table to only rotate when moving between the RAi or PRi subsystems. The design specifications for the manipulator are defined below:

a) Sample with mass up to 50 g;
b) Sample to be hosted on the rotary table with controlled motion;
c) Motion trajectory of the sample: controlled rotation of the rotary table;
d) Radial position: r = several radial positions from 10, 20, 30, 40, and 50 mm with radial positioning error of 5 μm;
e) Rotary angular motion: range 360° with positioning error of 7.5 arcsec;
f) Controlled positioning under AFM and nanomachining;
g) Maximum rotational speed: 10–30 deg/s;
h) Manipulation 3D within 20×20×20 mm³ with positioning error < 0.1 μm;
i) Force measurement up to 0.5 N with a resolution better than 1 μN;
j) Thermal drift kept within +/− 1°C.

4.8.2 Error Modeling and Experimental Testing

In Figure 4.21, the sample S is locked on the rotary table and assigned rotation only when moving between the functional areas or subsystems, such as manipulation or inspection nanomanufacturing represented by the corresponding referential $Ri(R1$ for nanomanipulator 1 to represent its coordinate system $(o1, x1, y1, z1)$ of this subsystem) (Fig.4.21). This figure shows the coordinate system of each subsystem with respect to the main coordinate system (O, xo, yo, zo) and, in particular, the coordinate system of rotary table $(O', x_{rt}, y_{rt}, z_{rt})$ with rigid body errors, to be discussed later.

Two fundamental relationships are written between the three end effectors of the robotic manipulators and the sample held on the rotary table and later on the other side under the AFM/lithography probes [8]. The total positioning error on the sample under consideration results

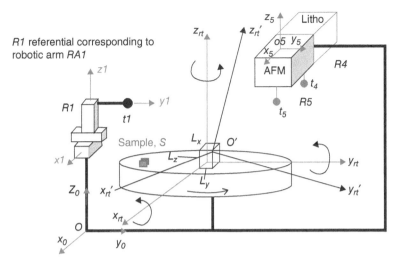

Figure 4.21 3D axis configuration of the nano-manipulator.

from the motion error of the sample on the rotary table combined with the AFM probe positioning error.

Hence, the vector position P of the AFM probe tip $t5$ with respect to the reference coordinate system is given in Eq. 4.51.

$$\{^{O}P_{t5}\}_{AFM} = [^{O}T_{t5}]\{^{05}P_{t5}\}_{AFM},$$ (4.51)

where:

$\{^{05}P_{t1}\}_{AFM} = [0,0,0,1]^{-1}$, with respect to the AFM reference. Similar relationship can be written with respect to lithography.

$[^{O}T_{05}]$ is the transformation motion error composed of $[^{05}T_{t5}]$ the AFM errors and $[^{O}T_{05}]$ the assembly errors of the AFM with respect to the reference origin O, respectively.

Similarly, the sample position error on the rotary table is given by Eq. 4.52:

$$\{^{O}P_{s}\}_{RT} = [^{O}T_{s}]\{^{O'}P_{s}\}_{RT},$$ (4.52)

where $[^{O}T_{s}]$ is the errors transformation matrix between the sample S with respect to the reference origin O.

The error models of the three subgroups will be described hereafter.

A) Error Modeling of the 3-axis Nanomanipulator Arm

Each robotic arm that operate in 3D axes x, y, z has an error matrix over its related axis. The end effector mounted on the z-axis shows an Abbé error taken into consideration while writing the overall motion of the arm, as shown in Figure 4.22.

The holders, for example, arms, have low weight and high bending stiffness. No deformation has been noticed during operations. The position sensor giving positions for each axis are known in Eq. 4.51. The robotic tool tip1, t_1, will move according to Eq. 4.53.

$$\{^{o1}P_{t1}\} = [^{o1}T_{z}]\{^{z}P_{t1}\},$$ (4.53)

where $\{^{z}P_{t1}\} = [0,0,0,1]^{-1}$ with respect to local coordinate origin; $o1$.

The overall three axes manipulator motion $[^{r}T_{z}]$ is defined by Eq. 4.54:

$$[^{o1}T_{z}] = [^{o1}T_{x}][^{x}T_{y}][^{y}T_{z}]$$ (4.54)

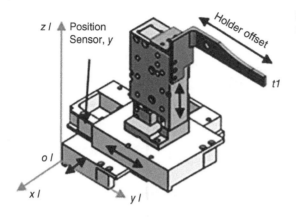

Figure 4.22 Configuration of the nanomanipulator.

Each axis is defined by the following Denavit-Hartenberg error matrix along the x-axis in Eq. 4.55.

$$^{01}T_x = \begin{bmatrix} 1 & -\varepsilon_{zx}(x) & \varepsilon_{yx}(x) & x + \delta_{xx}(x) + a_x \\ \varepsilon_{zx}(x) & 1 & -\varepsilon_{xx}(x) & \delta_{yx}(x) + b_x \\ -\varepsilon_{yx}(x) & \varepsilon_{xx}(x) & 1 & \delta_{zx}(x) + c_x \\ 0 & 0 & 0 & 1 \end{bmatrix}, \tag{4.55}$$

where:

- $\varepsilon_{xy}(x)$ is the roll error of x-axis;
- $\varepsilon_{yx}(x)$ is the pitch error of x-axis;
- $\varepsilon_{zx}(x)$ is the yaw error of x-axis;
- a_x, b_x, c_x are the constant offset in x, y, and z between the reference r and x-axis;
- $\delta_{xx}(x)$: linear displacement error of x-axis;
- $\delta_{yx}(x) = \delta'_{yx}(x) + \alpha_{xy}x$,
- $\delta_{zx}(x) = \delta'_{zx}(x) + \alpha_{xz}x$,
- $\delta'_{yx}(x)$: y straightness error of x-axis moving in x direction;

- α_{xy}: squareness error between x and y axes;
- x : nominal x-axis position that amplifies α_{xy} to yield Abbé error in y-direction.

Similarly, for motions in y- and z-axis, respectively:

$$^{y}T_z = \begin{bmatrix} 1 & -\varepsilon_{zz}(z) & \varepsilon_{yz}(z) & \delta_{xz}(z) + a_z \\ \varepsilon_{zz}(z) & 1 & -\varepsilon_{xz}(z) & \delta_{yz}(z) + b_z \\ -\varepsilon_{yz}(z) & \varepsilon_{xz}(z) & 1 & z + \delta_{zz}(z) + c_z \\ 0 & 0 & 0 & 1 \end{bmatrix}, \tag{4.56}$$

$$^{x}T_y = \begin{bmatrix} 1 & -\varepsilon_{zy}(y) & \varepsilon_{yy}(y) & \delta_{yy}(y) + a_y \\ \varepsilon_{zy}(y) & 1 & -\varepsilon_{xy}(y) & y + \delta_{yy}(y) + b_y \\ -\varepsilon_{yy}(y) & \varepsilon_{xy}(y) & 1 & \delta_{zx}(y) + c_y \\ 0 & 0 & 0 & 1 \end{bmatrix}. \tag{4.57}$$

To eliminate the effect of the manipulator Abbé errors on the sample, an error compensation has been carried out for a better understanding of the tip t_1 position. Figure 4.23 shows the motion of the tip t_1 before and after error compensation.

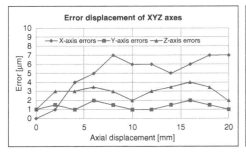
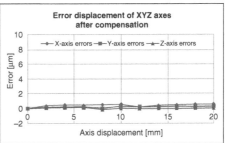

Figure 4.23 Manipulator's tip t_1 position error within the workspace before and after compensation.

B) Error Modeling of the Rotary Table

As in the robotic nanomanipulator, the error model of the rotary table is composed of 6 error components, 3 rotation errors (R_x, R_y, R_z) in which R_z is the actual angular position, and 3 translation errors (L_x, L_y, L_z), as shown in Figure 4.21. Assuming small angular errors and ignoring second-order terms, the transformation matrix [$^0M_{rot}$] written from Eq. 4.52 is as shown in Eq. 4.68.

$$[^0T_s] = {}^0M_{Rot}(\theta_z) = \begin{bmatrix} 1 & -R_z(\theta_z) & R_y(\theta_z) & L_x(\theta_z) \\ R_z(\theta_z) & 1 & -R_x(\theta_z) & L_y(\theta_z) \\ -R_y(\theta_z) & R_x(\theta_z) & 1 & L_z(\theta_z) \\ 0 & 0 & 0 & 1 \end{bmatrix}. \tag{4.58}$$

For any point on the rotary table, the positioning error is defined as $\delta P = P_A(\theta_z) - P_C(\theta_z)$, where P_A is the actual position, and P_C is the commanded position.

The error vector is then defined as the difference between the transformed point P_A and the commended Pc when the table rotates with an angle θ_z, as shown in Eq. 4.59:

$$E(\theta_z) = M_{rot}(\theta_z)P - P_C(\theta_z) = [e_1, e_2, e_3, 1]^T. \tag{4.59}$$

The adopted compensation procedure includes the following steps:

i) Current nominal position $P = [0, 0, 0, \theta_z]$;
ii) Actual position $P_A = M_{rot}(\theta_z)P$;
iii) Calculate the angle difference $\delta\theta$ between P_A and P_C;
iv) Calculate $\Delta P = M_{rot}(\theta_z - \delta\theta)P_A$
v) Calculate the residual $\varepsilon = \Delta P - P$ with ε (ε_x, ε_y, ε_z);
vi) If ε < tolerance, then stop;
vii) Otherwise, return to new position (ε_x, ε_y, ε_z, $\theta_z - \delta\theta$).

This is applied only to the rotary table to move samples to precision locations. Extensive measurements have been performed to obtain the complete quantified error model by identifying the rotational errors and the translational errors as explained next.

1) Measurement of Rotational Errors To measure the three rotational errors of the rotary table, a polygon mirror with 12 faces and a high precision autocollimator, from Taylor Hobson, were used in a standard way (Figure 4.24a).

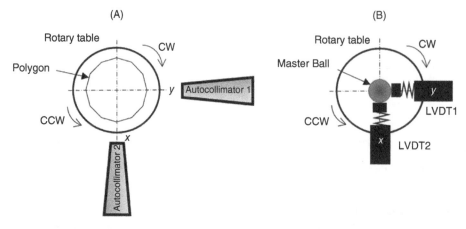

Figure 4.24 Schematic views of experimental a) rotational errors and b) translational errors.

2) Measurement of l_x and l_y   The rotary table error characterization follows BS3800 with actual experiments. The setup of the master ball on the rotary table has two types of eccentricity errors: eccentric setting of the ball center and D_x and D_y in the xy-plane. Added to this are the eccentric errors ε_x and ε_y in the xz- and yz-plane, respectively, due to the elevated height h of the master ball placed in the center of the rotary table.

$$\begin{Bmatrix} \varepsilon_x \\ \varepsilon_y \end{Bmatrix} = \begin{Bmatrix} h \sin R_y \\ -h \sin R_x \end{Bmatrix}. \tag{4.60}$$

The eccentric coordinates of the ball center can be written as:

$$\begin{Bmatrix} L_x \\ L_y \end{Bmatrix} = \begin{Bmatrix} M_x - D_x - x \\ M_y - D_y - y \end{Bmatrix}, \tag{4.61}$$

where M_x and M_y is translational measured values using LVDT sensor in x- and y-axis (Fig.4.13b) on the master ball including the setup errors. L_x and L_y can be extracted as shown in Eq. 4.58.

Figure 4.25 shows the L_x and L_y plots after the setup of error elimination has been applied. The error modeling and experimental error measurements have been carried out using 12 balls on the rotary table and using a CMM to locate its position. The results were averaged and taken in clockwise and counterclockwise directions. Both experimental and modelling data show good agreement in both directional errors of the rotary table, and the overall variation trend is mainly due to geometric errors of the balls inside the rotary table.

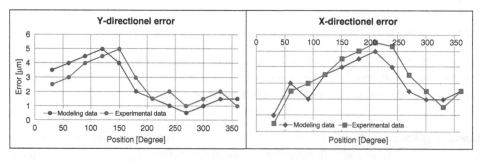

Figure 4.25 Error progression over one rotation in x and y directions.

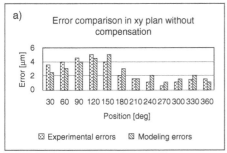
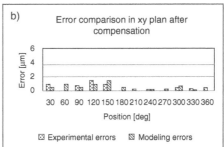

Figure 4.26 Eccentricity error before (a) and after compensation (b) of the rotary table in xy.

Error compensation was introduced to minimize the errors of the rotary table using the procedure previously described. The difference between a position recorded by CMM and the same position through compensation that should be acceptable was measured for verification. The resulting errors in the xy-plane from the experimental tests and modeling are shown in Figure 4.26a, demonstrating agreement and a good level of acceptance with an overall 1 μm positioning. After the compensation in Figure 4.26b, the error is systematically reduced in all 12 positions of the rotary table following the procedure from Eq. 4.59 and steps after that.

C) Atomic Force Microscopy (AFM) Error Model

As previously introduced, the nanomanufacturing zone is composed of a metrological AFM and a modified AFM for lithography designed to operate simultaneously to offer an in-process inspection. The AFM system, as opposed to previous mechanical systems, is based on a scanner that provides the lateral motion of the probing system. This motion is electronically controlled, that is, a voltage step is assigned to a relative displacement. Distortions are proportional to the measured range and usually less than 1%. They include linear and nonlinear scaling as well as cross-talk to include squareness errors. Comprehensive details are described in [8].

The systematic deviation of the instrument measured position from the real tip position is modeled as $x = C_{xx'} \cdot x'$, where $C_{xx'}$ is the calibration coefficient. Hence, the error transformation matrix for the 3D coordinate system referring to [8] is written as shown in Eq. 4.62. Because the laser directly reflects the laser beam from the probe tip's head, the Abbé error of the cantilever probe is negligible. The overall motion errors can be modeled using Eq. 4.62 in the first- and second-order Taylor series approximation. The current position of the probe tip is written with respect to the reference (x', y', z') as:

$$\begin{Bmatrix} x \\ y \\ z \end{Bmatrix} = \begin{bmatrix} C_{xx'} & C_{xy'} & C_{xz'} \\ C_{yx'} & C_{yy'} & C_{yy'} \\ C_{zx'} & C_{zx'} & C_{zz'} \end{bmatrix} \begin{Bmatrix} x' \\ y' \\ z' \end{Bmatrix} + \begin{bmatrix} C_{xx'^2} & C_{xy'^2} & C_{xz'^2} & C_{xx'y'} & C_{xx'z'} & C_{xy'z'} \\ C_{yx'^2} & C_{yy'^2} & C_{yz'^2} & C_{yx'y'} & C_{yx'z'} & C_{yy'z'} \\ C_{zx'^2} & C_{zy'^2} & C_{zz'^2} & C_{zx'y'} & C_{zx'z'} & C_{zy'z'} \end{bmatrix} \begin{Bmatrix} x'^2 \\ y'^2 \\ z'^2 \\ x'y' \\ x'z' \\ y'z' \end{Bmatrix} + \sigma(x'^2, y'^2, z'^2)$$

(4.62)

where x, y are the lateral positions and z the vertical position in the x', y', and z' reference superimposed to $(05, x5, y5, z5)$ AFM coordinate system with dimensions length. The coefficient C_{ijk} are calibration coefficients evaluated through incremental rate as shown in Eq. (4.63) or, alternatively,

Table 4.4 Current AFM specifications.

Nanite AFM scan head specifications	Variable
Maximum scan range (XY)[1]	110 µm
Maximum Z-range[1]	20 µm
XY-linearity mean error	< 0.6%
Z-measurement noise level (RMS, static mode)	typ. 350 pm (max. 500 pm)
Z-measurement noise level (RMS, dynamic mode)	typ. 90 pm (max. 150 pm)

1) Manufacturing tolerances are ±10 % for the 110 µm scan head

made available by the supplier company, for instance, Nanosurf.

$$C_{xx'} = \frac{\partial x}{\partial x'}, \quad C_{xx'^2} = \frac{\partial^2 x}{\partial x'^2}, \quad C_{xx'y'} = \frac{\partial^2 x}{\partial x' \partial y'}. \tag{4.63}$$

AFM uses standard tapping mode AFM probes in cantilever assembly with length ranging from 125 to 225 µm, with a bending stiffness range of 40–49 N/m. The specifications are described in Table 4.4. The AFM is calibrated using length standard reference (Figure 4.27). A couple of tests have been carried out to verify the AFM tip errors. The background noise of the probe tip was recorded as very low, that is, within +/− 0.2 nm, as shown in Figure 4.28. Table 4.4 indicates a typical value of 1.8 nm. This is improved due to the active damping optical table and the Perspex cover of the entire instrument. However, the drifts of the probe tip over 24 hours observation varies within +/− 1.5 nm for lateral motion x and y.

Table 4.5 contains an error budget divided into subsystems after discussing all possible errors governing this integrated nanomanipulator. The shown error budget is based on the connectivity rules that define the instrument behavior of its previously discussed subsystems and their interfaces. Tightening interfaces has been carried out using a torque meter to avoid excessive and nonuniform tightening. Hence, no structural deformation has been observed since all structural parts are made from granite material. Independent subsystems accuracy is an advantage. Each subsystem is represented by its local error definition through modeling and experimental measurements. The

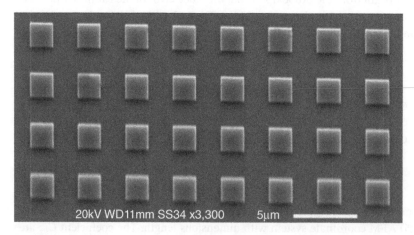

20kV WD11mm SS34 x3,300   5µm

Figure 4.27 Calibrating reference 2x2 µm squares.

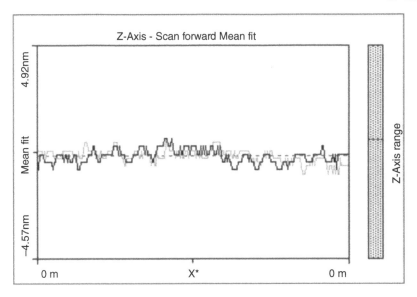

Figure 4.28 Background noise of the probe tip at 20°C.

Table 4.5 Errors budget table.

Variable	Range	Abbé Error (after compensation)	Positioning	Resolution	Repeatability	24H drift [nm]
Manipulators *R1*, *R2* & *R3*						
x-axis	20 mm	<1 µm	< 10 nm	1 nm	5 nm	
y-axis		0.5 µm				
z-axis		0.5 µm				
Rotary table, *RO'*						
θ-axis	360°		5 arcsec		±2 arcsec	
x-axis	40 mm	n/a	1 µm	0.5 µm	±0.5 µm	
y-axis	40 mm		1 µm	0.5 µm	±0.5 µm	
z-axis	-		0.1 µm	-	-	
AFM, *R4* & *R5*						
x-axis	110 µm		<0.6 %		-	1.5 nm
y-axis	110 µm	± 1 nm	<0.6 %	1 nm	-	1.5 nm
z-axis	20 µm		90pm		-	-
Manipulator/Rotary table						
X, Y, Z position			Locked within < 1 µm	Manipulator position error < 0.6 µm		
AFM/Rotary table						
X, Y, Z position			Locked within < 1 µm	AFM position error < 2 nm		

manipulation shows a positioning accuracy of 10 nm in 3D, the rotary table with a positioning accuracy of 1 μm in the *xy*-plane and the AFM with nanomachining at an accuracy of 2 nm. Since the rotary table has relatively lower accuracy compared to other subsystems, a programming loop is added to record the position of the sample once on the rotary table, using the nanomanipulator and its accurate position updated in the register [11–13].

4.9 Solved Problems

Problem 1

Table S4.1

Material	n	Material	n	Material	n
vacuum	1	crown Glass	1.52	fused quartz	1.4585
air	1.0003	salt	1.54	whale oil	1.460
water	1.33	asphalt	1.635	diamond	2.42
ethyl alcohol	1.36	heavy flint glass	1.65	lead	2.6

1) Light with a vacuum wavelength of 504 nm travels through whale oil. Use Table S4.1 to determine the wavelength of the light in the oil.
   Ans: 345 nm
2) As light travels from a sample of material into air, its wavelength changes from 300.0 nm to 726 nm. What is the material?
   Ans: Diamond

Problem 2

A ball is tossed straight up into the air with initial speed $v_0 = 4.0 \pm 0.2$ m/s. After a time $t = 0.60 \pm 0.60$ s, the height of the ball is $y = v_0 t - \frac{1}{2}gt^2 = 0.636$ m. What is the uncertainty of y? Assume $g = 9.80$ m/s^2 (no uncertainty in g).

Answer: Let's start by naming some things.

Let $a = v_0 t = 2.4$ m and and $b = \frac{1}{2}gt^2 = 1.764$ m

Then using the multiplication rule, we can get the uncertainty in a:

$$\delta a = |a|\sqrt{\left(\frac{\delta v_0}{v_0}\right)^2 + \left(\frac{\delta t}{t}\right)^2}$$

$$= v_0 t\sqrt{(0.05)^2 + (0.10)^2}$$

$$= (2.4\text{ m})(0.112) = 0.27\text{ m}.$$

For δb, we can use the power rule:

$$\delta b = 2b\left(\frac{\delta t}{t}\right)$$

$$= 2(1.764\text{ m})\,(0.10)$$

$$= 0.35\text{ m}.$$

Finally, $y = a + b$, so we can get δy from the sum rule:

$$\delta y = \sqrt{(\delta a)^2 + (\delta b)^2}$$
$$= \sqrt{(0.27 \ m)^2 + 0.35 \ m)^2} = 0.44 \ m$$

Thus, y would be properly reported as 0.6 ± 0.4 m.

Multiple Choice Questions of this Chapter

Multiple Choice Questions are given for each chapter with solutions in an online extension of this book. Please use link: www.wiley.com\go\mekid\metrologyandinstrumentation\

References

1 "International Standard ISO 5725-4, accuracy (trueness and precision) of measurement methods and results—Basic methods for the determination of the trueness of a standard measurement method," International Organization for Standardization, Geneva, 1994.

2 Blumenthal, L. M., 1926, "Concerning the remainder term in Taylor's formula," *Amer. Math. Monthly* 33, pp. 424–426.

3 "Guide to the expression of uncertainty in measurement," First Edition, International Organization for Standardization, Geneva, 1993.

4 S. Mekid, 2005, "Design strategy for precision engineering: second order phenomena," *J. Engineering Design*, 16(1).

5 Bryan, J.B., 1979, "The Abbé principle revisited: an updated interpretation," *Precision Engineering*, 1(3), pp. 129–132.

6 Donaldson, R.R., 1980, "Error budgets," *Technology of Machine Tools*, 5 (October 1980), Sec. 9.14, pp. 1–14.

7 Ni, J., 1987, "Study on online identification and forecasting compensatory control of volumetric errors for multiple axis machine tools," PhD dissertation, University of Wisconsin–Madison.

8 Mekid, S., 2020, "Integrated nanomanipulator with in-process lithography inspection," *IEEE Access*, 8, art. no. 9097243, pp. 95378–95389.

9 Lo, C.H., 1994, "Real-time error compensation on machine tools through optimal thermal error modelling," PhD dissertation, University of Michigan.

10 Slocum, A.H., 1992, *Precision Machine Design* (Prentice Hall, Englewood Hill).

11 Mekid, S., and Lim, B., 2004, "Characteristics comparison of piezoelectric actuators at low electric field: analysis of strain and blocking force," *Smart Materials and Structures*, 13(5), pp. N93–N98.

12 Mekid, S., and Shang, M., 2015, "Concept of dependent joints in functional reconfigurable robots," *Journal of Engineering, Design and Technology*, 13(3), pp. 400–418.

13 Mekid, S., 2010, "Spatial thermal mapping using thermal infrared camera and wireless sensors for error compensation via open architecture controllers," Proceedings of the Institution of Mechanical Engineers. Part I: *Journal of Systems and Control Engineering*, 224(7), pp. 789–798.

5

Measurement and Measurement Systems

"I often say that when you can measure what you are speaking about, and express it in numbers, you know something about it; but when you cannot measure it, when you cannot express it in numbers, your knowledge is of a meager and unsatisfactory kind; it may be the beginning of knowledge, but you have scarcely in your thoughts advanced to the state of Science, whatever the matter may be."

—Lord Kelvin, 1883. Public Domain

National Physical Laboratory (NPL) has defined six guiding principles to good measurement practice shown hereafter:

The Right Measurements: *Measurements should only be made to satisfy agreed and well specified requirements.*

The Right Tools: *Measurements should be made using equipment and methods that have been demonstrated to be fit for purpose.*

The Right People: *Measurement staff should be competent, properly qualified and well informed.*

Regular Review: *There should be both internal and independent assessment of the technical performance of all measurement facilities and procedures.*

Demonstrable Consistency: *Measurements made in one location should be consistent with those made elsewhere.*

The Right Procedures: *Well-defined procedures consistent with national or international standards should be in place for all measurements.*

5.1 Introduction

Measurement and quantification are the basic concepts of metrology. This considers explicit and internationally accepted definitions, principles, and standards. The fundamentals of metrology are valid for all science and engineering fields with constant search for the best design technologies that minimize direct involvement of humans in measurement processes. The purpose of any measurement system is to provide the user with a numerical value corresponding to the variable being measured by the system as simple as the one in Figure 5.1.

Metrologists consider measurement as a complex structured process in which a measurement task is carried out by applying theoretical and methodological principles and by executing agreed procedures and techniques (methods) for data generation.

The measurement task must be defined by specifying the objects under measurement, the property of interest, and more importantly, the targeted measurand. What is not measurement are some

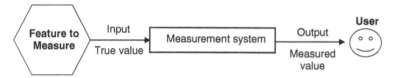

Figure 5.1 Measurement Process.

processes that appear to be measurements but are not, such as comparing two lengths or counting the number of balls in a bag. Also, any test requiring a yes or no answer or pass or fail result is not a measurement either. But measurement can lead to this type of decision making. As a result, the general model of the measuring system, or measurement methodology, must be specified, which includes the design of measurement procedures and the identification of measuring instruments, as well as proper explanations of how they enable capturing the value of the measurand.

Metrologists derive a specific model of the measuring system with respect to the planned measurands, such as length or height. After that, the scientists will define the specific parameters to be measured as well the measurement model comprising assumptions on their interrelations, including any calculation to produce the measurement results. Therefore, scientists design the process structures that enable empirical interactions with the value of the measurand. Measuring systems are based on the identification of systematic structural connections among properties or at least specific assumptions about such connection networks, which allow scientists to check measurement results via experimental cross-validation.

The generated results must be justified to be attributed to the objects and properties under measurement. The numerical assignments must be systematically connected through *unbroken documented chains of comparisons* to the measurand and a reference. This is the traceability of measurements. Every component of the chain implicates the possibility that the entities of the connected properties, that is, measurand and measurement unit, can be compared with one another regarding their quantities so that quantitative information from one property can be converted through sensing or transducing into quantitative information in another property (Figure 5.2).

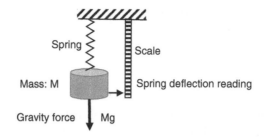

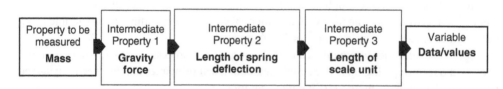

Figure 5.2 Unbroken chain of comparisons leading to quantitative information.

If direct comparison with psychological and social-science practices is considered, it is perhaps important to consider the simplest and historically oldest physical measuring instruments because, in them, involvement of human factors is still greater and more directly apparent than in today's sophisticated measuring technologies. These latter instruments build on knowledge gained from earlier instruments while involving more complex physical processes, many of which are imperceptible to humans [1].

5.2   What Can Be Standard in a Measurement?

Measurements have a standard ISO/IEC/IEEE 15939:2017, that applies to systems and software engineering measurement processes. This standard provides an elaboration of the measurement process from ISO/IEC 15288 and ISO/IEC 12207. The measurement process is applicable to system and software engineering and management disciplines. The process is described through a model that defines the activities of the measurement process that are required to adequately specify what measurement information is required, how the measures and analysis results are to be applied, and how to determine if the analysis results are valid. The measurement process is flexible, tailorable, and adaptable to the needs of different users.

ISO/IEC/IEEE 15939:2017 identifies a process that supports defining a suitable set of measures that address specific information needs. It identifies the activities and tasks that are necessary to successfully identify, define, select, apply, and improve measurement within an overall project or organizational measurement structure. It also provides definitions for commonly used measurement terms.

Measurement supports the management and improvement of processes and products. Measurement is a primary tool for managing system and software life cycle activities, assessing the feasibility of project plans, and monitoring the adherence of project activities to those plans. System and software measurement is also a key discipline in evaluating the quality of products and the capability of organizational processes. It is becoming increasingly important in two-party business agreements, where it provides a basis for specification, management, and acceptance criteria. Continual improvement requires change within the organization. Evaluation of change requires measurement. Change is not initiated by measurement. Measurement should lead to action rather than simply collecting data. Measurements should serve a specific purpose.

This document defines a measurement process applicable to system and software engineering and management disciplines. The measurement process defined in this document, while written for system and software domains, can be applied in other domains. The purpose of this document is to describe the activities and tasks that are necessary to successfully identify, define, select, apply, and improve measurement within an overall project or organizational measurement structure. It also provides definitions for measurement terms commonly used within the system and software disciplines.

This document neither catalogs measures nor provides a recommended set of measures for use on projects. It does identify a process that aids in the development of a suitable set of measures to address specific information needs. This document is intended to be used by suppliers and acquirers. Suppliers include personnel performing management, technical and quality management functions in system and software development, maintenance, integration, and product support organizations. Acquirers include personnel performing management, technical, and quality management functions in procurement and user organizations.

The following are examples of how this document can be used:

1) By a supplier to implement a measurement process to address specific project or organizational information requirements;
2) By an acquirer (or third-party agents) for evaluating conformance of the supplier's measurement process to this document;
3) By an acquirer (or third-party agents) to implement a measurement process to address specific technical and project management information requirements related to the acquisition;
4) In a contract between an acquirer and a supplier as a method for defining the process and product measurement information to be exchanged.

Scope of ISO/IEC/IEEE 15939:2017

The ISO 15939 document establishes a common process and framework for measurement of systems and software. It defines a process and associated terminology from an engineering viewpoint. The process can be applied to the project and products across the life cycle. The measurement process can be applied throughout the life cycle to aid the planning, managing, assessing, and decision-making in all stages of a system or software life cycle.

The ISO 15939 document also provides activities that support the definition, control, and improvement of the measurement process used within an organization or a project. The ISO 15939 document does not assume or prescribe an organizational model for measurement. The user of this document decides, for example, whether a separate measurement function is necessary within the organization and whether the measurement function should be integrated within individual projects or across projects, based on the current organizational structure, culture, and prevailing constraints.

The ISO 15939 document does not prescribe a specific set of measures, method, model, or technique. The users of this document are responsible for selecting a set of measures for the project and defining the application of those measures across the process, products, and other elements of the life cycle. The parties are also responsible for selecting and applying appropriate methods, models, tools, and techniques suitable for the project. The ISO 15939 document does not specify the name, format, explicit content, or recording medium of the information items to be produced. This document does not suggest that documents be packaged or combined in any way. These decisions are left to the user of this document. ISO/IEC/IEEE 15289 addresses the content for life cycle process information items (documentation).

The measurement process should be properly integrated with the organizational quality system. This document does not explicitly cover all aspects of internal audits and noncompliance reporting because they are assumed to be in the domain of the quality system.

The ISO 15939 document is not intended to contradict any existing organizational policies, standards, or procedures. Any conflict, however, should be resolved, and any overriding conditions or situations should be specified in writing as exceptions to the application of this document. A list of all ISO international standards in metrology is given in Appendix A1 and ASME standards in A2.

5.3 Definitions of Key Measurement Components

5.3.1 Measurement System

The measurement system is the method for associating numbers with physical quantities and phenomena (measurand). The imperial system and the International System of Units are two types of measurement systems (SI).

5.3.2 Measurement System Analysis

The measurement system analysis is a thorough assessment that can be applied to a measurement process. The purpose of this analysis is to identify the components of variation in the measurement process. Hence, a designed experiment may be dedicated to this activity. The available tools, so far, are calibrations analysis, analysis of components of variance, fixed effect of ANOVA, and gauge R&R. This process will go through the following components evaluation and approaches:

1) The measuring device;
2) The procedures and operators;
3) The measurement interactions;
4) Identifying the proper measurement and approach;
5) Determining the measurement uncertainty of individual measurement devices and/or measurement systems.

5.3.3 Measurement Process

According to the VIM, measurement is defined as the "process of experimentally obtaining one or more quantity values that can reasonably be attributed to a quantity" [2]. The process, however, is not well defined in VIM. The definition is included in ISO 10012 [3], but it is still ambiguous and may be incomplete because it focuses on the manufacturing process. The measurement process consists of more than just the measurement itself. It consists of the tools and other requirements required to obtain reasonable measurement results.

To summarize, the goal of any measurement system is to provide the user with a numerical value corresponding to the variable being measured by the system, which can be as simple as the one shown in Figure 5.2. It is well understood that the measured value is not always the variable's "true value." A "true value" would be obtained by a perfect measurement. It is also defined as the value that is approached by averaging an increasing number of measurements with no systematic errors. To characterize these measurements and the accompanying errors, a number of definitions and rules have been set and detailed in Chapters 4 and 6.

5.4 Physical Measurement Process (PMP)

One of the objectives of measuring is to determine physical properties of objects; hence, any measurement will require a measurand (object to measure) and a measuring standard in order to achieve a successful measurement. The measuring system is designed to produce one property of the measurand. In these cases, the process involves no change in the intrinsic properties of the matter while any surrounding environmental parameter, such as temperature, vibration, and light, can provide information that may be necessary to properly quantify the measurand. The direct measurements of physical measurands are, for example, temperature, mass, and length. A PMP covers the measurement of an ongoing parameter such as the displacement of an object driven by a force or the variation of the temperature of a plate under a heating process.

- If the environmental parameters are controlled and monitored, the measurement will be accurate: the process is controlled.
- If the environmental parameters have an effect on the measurand, we must include the effect if it is known.

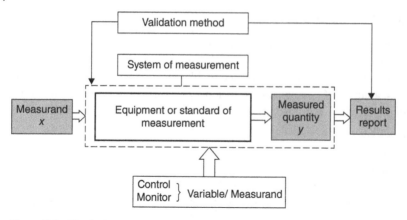

Figure 5.3 Physical measurement process.

- If a property to be measured needs multiple operation of the sample to reach it, the measurement system can be composed of multiple subsystems: $y = u + e$ [1] where, y: measurement result or a better estimation of the measurand, u: property of the sample intended to be measured, and e: measurement error.

Figure 5.3 shows the main components of PMP: the measurand variable and the measured value and reporting indicating the quality of the measurement as a whole process and the measurement system.

5.5 Difference between Number and an Analysis Model

The differences between number and an analysis model are:

1) A number derived from the result of a measurement process that meets the metrology requirements is a quantity expressed with a measurement unit;
2) A number with a measurement unit obtained through the proper application (manual or automatic) of its corresponding measurement method will have many more measurement qualities (in the metrology sense) than a number derived from an opinion only;
3) A number derived from a mix of mathematical operations without consideration of measurement units and scale types will still be a number, but it could be a meaningless one;
4) Practitioners may be happy with models that appear to take into account a large number of factors (as in many estimation models and quality models). However, feeling good does not add validity to mathematical operations that are inadmissible in measurement.

In practice, various types of quantitative models produce numbers in outputs (i.e., the outcomes of the models) that do not have the same qualities as numbers, which meet the requirements of metrology.

An estimation model provides a number as an estimate:

1) To every such estimated number is associated a (potentially large) range of variations, depending on the number of input parameters and their corresponding uncertainties, as well as on the uncertainties about the relationships across all such input parameters;
2) These estimated numbers are not meaningful without a knowledge (and understanding) of the corresponding uncertainties.

A quality model provides a number, which typically depends on:

1) A specific selection among a (potentially large) number of alternatives;
2) The assignment of a percentage to each contributing alternative, which is based on the opinion of one person (or a group of persons); and
3) Comparison of each contributing alternative with distinct threshold values, which are often defined by opinion as well.

5.6 Measurement Methods

5.6.1 Metrology and Measurement

The initial step before performing any measurement using an instrument is to read the specifications and conditions of use. The measurands expected values may be of very tight quantitative values. The measurement is required to be achieved with high accuracy. It is also required to provide report on the performance of the instrument and any influencing external parameter. As a final stage, it is required to check whether there is any compliance with initial targets; otherwise, adjustments will be in order to meet the specifications.

According to the *Vocabulaire international de métrologie* (VIM)[1], Metrology is a field of knowledge concerned with measurement. It includes all theoretical and practical aspects of measurement, including the measurement uncertainty and field of application. Measurement is the process of experimentally obtaining information about the magnitude of a quantity. It implies a measurement procedure based on a theoretical model. In practice, measurement presupposes a calibrated measuring system, possibly subsequently verified.

To figure out one micrometer, the diameter of a hair is about 40–80 µm (Figure 5.4), while one nanometer is almost as wide as a DNA molecule and 10 times the diameter of a hydrogen atom. It deals with how much your fingernails grow each second. It also relates to the thickness of a drop of water that spreads over a square meter. It is one-tenth the thickness of the metal film on your tinted sunglasses or your potato chip bag. The smallest lithographic feature on a Pentium computer chip is about 100 nanometers.

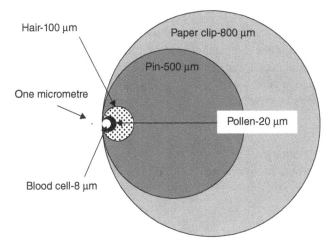

Figure 5.4 Relative scale of five items.

1 VIM draft version of April 2004.

Different methods of measurement are explained below in details.

a) Direct Method of Measurement (DMM)

In the direct method of measurement, the measured quantity is directly compared with the standard quantity. There are no mathematical calculations to find the results. The method is very accurate. The values of the physical quantities such as length, temperature, and pressure are expressed in numbers. Example 1: measuring a length using a graduated ruler.

b) Indirect Method of Measurement (IDM)

In most cases, the results of direct measurement are inaccurate in figures. Hence, the direct method is rarely preferred for measurement. In IDM, the physical parameters of the quantity are measured directly, and then the numerical value of the quantity is determined by the mathematical relationship as explained later. Example 2: measuring weight = length × breadth × height × density.

c) Fundamental Method of Measurement

This is the absolute method of measurement. It is based on the measurement of the base quantities used to define the quantity. Example 3: measuring a quantity directly in accordance with the definition of that quantity or measuring a quantity indirectly by direct measurement of the quantities linked with the definition of the quantity to be measured.

d) Comparison Method of Measurement

This method involves comparison with either a known value of the same quantity or another quantity that is a function of the quantity to be measured.

e) Substitution Method of Measurement

The quantity to be measured is done by direct comparison on an indicating device, by replacing the measuring quantity with some other known quantity that produces a similar effect on the indicating device. Example 4: determination of mass by the Borda method.

f) Transposition Method of Measurement

This measurement involves direct comparison in which the value of the quantity to be measured is first balanced by an initial known value A of the same quantity; next, the value of the quantity to be measured replaces the known value and is balanced again by a second known value B. When the balance arrow gives the same indication in both cases, the value of the quantity to be measured is V_{AB}. Example 5: determination of a mass by means of a balance and known weights, using the Gauss double weighing method.

g) Differential or Comparison Method of Measurement

This method involves measuring the difference between the given quantity and a known master of almost similar value. Example 6: determination of a diameter with master cylinder on a comparator.

h) Coincidence Method of Measurement

In this differential method of measurement, the very small difference between the given quantity and the reference is determined by the observation of the coincidence of scale marks. Example 7: measurement on Vernier caliper.

i) Null Method of Measurement

In this method the quantity to be measured is compared with a known source, and the difference between these two is made zero.

j) Deflection Method of Measurement

In this method, the value of the quantity is directly indicated by deflection of a pointer on a calibrated scale.

k) Interpolation Method of Measurement

In this method, the given quantity is compared with two or more known values of almost similar value, ensuring at least one smaller and one bigger than the quantity to be measured, and the readings are interpolated.

l) Extrapolation Method of Measurement

In this method, the given quantity is compared with two or more known smaller values, and then the reading is extrapolated.

m) Complementary Method of Measurement

This is the method of measurement by comparison in which the value of the quantity to be measured is combined with a known value of the same quantity and then adjusted such that the sum of these two values is equal to predetermined comparison value. For example, determination of the volume of a solid by liquid displacement.

n) Composite Method of Measurement

It involves the comparison of the actual contour of a component to be checked with its contours in maximum and minimum tolerable limits. This method enables the checking of the cumulative errors of the interconnected elements of the component, which are controlled through a combined tolerance. This method is most reliable to ensure interchangeability and is usually effected with composite "go" gauges. Example 8: checking of the thread of a nut with a screw plug "go" gauge.

o) Element Method

In this method, the several related dimensions are gauged individually, that is, each component element is separately checked. Example 9: In the case of thread, the pitch diameter, pitch, and flank angle are checked separately, and then the virtual pitch diameter is calculated. It may be noted that the value of virtual pitch diameter depends on the deviations of the above thread elements. The functioning of thread depends on virtual pitch diameter lying within the specified tolerable limits. In case of composite method, all the three elements need not be checked separately and is thus useful for checking the product parts. The element method is used for checking tools and for detecting the causes of rejections in the product.

p) Contact and Contactless Methods of Measurements

In a contact method, the measuring tip of the instrument is directly in contact with the surface to be measured. In such cases, arrangements for constant contact pressure should be provided in order to prevent errors due to excess contact pressure. Example 10: touch probes used in CMMs.

In a contactless method, the measurement is carried out without any contact. Example 11: instruments including tool-maker's microscope and projection comparator. For every method of measurement, a detailed definition of the equipment to be used, a sequential list of operations to be performed, and the surrounding environmental conditions and descriptions of all factors influencing accuracy of measurement at the required level must be prepared and followed.

5.6.2 Metrological Characteristics of Measuring Instruments

Every measurement device comes with a manual of use and a calibration certificate. Any considered measuring instrument has its metrological properties usually specified, such as the range of measurement, scale graduation value, scale spacing, sensitivity, and reading accuracy. The following metrological properties are defined [4].

a) *Range of measurement:* It indicates the size values between which measurements may be made on the given instrument.

b) *Scale range:* It is the difference between the values of the measured quantities corresponding to the terminal scale marks.

c) *Instrument range:* It is the capacity or total range of values that an instrument is capable of measuring. For example, a micrometer screw gauge with a 25 to 50 mm capacity has an instrument range of 25 to 50 mm, but the scale range is 25 mm.

d) *Scale spacing:* It is the distance between the axes of two adjacent graduations on the scale. Most instruments have a constant value of scale spacing throughout the scale. Such scales are said to be linear. In case of nonlinear scales, the scale spacing value is variable within the limits of the scale.

e) *Scale division value:* It is the measured value of the measured quantity corresponding to one division of the instrument, for example, for ordinary scale, the scale division value is 1 mm. As a rule, the scale division should not be smaller in value than the permissible indication error of an instrument.

f) *Sensitivity (amplification or gearing ratio):* It is the ratio of the scale spacing to the division value. It could also be expressed as the ratio of the product of all the larger lever arms and the product of all the smaller lever arms. It is the property of a measuring instrument to respond to changes in the measured quantity.

g) *Sensitivity threshold:* It is defined as the minimum measured value that may cause any movement whatsoever of the indicating hand. It is also called the discrimination or resolving power of an instrument and is the minimum change in the quantity being measured that produces a perceptible movement of the index.

h) *Reading accuracy:* The accuracy that may be attained in using a measuring instrument.

i) *Reading error:* It is defined as the difference between the reading of the instrument and the actual value of the dimension being measured.

j) *Accuracy of observation:* It is the accuracy attainable in reading the scale of an instrument. It depends on the quality of the scale marks, the width or the pointer/index, the space between the pointer and the scale, the illumination of the scale, and the skill of the inspector. For accurate reading of indications, the width of scale mark is usually kept one-tenth of the scale spacing.

k) *Parallax:* This refers to the apparent change in the position of the index relative to the scale marks when the scale is observed in a direction other than perpendicular to its plane.

l) *Repeatability:* It is the variation of indications in repeated measurements of the same dimension. The variations may be due to clearances, friction, and distortions in the instrument's mechanism. Repeatability represents the reproducibility of the readings of an instrument when a series of measurements in carried out under fixed conditions of use.

m) *Measuring force:* This is the force produced by an instrument that acts upon the measured surface in the direction of measurement. It is usually developed by springs whose deformation and pressure changes with the displacement of the instrument's measuring spindle.

5.7 Instrumentation for Measurement

5.7.1 Background

Metrology can be divided according to the type of the quantity under consideration, such as metrology of mass or metrology of time. The applications are divided into industrial metrology, medical metrology, and so forth.

Engineering metrology is restricted to the measurement of length, angles, and other quantities, which are expressed in linear or angular terms. A unit is assigned to the quantity to be measured. This is further accompanied by the need of a universal standard, and the various units for various parameters of importance must be standardized. The quantity measured should be presented with sufficient correctness and accuracy. This will depend on the method of measurement [5, 6]. A proper reporting of measurement should include the following aspects for referencing and archiving of measurements:

 i) The purpose of measurement;
 ii) Methods of measurement based on agreed units and standards;
 iii) The average measured quantity;
 iv) Associated errors of measurement;
 v) Units of measurement and their standards related to the establishment, reproduction, conservation, and transfer of units of measurement and their standards;
 vi) Measuring instruments and devices;
 vii) Accuracy of measuring instruments and their related maintenance;
 viii) Design, manufacturing, and testing of the samples used.

5.7.2 Measurement Instrumentations

The instrumentation used for measurement can be divided into two groups:

1) Portable measuring devices and manual gauge;
2) Digital instrument allowing for one or more measurements sometimes recorded with data post-processing.

The following instruments are popular mechanical gauges and useful tools with the main functions of measuring geometric parameters such as lengths or diameters. The scale unit is usually in mm (cm) or in inches.

5.7.3 Digital Measuring Device Fundamentals

In a digital measuring device, the value of the measured physical quantity is automatically represented by a number on a digital display. The digital measuring devices can be divided into digital measuring instruments and digital measuring transducers.

Digital measuring instruments are self-contained devices that automatically present the value of the measured quantity on a digital display. Digital measuring transducers are sensors lacking a digital display; the measurement results are converted into a digital code for subsequent transmission and processing in measuring systems. Laser interferometers, electronic temperature sensors, voltmeters, and so forth, are some examples of digital measuring devices. Pressure, speed, force, and so forth, can be measured through a voltage when this measurand is converted into an electrical quantity.

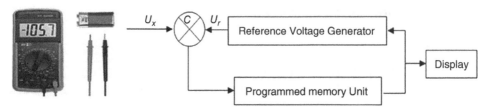

Figure 5.5 Schematic diagram of a digital DC voltmeter.

The operation of digital measuring devices is based on the digitization and coding of the value of the measured physical quantity. The converted signal is sent to a processing system and then to transmission or to display. The digital display shows the converted measurement result as a number expressed by a decimal number system as shown in many digital systems.

Depending on the application, the accuracy and range of operation relies on the type of conversion for the measured quantity. Three types of commonly used methods of analog-to-digital conversion include pulse count modulation, successive approximation, and flash converters, for ,instance, direct conversion. Devices that measure the rotation angle of a shaft generally utilize an encoding disk or drum that is fastened to the shaft. The encoding disk gives the value of the measured angle using a reader, which feeds the value, in the form of a coded signal to the display after being processed if needed. A schematic diagram of a digital DC voltmeter is shown in Figure. 5.5. In this figure, the voltage U_x is measured and sent to one of the inputs of the comparator C. There is a reference voltage U_r applied to the other input from a program-controlled reference voltage generator. The comparator will generate one of the two signals either $U_r > U_x$ or $U_r \leq U_x$. If the signal $U_r \leq U_x$ is received, the control memory unit triggers the reference voltage generator to increment U_r. If the signal $U_r > U_x$ is received, the control memory unit triggers the reference voltage generator to replace the last of the memory increments by a smaller increment. This process is repeated until the increase in U_r becomes equal to the smallest increment possible with the given pulse generator. The control memory unit then generates the code corresponding to the sum of all the increments, and the code is transmitted to the display.

5.8 Non-Portable Dimensional Measuring Devices

There are several non-portable equipment for length measurement and inspection. The most important ones for metrology length are laser interferometry and the coordinate measuring machine (CMM) in its various types.

5.8.1 Laser Interferometry, Application to CNC Machines

a) Definition
This is a contactless laser-based method used to measure small to large displacements and angles. The uncertainty reaches subnanometer scales. Considering possible bending of the laser for large displacement, large distances can reach 30 m and even 60 m.

A laser source (Figure 5.6) is emitted through a half-silvered mirror (beam splitter) reflecting light to two mirrors attached to two bodies. With one mirror as the reference, the other mirror as it

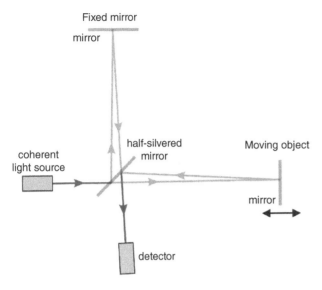

Figure 5.6 Michelson interferometer.

translates the laser feedback to the receiver can be compared to the laser reaching the fixed mirror, and hence the displacement can be measured.

Ultra-high resolution is provided by the use of heterodyne detection techniques. Two extremely stable frequency-shifted orthogonally polarized beams are emitted by a laser light source at much known frequencies (i.e., He:Ne laser head). The beam is divided into reference and measuring beam components (beam splitter). The detectors sense the intensities of the two components of the sampled beam, and their output is used to control the temperature of the laser tube. By adjusting the temperature of the laser tube, the distance between the ends can be controlled and the frequency stabilized to about 1 part in 10^8. The stability is then in the range of 1 part in 10^7. The resolution over the machine workspace is about one nanometer.

This instrument allows the possible measurements of geometric errors in CNC machine tools for example:

i) Displacement;
ii) Straightness;
iii) Angles;
iv) Flatness;
v) Squareness.

This type of measurement allows the following:

i) Machine tool calibrations according to standards of measurements (ISO, VDI, BS, JIS, ANSI, and NMTBA);
ii) Machine certification ISO 9000.

b) Principles of Measurement of Straightness, Displacement XYZ, and Angles

b-1) Linear Position Measurement The positioning precision is measured using the laser interferometry configuration described in Figure 5.7, where the interferometer optics are mounted on the moving table and the laser head is mounted on a stable tripod, while the reflector is on a fixed position. The inspection of the positioning follows a protocol established in ISO 230-1 to ISO 230-4.

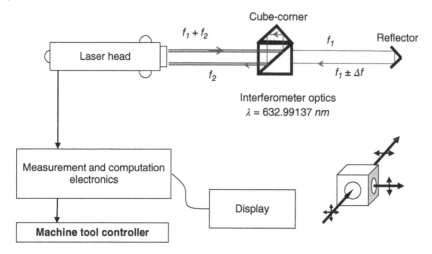

Figure 5.7 Laser interferometer configuration for displacement measurement.

b-2) Straightness Measurement Like positioning, the straightness configuration (Figure 5.8) uses interferometer optics but comprises a beam splitter and is in a fixed position, while the retrore-flector is mounted on the moving table depending on whether horizontal or vertical straightness is assessed. Tests are conducted following ISO 230-1to ISO 230-4. Table 5.1 gives an overview of the expected resolution and accuracy of the straightness, depending on the range of measurement and the used optics.

b-3) Angular Measurement In this configuration, the angular error is assessed using laser interfer-ometry. The angular retroreflector is mounted on a moving table. The optics is perfectly aligned,

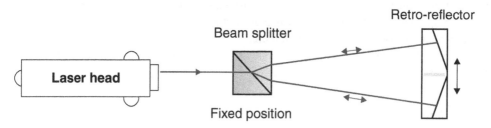

Figure 5.8 Laser interferometer configuration for displacement measurement.

Table 5.1 Standard scale of sizes.

Symbol	Corresponding scale
femto (f)	10^{-15}
pico (p)	10^{-12}
angstrom (Å)	10^{-10}
nano (n)	10^{-9}
micro (μ)	10^{-6}
milli (m)	10^{-3}

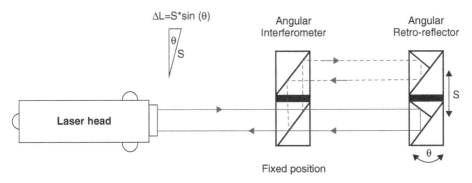

Figure 5.9 Laser interferometer configuration for angular measurement.

and the angular interferometer is composed of beam splitter with a periscope, while the angular reflector is composed of two retroreflectors separated by a distance S. The laser source beam is split into 2 arms, one going straightforward and the other one coming back. The change in angular motion as shown in Figure 5.9 is detected between optical path lengths in the two arms.

c) Effects of the Environment on Laser Measurements
The laser wavelength usually selected as He-Ne laser is 632.8 nm and evolves in air. The optical refraction of the index of air may change as it is affected by variations in temperature, pressure, and humidity.

The wavelength λ_n is constant in vacuum, but in any nonvacuum environment, it depends on the index of refraction of the environment. Since most laser interferometer systems operate in air, it is necessary to correct for the difference between λ_n and the wavelength in air λ_a. The index of refraction n of the air depends on the ratio of λ_n and λ_a as:

$$n = \lambda_n / \lambda_a. \tag{5.1}$$

The new speed of light in air, v, is calculated from the speed of light in vacuum, c, over the index of refraction in air, given as:

$$v = c/n. \tag{5.2}$$

The correct compensation factor C for the measurement conditions is calculated through Edlen's (or modified) equations (3) and (4).

$$C = \frac{10^6}{N + 10^6}, \tag{5.3}$$

$$N = 0.3836391P \times \left[\frac{1 + 10^{-6}P(0.817 - 0.0133T)}{1 + 0.0036610T}\right] - 3.033 \times 10^{-3} \times H \times e^{0.057627T}, \tag{5.4}$$

where:

P: air pressure;
T: air temperature;
H: relative humidity.

Example 12: For standard conditions $H = 50\%$, $P = 760$ mm Hg and $T = 20°C$, the compensation factor is then C= 0.9997287628. Recommended environmental conditions are shown in Figure 5.10.

| LABORATORY TYPE | FIELD | TEMPERATURE (DEGREES CELCIUS) | | RELATIVE HUMIDITY (%) | AIRBORNE PARTICLE COUNT[b] |
		SET POINT AND LIMITS	MAXIMUM RATE OF CHANGE (K/HOUR)[a]		
I	MECHANICAL/DIMENSIONAL	20 ± 1	0.5	30 - 55	50 000
	ELECTRIC/ELECTRONIC	23 ± 1	1.0	30 - 55	100 000
	OTHER	The environmental conditions for other fields of measurements should be developed and assessed with respect to appropriate applicable influencing factors.			
II	MECHANICAL/DIMENSIONAL	20 ± 2	1.0	30 - 55	150 000
	ELECTRIC/ELECTRONIC	23 ± 2	1.5	30 - 55	250 000[c]
	OTHER	The environmental conditions for other fields of measurements should be developed and assessed with respect to appropriate applicable influencing factors.			
III	GENERAL	Range of 18 - 28 with preferred setpoint of 23	1.5 (1.0 - for Mechanical/ Dimensional)	10 - 60	[d]

a) Certain measurement or comparison equipment may require more stringent control.
b) Measured in accordance with U.S. Federal Standard No. 209, "*Clean Room and Work Station Requirements. Controlled Environment*"
c) There should be no accumulation of particles on or under benches, cabinets, equipment, instrumentation, etc.
d) Careful housekeeping with no accumulation of particles on or under benches, cabinets, equipment, instrumentation, etc.

Figure 5.10 Recommended environmental conditions.

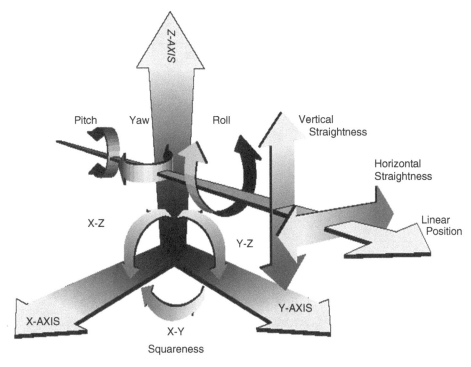

Figure 5.11 Associated errors for a 3-axis machine. [Courtesy of HP].

Source of errors

1) Intrinsic (laser wavelength accuracy, measurement resolution, optics nonlinearity);
2) Environmental (atmospheric compensation, material expansion, optical thermal drift);
3) Installation (alignment error, cosine error, Abbe error).

d) Inspection of Motion Errors in Various Applications

Inspection of CNC Machines and Coordinate Measuring Machines (CMM)   Laser interferometry is mainly used to assess the displacement (positioning and straightness) over the three axes of motion and the rotations errors around these axes. The overall errors in a 3-axis machine are shown below (Figure 5.11). The example of application of one axis is shown in this figure. The table is driven by a servo motor and may induce multiple errors as seen previously.

In a one-axis machine, the measurement is about displacement and position error in one axis, but the associated errors to this one axis are five errors; vertical and horizontal straightness, pitch, yaw, and roll, as shown in Figures 5.12–5.13. The principle of the Michelson interferometer is physically shown in Figures 5.14–5.16 by Renishaw equipment as an example. The standards ISO 230-1 to 230-4 are used to assess the errors in CNC and CMM machines [7–9].

Example of Configuration for Error Measurements   The old method used to measure straightness over the working length of a table involves using a straightedge mounted on the table as shown in Figures 5.15 and 5.16. The new method uses laser interferometry as shown in the same Figure 5.17. The method is very fast and more accurate compared to the previous straightedge ruler. It is still a good reference, as shown in the same figure above. Samples of error evaluations in positioning, straightness, and angular assessments have been carried out on a machine detail. Results are shown in Figure 5.18. More calibration techniques and maintenance of the precision will be discussed in Chapter 7.

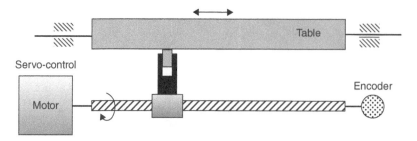

Figure 5.12 Principle of driving a table with servo control.

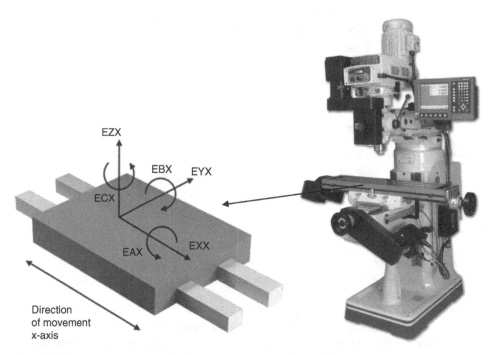

Figure 5.13 6 main errors associated to 1-axis moving table in CNC machine.

Figure 5.14 Corresponding physical Renishaw interferometer. *Source:* Courtesy Renishaw.

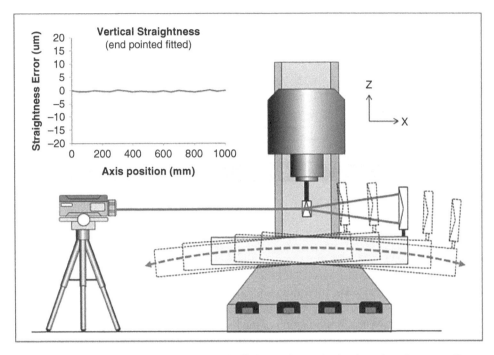

Figure 5.15 Vertical straightness measured in CNC machine tool using laser interferometry. *Source:* Courtesy Renishaw.

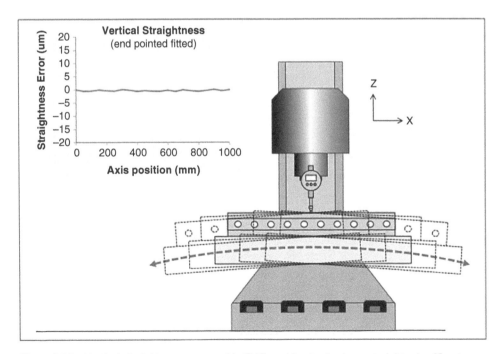

Figure 5.16 Vertical straightness measured in CNC machine tool using a straight edge [Courtesy Renishaw]. *Source:* Courtesy Renishaw.

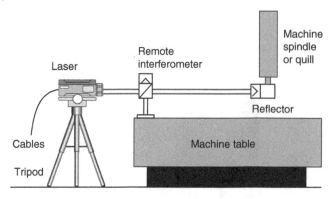

Figure 5.17 Positioning error of the machine spindle.

5.8.2 Coordinate Measuring Machine (CMM)

Description of a CMM

The coordinate measurement machines (CMM) have existed for about 60 years, starting since 1959 when the first CMM appeared at the International Machine Tool Exhibition in Paris. It was manufactured in 1951 by Ferranti, a British company that also developed the first commercial computer for a general audience. The current production of modern CMM industry is over 7,000 CMMs per year. Many of the old versions are also retrofitted and upgraded. These CMMs can be manual or driven automatically by a special program. Although CMMs are now universal in manufacturing, it is hard to imagine a world when quality control was dependent on hand measurements. Several big companies from the most developed countries of that time, including the USA, Japan, UK, Germany, and France, joined in the production of commercial CMMs during the 1960s.

The coordinate measuring machine market is expected to grow from $2.8 billion in 2018 to $4.1 billion by 2023, at a CAGR of 8.00% during the forecast period. The study involved two major activities in estimating the current size of the coordinate measuring machine (CMM) market. These are given below.

1) The CMM market, by application:
   - Quality control and inspection;
   - Reverse engineering;
   - Virtual simulation;
   - Others (profiling and performance assessment).
2) CMM market, by industry:
   - Aerospace and defense;
   - Automotive;
   - Medical;
   - Electronics;
   - Energy and power;
   - Heavy machinery;
   - Others (education, forensics, fashion and jewelry, and research).

According to BS 6808 Part 1, the coordinate measuring machine (CMM) is defined as a machine having a series of movable members, a sensing probe, and support member. The latter can be operated in such a way that the probe is brought to a fixed and known relationship with points on the work piece, and the coordinates of these points can be displayed or otherwise defined with respect

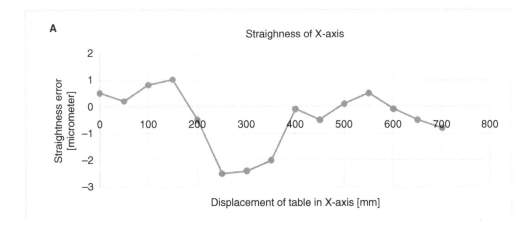

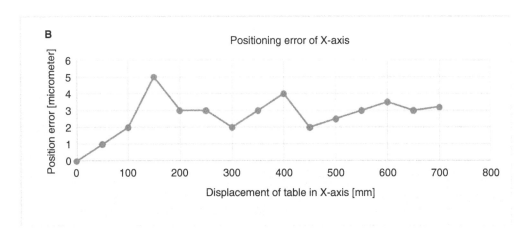

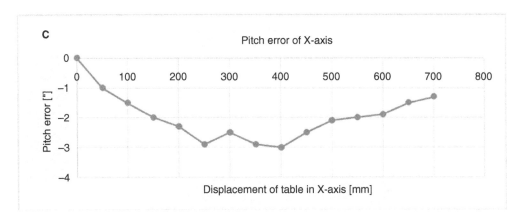

Figure 5.18 Measurement errors in axis a) Position, b) Straightness, c) Pitch error.

to the origin of the coordinate system. The coordinate system is a rectangular or Cartesian one but can also be polar or cylindrical.

The CMM is composed of two groups:

i) Mechanical System
- Structural elements (e.g., granite material);
- Kinematic supports;
- Bearing systems (e.g., air bearings);
- Guide ways (ceramic drives);
- Drive systems (actuators);
- Displacement transducers (e.g., optical encoders).

ii) Measuring System
- Probe head for all spatial directions (touch probe or noncontact probe);
- Optional remote control unit;
- Control unit (live interaction; point-point, contour path);
- PC and software to represent and to analyze the results.

Other features of CMM
- CMMs are programmable, flexible instruments used to collect and report dimensional data for virtually any type of manufactured component, for instance, engine blocks, door panels, camera bodies, and turbine blades;
- CMMs typically collect their data by touching a component with a calibrated probe as directed by an operator. The CMM records the location of these touch points;
- CMMs can also use machine vision to collect data without contacting the component (video probes and scanners);
- CMMs remove the requirement for dedicated gauging in manufacturing. Instead of buying one gage to measure 20 mm holes, another to measure 35 mm holes, and yet another to measure a surface's flatness (all which may require replacement if the component changes), the CMM can be programmed to measure any combination of features. The CMM is reprogrammed whenever the requirements change.

Types of CMM
There are several configurations of CMMs that exist already in the market and labs (Tables 5.2, 5.3, 5.4 and Figure 5.20). The question is which of the configurations is best. By best we mean precise and with 3D measurement coverage. The most accurate machines have a vertical spindle as a machine tool and here, a vertical arm with probe head in a CMM. This minimizes the gravitational influences. Accessibility to automatic loading and convenience of use could be important, and hence the horizontal arm may be better. However, the gravitational effects may have a significant influence on the geometry accuracy and may add Abbe errors. Other designs with horizontal arms are with extensions, and hence if counterbalanced, the bending moment of the column can be balanced (Figure 5.19). Some others do not have this feature and may result in errors due to increased bending moment. The recent robotic arms, for example, Faro and Romer arms, appeared as alternative portable CMM providing manufacturers with fast and easy verification of product quality. The current challenge in inspecting medium to large parts that cannot be hosted on the CMM table is that they can only be carried out by moving the instrument to the part that requires inspection. These arms can perform 3D inspection, tool certification, CAD comparison, dimensional analysis, and reverse engineering using both contact and noncontact probes. They are very

Table 5.2 Comparison between 3 different techniques and ranges measuring straightness.

Measurement	Range	Resolution	Accuracy
Straightness measurement (angular optics)	0–15 m	0.01 μm (for 100 mm base)	±0.2%
Straightness measurement (3D optics)	0–6 m	0.1 μm	±(10 + 10L) μm
Straightness measurement (Wollaston prism)	0.3–9 m (vertical range up to ±30 mm)	0.01 μm	±0.5% × L μm

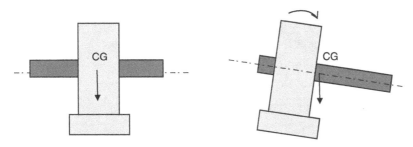

Figure 5.19 Horizontal arm inconvenience due to center of gravity CG.

Figure 5.20 Robot arms CMM (Faro and Romer). *Source:* Faro and Romer/Matt Findlay.

versatile and stable, having encoders in the joints to record any motion in 3D space through transformation motion between arms. At the moment, the measurement is only manual.

Coordinate System and Probing System

The CMM has a reference coordinate system linked to the machine itself. This is called the datum. The probe moves in 3D axis with respect to the machine reference (Figure 5.21). The contact probe gathers data by physically touching the specimen directly, which can be classified into two specific families of manual hard probes and touch trigger probes.

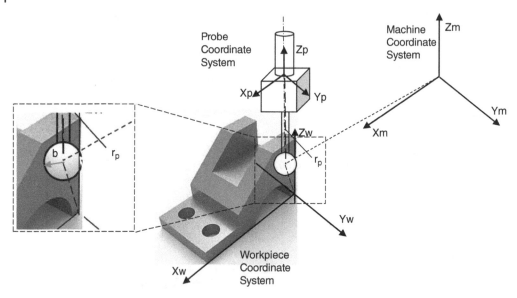

Figure 5.21 Configuration of CMM in coordinate measuring system.

The hard probes are available in a variety of configurations and are used with manual CMMs for low and medium accuracy requirements. Their repeatability quality depends on their operator touch. This hard type probe is not commonly used in mass production, which requires high-level accuracy. Touch trigger probe is the commonly used type in CMM; it has a precision-built-in and

Table 5.3 Types of CMM configuration -1

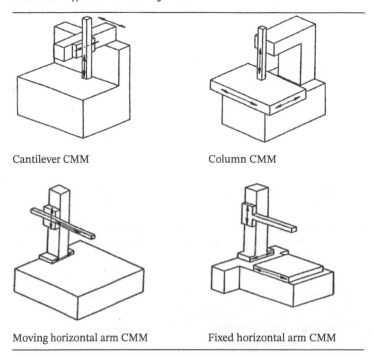

| Cantilever CMM | Column CMM |

| Moving horizontal arm CMM | Fixed horizontal arm CMM |

Table 5.4 Types of CMM configuration -2

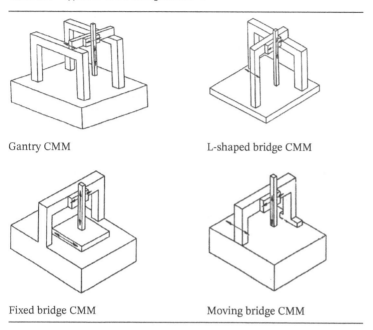

Gantry CMM   L-shaped bridge CMM

Fixed bridge CMM   Moving bridge CMM

touch-sensitive device that generates an audible signal through probe tip contact (Figure 5.22) with a visual LED light. The probe can be rotated in 3D axes and has multiple add-ons optional lengths depending on the features to inspect. The CMM is very versatile, and flexible data can be recorded. The probes can also be fitted in a CNC for in-process inspection.

A part is manufactured in a CNC machine and can be very complex in terms of the number of features that are functional or nonfunctional and the number of tolerances that are also functional or nonfunctional. As seen previously, the machine tool has variations over time for various reasons, and hence the form (shape), its position, size, and tolerance can be affected. There is a need for a

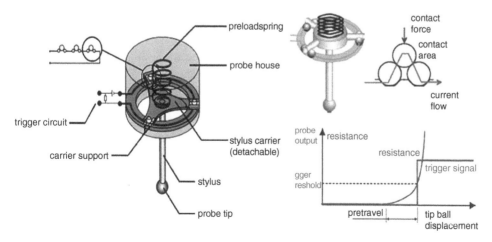

Figure 5.22 Touch probe system.

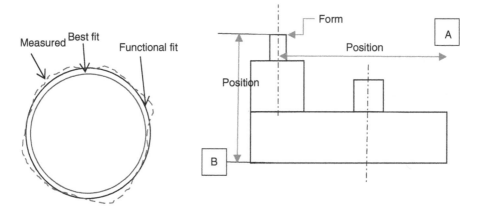

Figure 5.23 Features to inspect.

range of measurement techniques to detect these in one easy process. The detection can be split into two processes.

1) Inspection of Size and Position/Location Since specific sizes are needed for inspection, discrete point measurement can be carried out. The data is rated between 1 and 2 points per second. Some conditions on the probe and the stylus need to be observed, such as the effect of length and wear.

2) Inspection of Form (Feature or Geometry) Here, multiple points are needed to build and inspect the form, hence the need for scanning.

2.1. Form needs clearance and position/location: Only discrete points are required and hence measured. A part can be inspected in 3D dimensions in a CMM.

2.2. The functional fits become important, so the form is critical and hence requires scanning to collect points and detect fitness. There is a need to define the touch probes necessary for the various features to be inspected on the part, for example, long cylinder or hidden, as they can be far to reach. Based on touch probing experience, errors can emerge when collecting several points and forming a contour, as shown in Figure 5.23.

3) Probing Requirements Various set of probes exist, as not all parts can be inspected with only one probe. In the range of the various features in parts, the inspection requires a range of sensors; it is recommended to have sensors changing system from which the appropriate probe/sensor is selected automatically when needed without requalification. The changing system allows automatic switching, sensor recognition, electrical connections, and alignment, all automated. Key differences exist and are defined here.

Touch-trigger probe: This probe is suitable for inspection of discrete points, size, and position of features (Figure 5.24). They are affordable, small sized, and very versatile. They are used for inspecting 3D prismatic parts with standard surfaces, but less information is given on the surface as it assumes a flat surface with 3 points. Dynamic performance does not improve the measurement accuracy. They can have various heads and long extensions, depending on the feature to measure.

The probe tip, that is, the tip of the probe length, has coordinates with respect to the probe coordinate system. Hence, any touch to the part will be registered with respect to the CMM

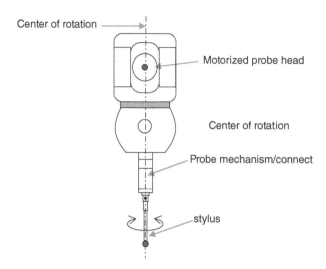

Figure 5.24 Fundamental touch probe.

reference system. The radius of the tip is taken into consideration. To minimize wear during contacts, the tip is made of a hard material.

A coordinate measurement machine (CMM) typically uses probes to sense a part. The latter is made of multiple points. The CMM senses the position of a point in space and measures it based on its distance from a three-dimensional reference position. CMMs are often used to ensure a part or assembly falls within a specified range (tolerance) of the design intent.

The probe gathers the points through a human operator or via automation through a process called direct computer control (DCC). The DCC CMMs can assess identical objects repeatedly through programming.

However, it happens that the probe ball tip may have an error due to the finite size of the ball tip, and hence the contact point on a cylindrical surface will be along the stylus axis, but alternatively, this can be at some point on the side of the ball where the test surface and the stylus tip ball slope match horizontally. The ball does not touch the test artifact specimen along the same stylus slope angle and hence an error E (Figure 5.25) exist in the measured length for any measurement point where the test part surface slope (θ) is not zero degree, that is, when the tip ball is at the bottom of the test artifact.

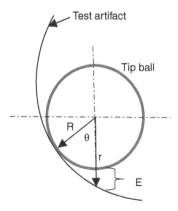

Figure 5.25 Error E of the probe at angle θ.

Usually, this form of error is removed through calibration. A typical probing error according to ISO 10360-2 is established using stylus with a grade 5 ball, for instance, sphericity at 0.13 µm. Probing using CMMs is also used for reverse engineering to rebuild the part shape and hence ultimately issue a technical drawing.

Scanning Probes: These are ideal for features where form/geometry is important and support digitizing of contoured surfaces. They can be passive (for quick probing) and active probes that are slow while settling at a target force to capture a reading. The data is captured at an average of 500 points per second and may degrade the stylus over time, for instance, because of wear. Large data can be collected for better representation of forms and building of datum stability. Digitizing leads to collection of large number of points to build the surfaces and shapes and help reverse the engineering parts. This operation can also be carried out in a machine tool. These sensors are relatively costly.

4) Performance of Scanning The modern CMMs are becoming very quick in 3 axes motion, yet the conventional scanning is performed in an average low speed of 15 mm/s. The mechanical reason behind it is that it induces dynamic forces that affect the precision of measurement. The dynamic errors are usually not only attributed to accelerations but also to inertial forces and vibration, for instance, the ground. According to Figure 5.21 an increase in acceleration increases the error.

In dynamic measurement at high speed, extra errors are generated. Figure 5.22 shows an increase in errors of form to almost 8 µm when measuring a ring of 60 mm diameter. It is worth mentioning that the effect of the probe dynamics is minor compared to machine dynamics. This is why the structure and material selection for the CMM are fundamental.

Why This Is Important? The scanning will result in a mesh with a resolution that is important for the accuracy of measurement in digital twin metrology or in reverse engineering.

The refinement of the mesh of any scanned feature is an art that depends on the probe (contact or contactless), the type of the surface (shiny or mat for optical scanning), the scanning resolution, and the mesh technique programmed in the post processing software. Figure 5.29 shows two different meshes quality of a feature.

Optical probe: They are suitable for noncontact probing to capture multiple external features that are not very deep, such as printed circuit boards. The optical probe is equipped with a combined video, optics, and laser technology (Fig. 5.26). Those with a laser line can measure coordinates at discrete points on the work piece, similar to touch-trigger contact probes but with a laser beam instead of a stylus and ball. The coordinates are calculated through a triangulation method. A large amount of data is quickly collected, boosting throughput and efficiency. It was noticed that drawbacks included reduced accuracy due to lack of discrete physical measurements; if the surface finish is smooth and shiny, reflections from noncontact sensors create inaccurate data. An alternative solution is to spray the surface with a very thin layer of mat paint usually used for reverse engineering. The increase in cost of equipment is becoming effective with mass production. An example of a CMM scanning an engine is shown in Figure 5.30.

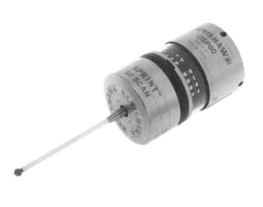

Figure 5.26 Measurement touch probe –OSP60 (Renishaw) and Metris scan probe. *Source:* Renishaw.

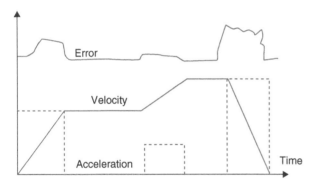

Figure 5.27 Source of dynamic errors in touch probes.

CMM Performance Evaluation Methods

The objective is to minimize the effects of the operator and testing equipment. Several international standards are available to manage this performance and are given below:

a) ISO 10360-International
 i) Definition and application of the fundamental geometric principles;
 ii) Performance assessment of CMMs.

b) BS 6808-British Standard
 i) Define relative terms to CMM;
 ii) Methods for verifying the performance of CMM;
 iii) Code of practice: environmental test, choice of artifact and machines with rotary tables.

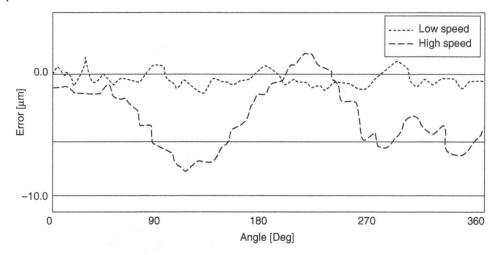

Figure 5.28 Form error in low and high measurement speed probing.

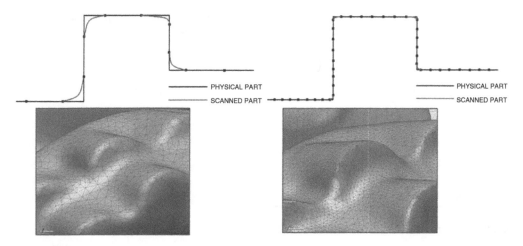

Figure 5.29 Resolution of the feature depending on the resolution of the scanned mesh. *Source:* Courtesy of CREAFORM.

c) VDI/VDE 2617-German Standard

i) Generalities defining accuracy of CMM and required testing conditions;
ii) Methods for checking manufacturer's specifications;
iii) Components of measurement deviation of the machine.

d) ANSI/ASME B89.1.12-American Standard

– Requirements and methods for specifying and testing performance of CMMs. (3 axis and rotary table). Large CMMs.

e) JIS B7440-Japanese Industrial Standard

– Measurement accuracy and motion accuracy of each axis.

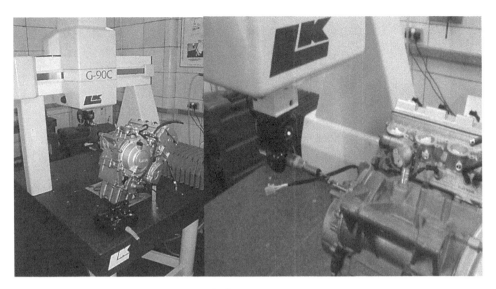

Figure 5.30 A) CMM inspecting a motor. B) Close view to the probe.

Types of Errors in CMMs

1) Geometric Errors Caused by the structural elements and include 21 parametric error components of CMM:

– 3 (scale + 2 straightness);
– 3 rotations;
– 3 out-of-squareness.

2) Kinematic Errors Positioning problems: probing speed, probing directions, motors, and so forth.

3) Thermal Error Thermal effects [10]are the largest single source of apparent non-repeatability and inaccuracy in CMMs with their complex, nonlinear nature; it has been widely addressed for CMMs. Based on ISO 10360-series, the inspection of 21 geometrical error sources are available in the following typical checklist:

a) Inspect and clean X, Y, and Z scales. All calibrations must start with a thorough cleaning and inspection of the scales.
b) Check and adjust all reader heads. Reader heads are checked with an oscilloscope and adjusted to the proper output.
c) Check X, Y and Z fine adjustment. The fine adjustment mechanisms are checked and rechecked for wear and tightened as necessary.
d) Inspect and adjust beam/plate parallelism. Beam parallelism is checked to the CMM's working surface [11, 12].
e) Inspect and adjust perpendicularity. Perpendicularity is checked in the XY, ZX, and ZY planes and adjusted to the best possible accuracy.
f) Inspect and adjust linearity. Linearity is checked in the X-, Y-, and Z-axis and is adjusted as required.
g) Check repeatability in X-, Y-, and Z-axis. Each axis is checked using your electronic touch probe when applicable.
h) Issue an official calibration certificate, accredited to standards.

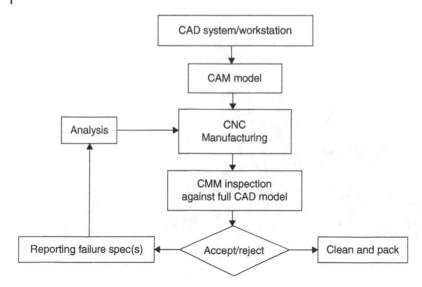

Figure 5.31 Digital manufacturing and inspection.

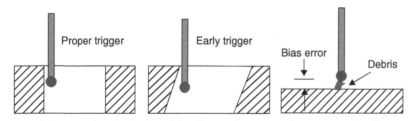

Figure 5.32 Stylus qualification and Probing strategy.

CMMs are widely used in manufacturing for inspection services, where the produced parts are subject to inspection and verification of design specifications before they are accepted (Fig. 5.31).

It is necessary to know the accuracy of the CMM equipment in use and also to confirm that it meets the initial requirements with respect to tolerances in the machining specifications. This is usually carried out for mass production, and to secure the accuracy, series standards ISO 10360 are used for this purpose.

CMMs are usually used in serial processes, that is, when the parts are machined, they are then cleaned and tagged to go for inspection in CMMs and other inspection machines for any specific feature. If they are rejected, there is a need to write a report on the failing specification for statistical analysis and to find out why this particular feature has failed to meet specification, especially if this failure was repetitive. Attention is to be paid to the faulty techniques used to inspect features, as shown in Figure 5.32.

Principle of Coordinate Measurement
The process of measuring starts by calibration or qualification of the measuring probe with the use of a known artifact, for instance, a reference ball (Fig. 5.33b). This part of the qualification is also available in the software to manage readings. After that, the center of the stylus is known with

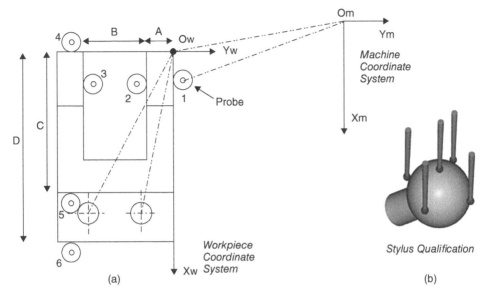

Figure 5.33 Probing most common mistakes to avoid.

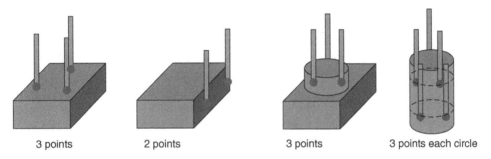

Figure 5.34 Basic probing for features.

respect to the zero of the CMM and the diameter of the tip (Fig. 5.33a). The probe moves over five points on the sphere while the sphere is mounted firmly on the table and the right name of the probe is entered correctly into the software.

The previous Figure 5.33 shows the probing strategy. After probe qualification, the part is located on the table of the CMM, with an overall probing used to determine the origin of the part Ow (Fig. 5.34) with respect to the CMM origin. Then any measured feature will encompass shape, size, and location with respect to the origin of the part and by default with respect to the CMM origin. Manually, probe position 1 to probing 2 will give dimension A by calculating the difference between the two points in the movement axis, then probing 3 to 4 will give B, and so on. When the feature is selected, for example, circle, to measure its diameter, the software will expect 3 points, if it is a planar surface, also 3 points are required to determine the position of the plane. If lateral points are added, then a 3D position is defined.

This type of inspection can be done automatically by direct computer control (DCC) that can be programmed to repeatedly measure identical parts. This is usually carried out by specialized software that also includes GD&T-based inspection.

Qualification

a) Part-to-CAD Inspection Technique   The quality inspection is needed for all parts manufactured in a production line. The method is carried out either using the systematic method or by random sampling. The quality control is comprehensive and requires a dense sampling of comparison points. The method becomes lengthy on free-form or complex shapes. Hence, the technique used is the superimposition of CAD of the part from scanning and the real CAD with the specification in tolerance and GD&T. This is called the part-to-CAD comparison, and then a 3D deviation over the whole CAD can be analyzed with color maps and user-defined comparison points based on all the measurements taken by the 3D scanner. The inspector can easily visualize and document a complete inspection of the part on all its surfaces in an official document using the high-density data provided by the scanner.

Alternatively, CMMs make use of the advantages offered by scanning technology. They are equipped with a scanning probe as contact or noncontact (Fig. 5.26) with special software modules to digitize complex profiles and shapes. The scanning option allows users to quickly collect a large number of data points by moving a hard probe, for example, along the surface of a work-piece feature, maintaining contact with the work piece throughout the operation. These systems can be used to digitize contoured components and prismatic parts for dimensional verification purposes. This process goes through the collection of scanned points to form a cloud of points. Special treatment is carried out to convert 3D scanned data into a parametrically accurate 3D CAD model; this is called reverse engineering.

Two applications result from this scanning: full CAD inspection carried out in superposition of the resulted scan on the original CAD file, and hence a full inspection report is generated. The second application is a consequence of the first in order to obtain a CAD through full scanning for reverse engineering. Usually, this is executed for obsolete parts that lack CAD files.

b) Full Scanning   This begins from a physical part, shown in Figure 5.35, to obtain a cloud of points and then finding the best mesh and forming the virtual current part. The CAD file will be superimposed on this constructed file to compare and determine the quantified differences. The full inspection option is very important depending on the features to be inspected (Fig. 5.36).

3D overview on the part inspection can be also generated to check specific details (Fig. 5.36 and Fig. 5.37) showing 3D inspection with range of variation. This part CAD inspection is a very genuine and useful method, whose process is shown in Figure 5.38. The scan is made on the part, and the virtual 3D part is built and superimposed to the theatrical one. This comparison shows immediately

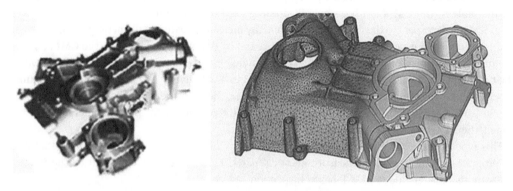

Figure 5.35   Scan-to-CAD of the part. *Source:* Courtesy of CREAFORM.

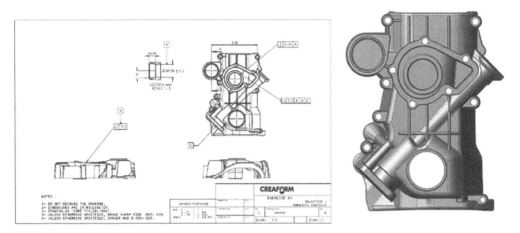

Figure 5.36 CAD file with GD&T and 3D scan of the part. *Source:* Courtesy of CREAFORM.

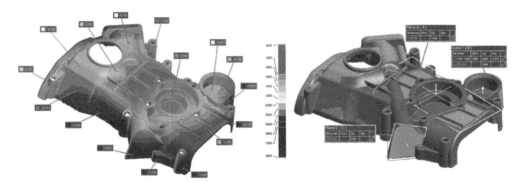

Figure 5.37 Sample part with full inspection in distributed errors. *Source:* Courtesy of CREAFORM.

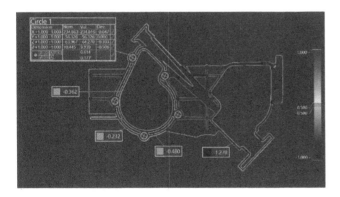

Figure 5.38 Partial inspection report (courtesy of CREAFORM). *Source:* Courtesy of CREAFORM.

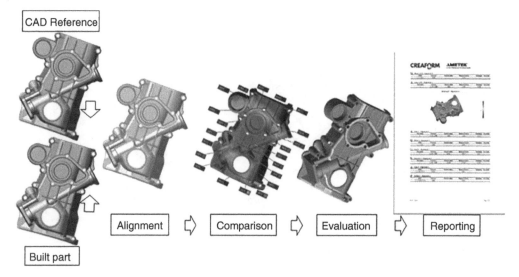

Figure 5.39 Part to CAD inspection.

the difference between the expected 3D part and the one obtained from the scanning after the part has been manufactured. An inspection report is issued for reference (Fig. 5.39).

c) How Is It Possible to Obtain Fast Scans from Reverse Engineering? The recent equipment has made it very simple and fast to acquire large files of scanned parts. The purpose of this reverse engineering is twofold: one is to obtain a CAD file from an obsolete part without CAD file and re-CAD it with strength analysis based on new needs. The second is used for inspection or part-to-CAD comparison, as introduced previously. Hence, reverse engineering will allow re-manufacturing and full inspection. The equipment is divided based on the size and application as shown in Figures 5.40–5.41.

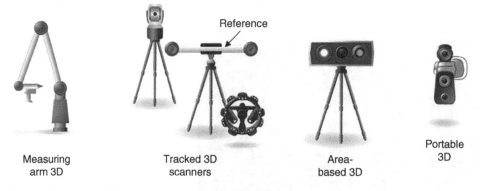

Figure 5.40 Various 3D scanner categories from CREAFORM.

Figure 5.41 Tracked 3D scanner from CREAFORM *Source:* Courtesy of CREAFORM.

d) Cases Studies

Case study: CMM and Reverse Engineering

One of the great applications is reverse engineering when parts are obsolete or when CAD drawings are missing. Starting from an existing part (Fig. 5.42) and using a scanner (handy or linked to a reference, e.g., CMM), a 3D model can be built out of a cloud of points (Fig. 5.43). This can result in a CAD file for inspection or other postprocessing such as re-CAD.

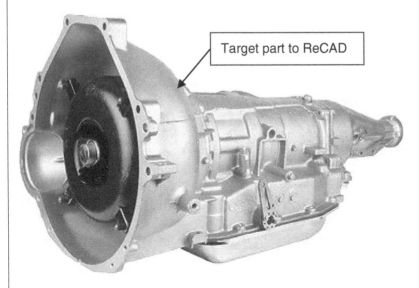

Target part to ReCAD

Figure 5.42 Target part to ReCAD.

(Continued)

(Continued)

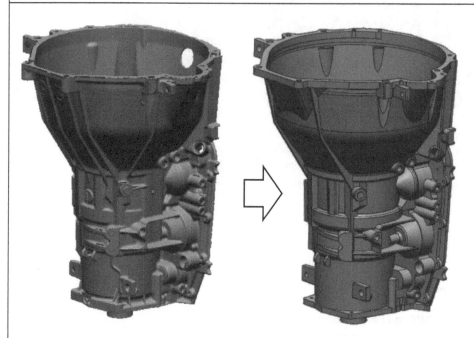

Figure 5.43 3D scan and model RE-CAD for the purpose. *Source:* Courtesy of CREAFORM.

Case study: CMM and Digital Inspection

i) Quality Control and Dimensional Inspection

There are various techniques used in industry to carry out quality checks and inspection of parts. It is expected that 3D measurement technology is seamlessly merged in the work process with efficacy.

Several scenarios exist already as shown below:

i) CMM with full report software to carry out inspection, and the software issues a full report, such as in Figure 5.44;
ii) Automating the full inspection of the part;
iii) Possibility of in-process inspection during machining;
iv) Integration of metrology and scanning in engineering processes and production;
v) Development of customized applications and interfaces for the user.

Figure 5.44 shows the use of both manufactured part and deviation file in CAD format file in order to locate all deviation with respect to tolerances inspected via CMM. The second file is the full report needed by engineers. The legend shows the spectrum of deviation and hence each inspected feature is labeled with a color.

ii) Inspection of a part

A handy probe is used to inspect the part shown in Figure 5.45. This is a quality dimensional control of the part where 21 points (Table 5.5, Fig. 5.46–47) are checked against assigned tolerances for each point, and a deviation is estimated to decide to pass or fail the point of inspection.

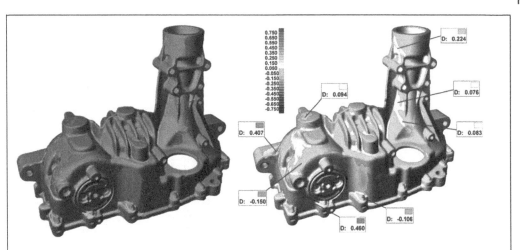

Figure 5.44 Inspection report (courtesy CREAFORM) Future of CMM with robotics and 3D scanners. *Source:* Courtesy of CREAFORM.

Figure 5.45 Reference frame 1 (courtesy CREAFORM).

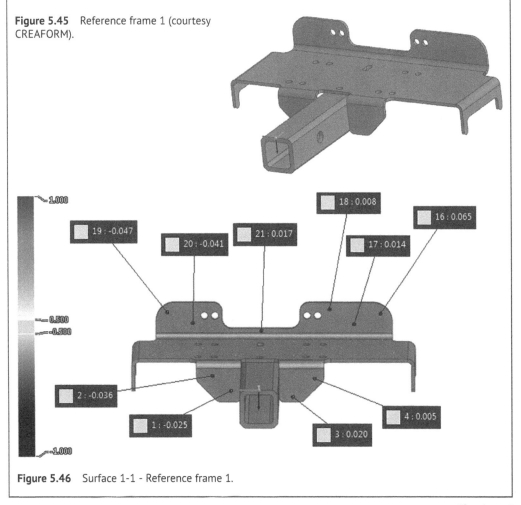

Figure 5.46 Surface 1-1 - Reference frame 1.

(Continued)

(Continued)

The inspection process uses the following methodologies (Table 5.7):

The inspection process will start by importing the reference model, then create control features on the reference, acquire data on the physical object by probing or scanning, aligning data vs reference(s), extract measurements of control features, and finally report the inspection.

Table 5.5

Surface 1 - Reference Frame 1

Probed Points	Tolerance	X	Y	Z	Deviations	Out of Tol.
1	±0.500	39.032	153.525	34.551	−0.025	
2	±0.500	63.897	153.536	14.258	−0.036	
3	±0.500	−44.468	153.480	42.900	0.020	
4	±0.500	−73.544	153.495	17.863	0.005	
5	±0.500	144.860	184.863	−6.057	0.057	
6	±0.500	135.814	262.704	−6.070	0.070	
7	±0.500	77.978	227.348	−6.078	0.078	
8	±0.500	23.645	275.209	−6.017	0.017	
9	±0.500	33.386	184.441	−6.048	0.048	
10	±0.500	−3.337	223.273	−6.033	0.033	
11	±0.500	−43.630	271.762	−5.944	−0.056	
12	±0.500	−44.681	184.314	−6.001	0.001	
13	±0.500	−107.965	231.765	−5.930	−0.070	
14	±0.500	−162.668	278.685	−5.888	−0.112	
15	±0.500	−178.660	202.249	−5.937	−0.063	
16	±0.500	−160.475	298.685	−37.790	0.065	
17	±0.500	−124.904	298.736	−23.187	0.014	
18	±0.500	−92.319	298.742	−43.466	0.008	
19	±0.500	126.249	298.797	−39.555	−0.047	
20	±0.500	91.773	298.791	−24.905	−0.041	
21	±0.500	−0.921	298.733	14.499	0.017	

GD&T	Tolerance		Measured Values			Out of Tol.
⌓ 1.000 A B C	1.000		0.224			

Circle 1 - Reference Frame 1

Dimensions	Tolerance	Nominal Values	Measured Values	Deviations	Out of Tol.
X	±1.000	0.000	0.002	0.002	
Y	±1.000	182.250	182.258	0.008	
Ø	±1.000	9.250	9.226	−0.024	
GD&T	Tolerance		Measured Values		Out of Tol.
⊕ Ø1.000 A B C	1.000		0.007		

Table 5.6 Actual Size = 15.0000 mm; Tolerance = 0.05.

Sample #	Measured feature [mm]					Average size	Deviation					Average deviation	Deviation from given tolerance value		Given tolerance value		
	1	2	3	4	5		1	2	3	4	5		Positive	Negative	Over size	Correct size	Under size
1	14.9536	14.9536	14.9622	14.9597	14.9585	14.9575	−0.0464	−0.0464	−0.0378	−0.0403	−0.0415	−0.0425		−0.0075		P	
2	14.9146	14.9158	14.9146	14.9170	14.9194	14.9163	−0.0854	−0.0842	−0.0854	−0.0830	−0.0806	−0.0837		−0.0337		P	
3	15.0171	15.0171	15.0146	15.0159	15.0171	15.0164	0.0171	0.0171	0.0146	0.0159	0.0171	0.0164	0.0329			P	
4	14.8035	14.8035	14.7937	14.8120	14.8157	14.8057	−0.1965	−0.1965	−0.2063	−0.1880	−0.1843	−0.1943		−0.1443			NP
5	14.9585	14.9487	14.9487	14.9487	14.9524	14.9514	−0.0415	−0.0513	−0.0513	−0.0513	−0.0476	−0.0486		−0.0014		P	
6	15.0134	15.0122	15.0122	15.0073	15.0085	15.0107	0.0134	0.0122	0.0122	0.0073	0.0085	0.0107	0.0415			P	
7	14.9915	14.9939	14.9915	14.9915	14.9927	14.9922	−0.0085	−0.0061	−0.0085	−0.0085	−0.0073	−0.0078		−0.0422		P	
8	14.9890	14.9841	14.9841	14.9878	14.9854	14.9861	−0.0110	−0.0159	−0.0159	−0.0122	−0.0146	−0.0139		−0.0361		P	
9	14.9976	14.9817	14.9854	14.9768	14.9902	14.9863	−0.0024	−0.0183	−0.0146	−0.0232	−0.0098	−0.0137		−0.0363		P	
10	15.0806	15.0781	15.0793	15.0745	15.0732	15.0771	0.0806	0.0781	0.0793	0.0745	0.0732	0.0771	0.0232			P	

(Continued)

(Continued)

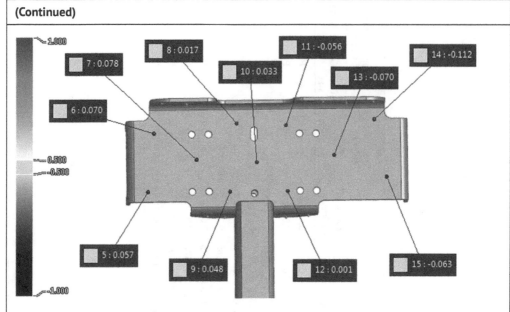

Figure 5.47 Surface 1-2 - Reference frame 1.

Table 5.7 Inspection methodologies as discussed with Creaform.

Inspection Methodologies	Impact on Workflow	Decisive Factor
1. With a CAD model as reference	Surface comparison and control feature creation directly on the CAD model are relatively simple	Having a CAD model available
2. With a 3D scan as reference	General surface comparison is simple, while control feature creation is more complicated	No CAD model is available or a before-after comparison is necessary
3. With a 2D drawing as reference	Surface comparison are impossible. Controls are specified, but creation of the control feature is significantly more complicated.	No CAD model is available and specific feature inspection is required

5.9 Metrology Laboratory Test for Students

This test is about dimensional verification of parts that have been manufactured by the same CNC machine according to given dimensional specifications. The set objectives for this exercise are as follows:

1) To find the average size of the component;
2) To determine the variation of the part dimension with respect to given tolerance;
3) To segregate the component under the headings: oversize, undersize, and correct size.

Several manufactured components from aluminum are given to students to check a one-dimensional size in each part. The equipment and instrumentation used for this exercise are described hereafter (Fig. 5.48). The protocol interface module (PIM) is a PLC compatible

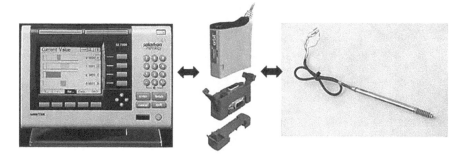

Figure 5.48 Orbit system comprising monitor, interface module and LVDT sensor.

EtherNet/IP to ORBIT®3 interface module to allow immediate connections between any Ether-Net/IP enabled controller and Solartron's flexible ORBIT®3 digital measurement system. This PIM interfaces most LVDT, rotary encoders, and temperature and pressure sensors. The linear velocity differential transducer (LVDT) is a common type of electromechanical transducer that can convert the rectilinear motion of an object to which it is coupled mechanically into a corresponding electrical signal.

A linear encoder is a sensor, transducer, or read-head paired with a scale that encodes position. The sensor reads the scale in order to convert the encoded position into an analog or digital signal, which can then be decoded into position by a digital readout (DRO) or motion controller. Solartron metrology linear encoders are highly accurate optical gauges designed for use in applications where consistent submicron measurement is required [13–15]. Accuracy is maintained along the entire measurement range. The LVDT sensor used here has a stroke length of 26 mm and a resolution of 0.05 μm.

A magnetic stand is used to hold the instrument on a bench (Fig. 5.49). It is a magnetic fixture based on a magnet that can effectively be turned on and off at will; they are often used in optics and metalworking, for instance, to hold a dial indicator.

An optical table is a vibration control platform that is used to support systems used for laser- and optics-related experiments, engineering, and manufacturing. A granite or reference table if well leveled can also be used. The instrument used to show and record data is the SI 7000 digital readout, 16 Channel Advanced Readout System (Fig. 5.50). This is for performing single or multi-point gauge measurements at very high levels of precision and accuracy. Dimensional inspection

Figure 5.49 Magnetic holder and Optical Table.

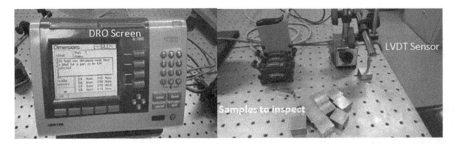

Figure 5.50 Experimental measurement.

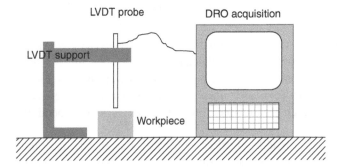

Figure 5.51 Schematic of the test.

of components can be made using a range of orbit transducers, encoders, and other sensors as part of in-line production activities or final quality inspection.

Measurements can be conducted fully under the control of the operator control or can be semi-automated and conducted in conjunction with a multipoint fixture gauge system. 10 parts made of steel by a CNC machine are inspected using an LVDT sensor for dimensional measurement having a resolution of one micrometer and recently calibrated (Fig. 5.51). Students are requested to go through the references [16–21]

Appendix A: A1- List of International Standards in Metrology

ISO Metrology and measurement including measuring instruments in general reference numbers, standard measures, general aspects of reference materials, etc.

Standard	Title	Technical committee
ISO 3:1973	Preferred numbers — Series of preferred numbers	ISO/TC 19
ISO 17:1973	Guide to the use of preferred numbers and of series of preferred numbers	ISO/TC 19
ISO/IEC Guide 98-1:2009	Uncertainty of measurement — Part 1: Introduction to the expression of uncertainty in measurement	ISO/TMBG
ISO/IEC Guide 98-3:2008	Uncertainty of measurement — Part 3: Guide to the expression of uncertainty in measurement (GUM:1995)	ISO/TMBG
ISO/IEC Guide 98-3:2008/Suppl 1:2008	Uncertainty of measurement — Part 3: Guide to the expression of uncertainty in measurement (GUM:1995) — Supplement 1: Propagation of distributions using a Monte Carlo method	ISO/TMBG

Standard	Title	Technical committee
ISO/IEC Guide 98-3:2008/Suppl 1:2008/Cor 1:2009	Uncertainty of measurement — Part 3: Guide to the expression of uncertainty in measurement (GUM:1995) — Supplement 1: Propagation of distributions using a Monte Carlo method — Technical Corrigendum 1	ISO/TMBG
ISO/IEC Guide 98-3:2008/Suppl 2:2011	Uncertainty of measurement — Part 3: Guide to the expression of uncertainty in measurement (GUM:1995) — Supplement 2: Extension to any number of output quantities	ISO/TMBG
ISO/IEC Guide 98-4:2012	Uncertainty of measurement — Part 4: Role of measurement uncertainty in conformity assessment	ISO/TMBG
ISO/IEC FD Guide 98-6	Uncertainty of measurement — Part 6: Developing and using measurement models	ISO/TMBG
ISO/IEC Guide 99:2007	International vocabulary of metrology — Basic and general concepts and associated terms (VIM)	ISO/TMBG
ISO 497:1973	Guide to the choice of series of preferred numbers and of series containing more rounded values of preferred numbers	ISO/TC 19
ISO 2533:1975	Standard Atmosphere	ISO/TC 20/SC 6
ISO 2533:1975/Add 1:1985	Standard Atmosphere — Addendum 1: Hypsometrical tables	ISO/TC 20/SC 6
ISO 2533:1975/Add 2:1997	Standard Atmosphere — Addendum 2: Extension to - 5000 m and standard atmosphere as a function of altitude in feet	ISO/TC 20/SC 6
ISO 5479:1997	Statistical interpretation of data — Tests for departure from the normal distribution	ISO/TC 69
ISO 5725-1:1994	Accuracy (trueness and precision) of measurement methods and results — Part 1: General principles and definitions	ISO/TC 69/SC 6
ISO 5725-1:1994/Cor 1:1998	Accuracy (trueness and precision) of measurement methods and results — Part 1: General principles and definitions — Technical Corrigendum 1	ISO/TC 69/SC 6
ISO 5725-2:2019	Accuracy (trueness and precision) of measurement methods and results — Part 2: Basic method for the determination of repeatability and reproducibility of a standard measurement method	ISO/TC 69/SC 6
ISO 5725-3:1994	Accuracy (trueness and precision) of measurement methods and results — Part 3: Intermediate measures of the precision of a standard measurement method	ISO/TC 69/SC 6
ISO 5725-3:1994/Cor 1:2001	Accuracy (trueness and precision) of measurement methods and results — Part 3: Intermediate measures of the precision of a standard measurement method — Technical Corrigendum 1	ISO/TC 69/SC 6
ISO/WD 5725-3	Accuracy (trueness and precision) of measurement methods and results — Part 3: Intermediate precision and alternative designs for collaborative studies	ISO/TC 69/SC 6
ISO 5725-4:2020	Accuracy (trueness and precision) of measurement methods and results — Part 4: Basic methods for the determination of the trueness of a standard measurement method	ISO/TC 69/SC 6

Standard	Title	Technical committee
ISO 5725-5:1998	Accuracy (trueness and precision) of measurement methods and results — Part 5: Alternative methods for the determination of the precision of a standard measurement method	ISO/TC 69/SC 6
ISO 5725-5:1998/Cor 1:2005	Accuracy (trueness and precision) of measurement methods and results — Part 5: Alternative methods for the determination of the precision of a standard measurement method — Technical Corrigendum 1	ISO/TC 69/SC 6
ISO 5725-6:1994	Accuracy (trueness and precision) of measurement methods and results — Part 6: Use in practice of accuracy values	ISO/TC 69/SC 6
ISO 5725-6:1994/Cor 1:2001	Accuracy (trueness and precision) of measurement methods and results — Part 6: Use in practice of accuracy values — Technical Corrigendum 1	ISO/TC 69/SC 6
ISO 10012:2003	Measurement management systems — Requirements for measurement processes and measuring equipment	ISO/TC 176/SC 3
ISO 11095:1996	Linear calibration using reference materials	ISO/TC 69/SC 6
ISO 11843-1:1997	Capability of detection — Part 1: Terms and definitions	ISO/TC 69/SC 6
ISO 11843-1:1997/Cor 1:2003	Capability of detection — Part 1: Terms and definitions — Technical Corrigendum 1	ISO/TC 69/SC 6
ISO 11843-2:2000	Capability of detection — Part 2: Methodology in the linear calibration case	ISO/TC 69/SC 6
ISO 11843-2:2000/Cor 1:2007	Capability of detection — Part 2: Methodology in the linear calibration case — Technical Corrigendum 1	ISO/TC 69/SC 6
ISO 11843-3:2003	Capability of detection — Part 3: Methodology for determination of the critical value for the response variable when no calibration data are used	ISO/TC 69/SC 6
ISO 11843-4:2003	Capability of detection — Part 4: Methodology for comparing the minimum detectable value with a given value	ISO/TC 69/SC 6
ISO 11843-5:2008	Capability of detection — Part 5: Methodology in the linear and non-linear calibration cases	ISO/TC 69/SC 6
ISO 11843-5:2008/Amd 1:2017	Capability of detection — Part 5: Methodology in the linear and non-linear calibration cases — Amendment 1	ISO/TC 69/SC 6
ISO 11843-6:2019	Capability of detection — Part 6: Methodology for the determination of the critical value and the minimum detectable value in Poisson distributed measurements by normal approximations	ISO/TC 69/SC 6
ISO 11843-7:2018	Capability of detection — Part 7: Methodology based on stochastic properties of instrumental noise	ISO/TC 69/SC 6
ISO/DIS 16269-3	Statistical interpretation of data — Part 3: Tests for departure from the normal distribution	ISO/TC 69
ISO/TS 17503:2015	Statistical methods of uncertainty evaluation — Guidance on evaluation of uncertainty using two-factor crossed designs	ISO/TC 69/SC 6
ISO/TR 21074:2016	Application of ISO 5725 for the determination of repeatability and reproducibility of precision tests performed in standardization work for chemical analysis of steel	ISO/TC 17/SC 1

Standard	Title	Technical committee
ISO 21748:2017	Guidance for the use of repeatability, reproducibility and trueness estimates in measurement uncertainty evaluation	ISO/TC 69/SC 6
ISO/TS 21749:2005	Measurement uncertainty for metrological applications — Repeated measurements and nested experiments	ISO/TC 69/SC 6
ISO/TR 22971:2005	Accuracy (trueness and precision) of measurement methods and results — Practical guidance for the use of ISO 5725-2:1994 in designing, implementing and statistically analyzing interlaboratory repeatability and reproducibility results	ISO/TC 69/SC 6
ISO/DIS 23131	Ellipsometry — Principles	ISO/TC 107
ISO/TS 23165:2006	Geometrical product specifications (GPS) — Guidelines for the evaluation of coordinate measuring machine (CMM) test uncertainty	ISO/TC 213
ISO/TS 28038:2018	Determination and use of polynomial calibration functions	ISO/TC 69/SC 6

A-2 ASME standards for dimensional metrology, calibration of instruments and uncertainty evaluation.

Code	Title
B89.1.14	Calipers
B46.1	Surface texture (surface roughness, waviness, & lay)
B89.1.13	Micrometers
B89.3.7	Granite Surface Plates
B89.1.10M	Dial Indicators (for Linear Measurements)
B89.6.2	Temperature and Humidity Environment for Dimensional Measurement
B89.7.3.1	Guideline for Decision Rules: Considering Measurement Uncertainty in Determining Confirm to Specifications
B89.1.5	Measurement of Plain External Diameters for use as Master Discs or Cylindrical Plug Gages
B89.1.9	GAGE BLOCKS
B89.1.17	Measurement of Thread Measuring Wires
B89.1.6	Measurement of Plain Internal Diameters for Use as Master Rings or Ring Gages
B89.7.3.2	Guidelines for the Evaluation of Dimensional Measurement Uncertainty (Technical Report)
B89.4.19	Performance Evaluation of Laser-Based Spherical Coordinate Measurement Systems
B89.7.2	Dimensional Measurement Planning
B89.4.10	Methods for Performance Evaluation of Coordinate Measuring System Software
B89.4.10360.2	Acceptance Test and Reverification Test for Coordinate Measuring Machines (CMMs) Part 2: CMMs Used for Measuring Linear Dimension (Technical Report)
B89.4.22	Methods for Performance Evaluation of Articulated Arm Coordinate Measuring Machines

Code	Title
B89.1.2M	Calibration of Gage Blocks by Contact Comparison Methods, Through 20in. and 500mm
B89.7.4.1	Measurement Uncertainty and Conformance Testing: Risk Analysis (An ASME Technical Report)
PDS-1.1	Dimensioning, Tolerancing, Surface Texture, and Metrology Standards-Rules for Drawings with Incomplete Reference to Applicable Drawing Standard
B89.7.5	Metrological Traceability of Dimensional Measurements to the SI Unit of Length (An ASME Report)
B89.1.7	Performance Standard for Steel Measuring Tapes
B89.1.8	Performance Evaluation of Displacement-Measuring Laser Interferometers
B89.7.3.3	Guidelines for Assessing the Reliability of Dimensional Measurement Uncertainty Statements
B89.3.1	Measurement of Out-of-Roundness
B89.7.1	Guidelines for Addressing Measurement Uncertainty in the Development and Application of ASME B89 Standards [Technical Report]
B89.3.4	Axes of Rotation: Methods for Specifying and Testing
B89 REPORT	Technical Report 1990, Parametric Calibration of Coordinate Measuring Machines
B89.4.21.1	Environmental Effects on Coordinate Measuring Machine Measurements [Technical Report]
B89.7.6	Guidelines for the Evaluation of Uncertainty of Test Values Associated with the Verification of Dimensional Measuring Instruments to Their Performance Specifications
B89	Technical Paper 1990, Space Plate Test Recommendations for Coordinate Measuring Machines
B89.4.23	X-Ray Computed Tomography (CT) Performance Evaluation

Multiple Choice Questions of this Chapter

Multiple Choice Questions are given for each chapter with solutions in an online extension of this book. Please use link: www.wiley.com\go\mekid\metrologyandinstrumentation\

References

1 Uher, J., 2020, "Measurement in metrology, psychology and social sciences: data generation traceability and numerical traceability as basic methodological principles applicable across sciences," *Qual Quant*, 54, pp. 975–1004.

2 JCGM, 2012, "International vocabulary of metrology—basic and general concepts and associated terms (VIM)," *JGCM 200*, 3rd ed., www.bipm.org.

3 ISO, "Measurement management systems—requirements for measurement processes and measuring equipment," *ISO 10012*, 2003.

4 Mekid, S., 2008, *Introduction to Precision Machine Design and Error Assessment*, (ISBN13: 9780849378867), CRC Press.

5 Mekid, S., 2010, "Effects of miniaturisation on electromagnetic motors for micro mechatronic systems," IMECE2010-38651, ASME International Mechanical Engineering Congress Conference 2010, Vancouver, Canada.

6 Mekid, S., "Enhanced deterministic design: application to a micro CNC machine design," IMECE2010-38171, ASME International Mechanical Engineering Congress Conference 2010, Vancouver, Canada.

7 Sued, M.K., and Mekid, S., 2009, "Dimensional inspection of small and mesoscale components using laser scanner," International Conference on Advances in Mechanical Engineering, ICAME 2009, Malaysia.

8 Ogedengbe, T., Mekid, S., and Hinduja, S., 2008, "An investigation of influence of machining conditions on machining error," Proceedings of the 5th Virtual International Conference on Intelligent Production Machines and Systems, 2008, Elsevier, Oxford.

9 Ogedengbe, T., and Mekid, S., 2008, "Machine scaling optimization for on demand precision machining." 10th EUSPEN Conference, 2008, Switzerland.

10 Mekid, S., 2009, "Spatial thermal error compensation using thermal stereo via OAC controller in NC machines," Proceedings of the 5th International Conference on Intelligent Production Machines and Systems, 2009, Elsevier, Oxford, Cardiff.

11 Khalid, A., and Mekid, S., 2010, "Characteristic analysis of parallel platform simulators with different hardware configurations," Proceedings of International Bhurban Conference on Applied Sciences & Technology Islamabad, Pakistan, January 11–14, 2010.

12 Khalid, A., and Mekid, S., 2010, "Design synthesis of a 3-SPS parallel manipulator," 36th International MATADOR Conference 2010, Manchester, UK.

13 Mekid, S., Khalid, A., and Ogedengbe, T., 2008, "Common physical problems encountered in micro machining," 6th CIRP International Seminar on Intelligent Computation in Manufacturing Engineering—CIRP ICME '08, July 2008, Italy, (ISBN 978-88-900948-7-3), pp. 637–642.

14 Vacharanukul, K., and Mekid, S., 2007, "In-process inspection of dimensional measurement and roundness," International Conference on Manufacturing Automation. National University of Singapore, May 2007.

15 Vacharanukul, K., and Mekid, S., 2007, "Differential laser-based probe for roundness measurement," EUSPEN Conference 2007, Germany.

16 Mekid, S., and Vaja, D., 2007, "New expression for uncertainty propagation at higher order for ultra-high precision in calibration," 9th EUSPEN Conference 2007, Germany.

17 Khalid A., and Mekid, S., 2006, "Design of precision desktop machine tools for meso-machining," Proceedings of the 2nd Virtual International Conference on Intelligent Production Machines and Systems, Elsevier, Oxford.

18 Khalid, A., and Mekid, S., 2006, "Design & optimization of a 3-axis micro milling machine," 6th Int. Conf. European Society for Precision Engineering and Nanotechnology, May 2006, Baden, Austria.

19 Mekid, S., and Khalid, A., 2006, "Robust design with error optimization analysis of a 3-axis CNC micro milling machine," 5th CIRP International Seminar on Intelligent Computation in Manufacturing Engineering—CIRP ICME '06, July 25–28, 2006, Ischia, Naples, Italy.

20 Owodunni, Hinduja, S., and Mekid, S., 2005, "Towards rapid sheet metal forming," Proceedings of the 1st Virtual International Conference on Intelligent Production Machines and Systems, (ISBN 0-080-44730-9), Elsevier, Oxford.

21 Mekid, S., Bonis, M., Glentzlin, A., and Sghedoni, M., 1995, "High precision optical delay line for stellar interferometers," International Precision Engineering Seminar 8th, Compiegne, May 1995, Elsevier, pp. 495–499.

6

Tolerance Stack-Up Analysis

"If you know what you want, you know your acceptance margin!"

—Samir Mekid

6.1 Introduction

Tolerance stacks or tolerance analysis methods are very old, back to 1925 by Gramenz [1], and many references and books have been published over the last century. This is to demonstrate that this type of tolerances and their related analysis are extremely important for manufacturing. The published methods are very different, and the used notations are not common. However, the objectives are similar.

It is interesting to note that many set methods were written as internal notes for private companies, such as Wade (1967) for IBM [2], Harry and Stewart (1988) [3] for Motorola, and Griess (1990) for Boeing [4]. With this, it was indispensable to have a standard unifying the rules and practices, and hence ASME Y14.5M-1994 [5] standard appeared to sort out this topic. Its scope was defined as follows: This standard establishes uniform practices for stating and interpreting dimensioning, tolerancing, and related requirements for use on engineering drawings and in related documents. For a mathematical explanation of many of the principles in this standard, see ASME Y14.5.1.

Tolerance analysis is a name given to a number of approaches used today in product design to understand how imperfections in parts as they are manufactured, and in products as they are assembled, affect the capability of a product to meet customer expectations. It is a way of understanding how sources of variation in part dimensions and assembly constraints propagate across parts and assemblies and how that total variation affects the capability of a design to achieve its design requirements within the process capabilities of manufacturing organizations and supply chains. Tolerances can be essential to ensuring that products assemble correctly, operate safely, and are profitable to manufacture.

This chapter contains a brief introduction to geometric dimension and tolerancing followed by tolerance stack-up rules and analysis with examples.

6.1.1 Importance of Tolerance Stack-Up Analysis

It is important for the designer to consider tolerance analysis thoroughly in the early stages of the product development for optimal design of assembly and to illustrate the problems associated with 2D tolerance analysis.

Tolerance stack-ups are vital to address mechanical fit and mechanical performance requirements. Mechanical fit is simply answering the question, "Do the parts that make up the assembly always go together?" Product manufacturers utilize an organized flow of information to translate customer requirements into product requirements. Mechanical performance requirements would include the performance of mechanisms, such as switches, latches, and actuators. Other performance requirements could include optical alignments or motor efficiency.

Tolerance stack-up calculations represent the cumulative effect of part tolerance with respect to an assembly requirement. The idea of tolerances "stacking up" would refer to adding tolerances to find total part tolerance and then comparing that to the available gap or performance limits in order to see if the design will work properly. This simple comparison is also referred to as worst-case analysis. Worst-case analysis is appropriate for certain requirements where failure would represent catastrophe for a company. It is also useful and appropriate for problems that involve a low number of contributing dimensions.

Many companies utilize a statistical method for tolerance analysis. One approach involves a simple calculation using the RSS method, root-sum-squared. Instead of summing tolerances, as in worst-case analysis, statistical analysis sums dimension variation, mathematically defined as the square of the standard deviation of the distribution of the dimension values. It is important to understand that the input values for a worst-case analysis are design tolerances, but the inputs for a statistical analysis are process distribution moments (e.g., standard deviation). Worst-case analysis can be used to validate the absolute extreme conditions permitted by a design. Statistical analysis can be used to predict the variation of an assembly based on the assumptions of variation of the part dimensions.

Assembly variation analysis provides insight required to identify the key part characteristics (KPCs) that must be controlled in order to produce a product that meets the expectation of the customer. The product development process should then become focused on defining and validating part manufacturing and assembly processes that can achieve high producibility levels. Goals of Cpk = 1.67 (process capability index) for key features and Cp = 1.33 (process potential index) for non-key features are commonly quoted. Utilizing the insight for variation analysis allows design engineers to allocate tolerance budgets strategically. Critical features will be held to tighter tolerances. Looser tolerance can be applied to less important features. These decisions not only ensure product quality and performance but also ensure manufacturability at the right price. The impact on the product development process can be huge.

Why Use Tolerance Analysis? With shorter product life cycles, faster time to market, and tighter cost pressures, the characteristics that differentiate a product from its competitors are now down in the details of a design. Engineers are going to perform a statistical tolerance analysis over a tolerance stack-up in order to improve cycle time and quality and to reduce costs. They are looking more closely at why they did not get the exact part and assembly dimensions they expected from manufacturing and then are attempting to optimize the tolerances on the next version of the product. Tolerance optimization during design has a positive impact on the yields coming out of manufacturing, and better yields directly affect product cost and quality. Analyzing tolerances and variations before trying to produce a product also helps engineers avoid time-consuming iterations late in the design cycle. Many companies have started to rely on tolerance analysis to gain competitive advantage in their industry.

Quality and profitability go hand in hand. Companies define quality in terms of production yield and reliability, which reflect the probability of defects given an overall number of units produced. When engineers perform tolerance analysis during design, they essentially are converting design intent into a statistical or probability-based design model. The model predicts the probable yield loss for the design, or the number of rejected parts per total produced that will cause a customer to be dissatisfied. Proper tolerancing can help designers pinpoint and avoid the problems that lead to rejected parts and thereby eliminate the non-value-added cost that damages profitability.

The Immediate Need Is for the Following:
- To ensure all parts fit together in assembly and the overall dimension is within specification;
- To measure the allowable value in form due to the variation in manufacturing;
- To select the relationship to manufacturing process for the tolerance needed and to observe the effects on fabrication and assembly;
- To make sure that no unnecessary tight tolerance is selected to be costly in manufacturing;
- To understand the worst-case fit and form of all the parts at the time of assembly;
- To manufacture the parts by securing the allowable variation in size and location.

Associated problems due to tolerance stack-ups:
- Failure to assemble;
- Interference between parts;
- Failure of parts to engage;
- Failure to the intended function of the assembly or system.

6.1.2 Need for Tolerance Stack-Up Analysis in Assemblies

Tolerance analysis is used to predict the effects of manufacturing variation on finished products. Either design tolerances or manufacturing process data may be used to define the variation. Current efforts in tolerance analysis assume rigid body motions. This presents a method of combining the flexibility of individual parts, derived from the finite element method, with a rigid body tolerance analysis of the assembly. These results can be used to predict statistical variation in residual stress and part displacement. This will show that manufacturing variation can produce significant residual stress in assemblies. It will demonstrate two different methods of combining tolerance analysis with the flexibility of the assembly.

6.1.3 Manufacturing Considerations in Engineering Design

In engineering design process [6–10], it is of primary interest to include tolerance assignment after tolerance analysis on parts for manufacturing. Therefore, manufacturing requirements will be set and related cost known in advance. Decision making at this stage is very important partly for cost analysis as this can end up being very high if high-precision manufacturing is needed.

Figure 6.1 shows how the assigned tolerance can relate both manufacturing and engineering design. The figure shows why the tolerance analysis is part of the design for assembly (DFA) and the design for manufacturing (DFM). Hence, it builds a strong link between design and manufacturing processes.

Mechanical parts are manufactured in CNC machine tools using usually a G-Code file to go automatically through all steps from a bland piece to a final shape of the part. Assuming, the material was well selected, the environment of temperature and vibration are accepted and the CNC machine is well calibrated, the expected part with the dimensions and tolerances should be accepted. It happens that some parts do not pass the quality check (QC) because of missing tolerance, and this may be due to any of the parameters announced previously.

On the other side, the tolerance in an assembly can vary from three known sources:

- Size variation of one or several parts;
- Geometric variation;
- Kinematic variation due to possibly size and geometric variations.

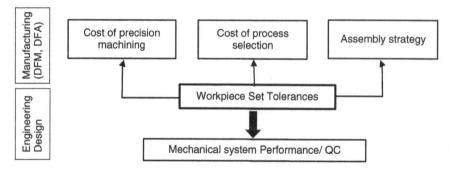

Figure 6.1 Effect of tolerances on manufacturing and specifications of the mechanical system.

The sources of variation in manufacturing are as follows:

a) **Tooling:**
This takes into consideration the variation in setting up a fixture for the part, the part location with respect to the tool, and the tool wear process with time dependence in mass production.

b) **Material:**
The material can be nonuniform with internal stress relaxation from prior or ongoing processes and dimensional variation of surfaces.

c) **Operator:**
The operator is variable as human with inconsistent setups and manual control of the process that can all affect the dimension and the tolerances.

d) **Equipment:**
The machines are the main source of variation to include a level of repeatability of motions, stiffness of the machine, and variation of the process parameters during manufacturing.

e) **Environment:**
The temperature and humidity have a usual effect on the parts. The temperature will let the part to expand by a few microns in one degree Celsius rise depending on length.

The overall actions to reduce these effects are:

1) Minimize the number of setups;
2) Tighter process controls;
3) Tighter environment controls;
4) Improve materials quality;
5) Operator training.

6.1.4 Technical Drawing

Technical drawing is a document that communicates precise description of a part. The description consists of pictures, words, numbers, and symbols, and in particular the following:

- Geometry of the part;
- Critical functional relationships;
- Tolerances;
- Material, heat treat, surface coatings;
- Part documentation such as part number and drawing revision number.

Figure 6.2 Cost of technical drawing error

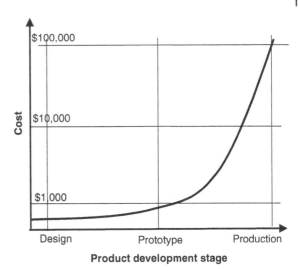

The issuer of a technical drawing should understand that this document with satisfactory dimensioning contributes to utility, transfer of information, facilitates mass production, and simplifies inspection of finished parts.

This document is a legal document; hence, it is important to notify clear details and high precision. The tolerances are particularly a criterion of acceptance of parts in a production line. Any mistake will cost high depending on the stage of product development, as shown in Figure 6.2. The drawing errors cost includes money, time, material, and customer satisfaction. The figure shows that highest damage happens when mistakes in drawings are made at the production stage. To avoid all types of mistakes beside calculations, symbols are standardized in ASME Y14.5 and in GD&T techniques.

Several design software exist to consider tolerance analysis, such as Solidworks, with a check of the effect of tolerances on parts and assemblies to ensure consistent fit of components and to verify tolerancing schemes before the product goes to production. Moreover, the tolerance analysis package can be coupled with FEM software, for instance, NASTRAN, to predict assembly stresses in flexible parts due to tolerance stack-up.

Example of Unnecessary Cost: With known nominal tolerance for a part made of steel, for example, if you choose tighter tolerances, then this generates high cost of manufacturing. What is the reason for this, any compromise or analysis to decide tighter tolerances?

If a tolerance is calculated and known to be fit for the expected operation, then there is no need to specify a tighter tolerance than necessary; otherwise, it becomes expensive (Fig. 6.3).

6.1.5   Definitions, Format, and Workflow of Tolerance Stack-Up

A tolerance is defined as the total amount by which a given dimension may vary, or the difference between the limits of the dimension of the feature in consideration [ASME Y14].

The parts are always manufactured with a variation of the dimensions within a certain limit. Hence, the assembly may be either too lose or too tight and sometimes impossible to assemble if the tolerances were not respected against specifications.

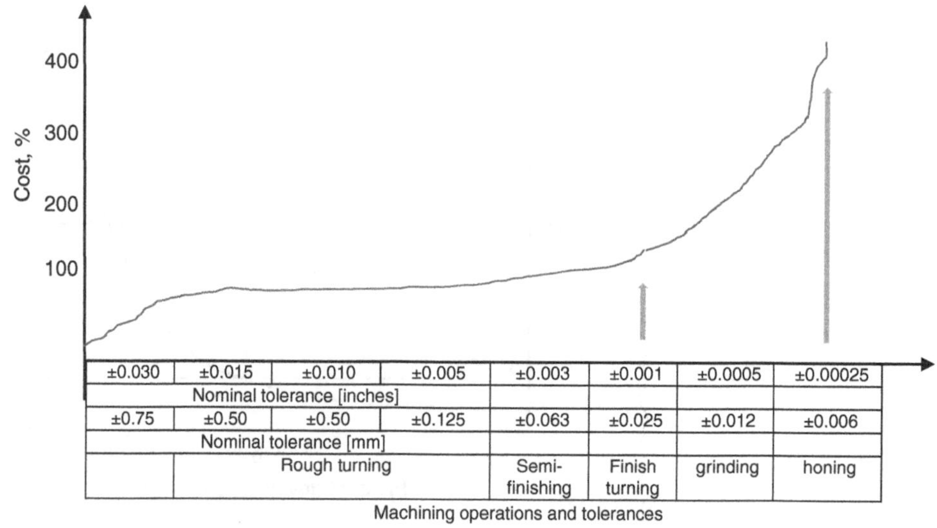

±0.030	±0.015	±0.010	±0.005	±0.003	±0.001	±0.0005	±0.00025
Nominal tolerance [inches]							
±0.75	±0.50	±0.50	±0.125	±0.063	±0.025	±0.012	±0.006
Nominal tolerance [mm]							
Rough turning				Semi-finishing	Finish turning	grinding	honing

Machining operations and tolerances

Figure 6.3 Manufacturing cost depending on the nominal tolerance for steel.

Tolerance analysis is used to predict the effects of manufacturing variation on finished products. Either design tolerances or manufacturing process data may be used to define the variation. Current efforts in tolerance analysis assume rigid body motions.

Workflow of Tolerance Stack-Up Introducing tolerance analysis at the correct points in your workflow can help maximize its effectiveness. Typically, the earlier you can introduce it to your process, the better. Let's use this sanitary needle assembly (Fig. 6.4) as an example of how integrating tolerance analysis can be beneficial at many stages in the manufacturing workflow.

1) During your initial design and production planning, introducing tolerance analysis can help guide design decisions and avoid undesirable interactions between parts. Having live feedback on how assemblies interact and whether they meet the stated criteria can drastically cut down on the amount of time spent in design reviews.

   An example of this would be using software such as CETOL 6s to identify whether the lid clearance (green arrows) on this sanitary needle design are adequate to ensure full engagement of the clasp (red arrow).

2) Once a design has been finalized and production planning begins, tolerance analysis can also assist in studying manufacturing feasibility and check fixture design. Ensuring that the correct features in your product are identified as critical can help manufacturing groups ensure that their processes yield products that will pass inspection.

   Our needle example being a medical device, the criteria for dimensional accuracy will be very strict. Having these dimensional checks be as accurate and well designed as possible can ensure that all parts that pass inspection truly meet the stated criteria for a successful assembly.

3) After a product has been through the design and manufacturing process, tolerance analysis can still play an important role in future production. Going back and feeding real production data into your tolerance analysis can drive process improvements, avoid scrap, and reduce down time for production line issues.

Figure 6.4 Sanitary needle. *Source:* Courtesy of Sigmetrix.

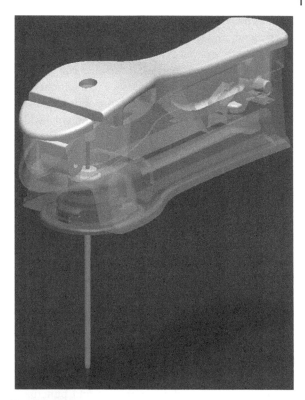

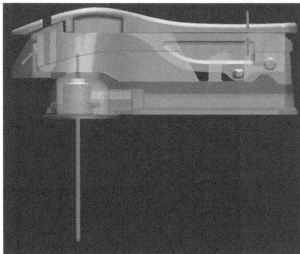

Figure 6.5 Sanitary needle- side view. *Source:* Courtesy of Sigmetrix.

If, for instance, we notice that the materials are behaving differently after the manufacturing process than we anticipated with our initial data, that information can be fed back into the tolerance analysis software to ensure that we have an accurate representation of our assembly digitally and that our tolerance controls we have in place are necessary. This can be useful for parts with designed deflection (like our needle's clasp in our example) or components that used intentional interference to close and seal completely.

Recommendations and Looking Forward   The future of manufacturing is one that will heavily rely on efficiency and streamlined process. Eliminating rework and redesign immediately translates to money and time saved.

By using tolerance analysis, companies can integrate process and design improvements earlier and take advantage of the benefits we've mentioned previously. Software programs such as CETOL 6s are making tolerance analysis more approachable for users of varying skill levels, and in doing so making tolerance analysis more common in a wide variety of industries.

6.2   Brief Introduction to Geometric Dimensioning and Tolerancing (GD&T)

The design of any mechanical system should lead to the manufacturing of components [11]. GD&T is a system and a language for refining and communicating design intent and engineering tolerances that helps engineers and manufacturers to optimally control variations in manufacturing processes. It allows to indicate what is functionally important and ultimately how to establish an inspection setup. It is a better accurate process to define tolerances and lead to more efficient manufacturing without necessarily tightening the tolerances that can be very expensive. This must be followed by tolerance stacks analysis.

GD&T is symbols and standards language utilized by engineers on technical drawings and blueprints to describe the design of products in three dimensions and be able to communicate all this information to explain the design intent (Table 6.3 and appendix A). GD&T is a high precision language with all symbols, grammar, and punctuation rules to be learned properly. A proper training is needed to master this topic. The chapter gives a short introduction.

When Should GD&T Be Used?   Designers should tolerance parts with GD&T when:

- Drawing delineation and interpretation need to be the same;
- Features are critical to function or interchangeability;
- It is important to stop scrapping perfectly good parts;
- It is important to reduce drawing changes;
- Automated equipment is used;
- Functional gaging is required;
- It is important to increase productivity;
- Companies want across-the-board savings.

International standards are employed to establish uniform rules, principles, and methods in dimensioning and tolerancing used on engineering drawings that apply coordinate and geometric dimensioning methods. Standard dimensioning will be introduced followed by GD&T to convey more information that cannot be shown by a simple dimensioning.

6.2.1   Notation and Problem Formulation

In manufacturing parts, various dimensional and geometric errors may take place during and after machining has finished. It is important to highlight the feature(s) that may cause assembly issue and affect the function and operation of the parts. Hence, on many occasions, functional tolerances participate directly in the operations.

6.2.2 Dimension Types

The dimensions are added to technical drawings to specify the nominal form, size, orientation, and location of part features. Every feature on a part must be dimensioned. Currently, all drawings are generated using 3D CAD solid modeling software with dimensions and tolerances.

The dimensions are classified under the following types:

- **Functional dimensions**: These are essential to the function of the product. They must be shown on the drawing and referred to a datum.
- **Nonfunctional dimensions**: These are not essential to the function of the product.
- **Redundant and auxiliary dimensions**: any redundant dimension should not be dimensioned unless they provide a useful information; they can be added as auxiliary dimension in brackets and not toleranced.
- **Reference dimensions**: this is usually a dimension without tolerance used for information only and within parentheses. It is often a repeat of another dimension for clarification.
- **Basic dimensions**: This is considered as a theoretically perfect dimension. They are used to describe the exact size, profile, orientation, and location of a feature. They are toleranced. They also indicate the datums and references.

CAD files define mathematically the part geometry and can form 3D object for various purposes, such as inspection, reverse engineering, and manufacturing details. This allows reduction of unnecessary dimensions and opens the possibility to queries rather than appearing permanently on screen.

There are several types and formats of dimensions. They are standardized and detailed in ASME Y14.5. Figures 6.6 and 6.7 show examples of drawings with geometric dimensions in linear, chain, angular, and angular repetition. The methods shown are equivalent. There is no difference in the legal interpretation for these methods except their format.

Some dimensions with ± tolerances are also used in drawing, but only to define the nominal size and size tolerance for features of size. Letters as tolerance are used too. This is common practice and will be explained later. Tolerances can be used under GD&T rules as shown in both Figures 6.8 and 6.9. The latter shows positional tolerancing with datum references.

Dimensions can also be presented in rectangular coordinate dimensioning method as shown in Figure 6.8, where all dimensions are referred to two perpendicular datum. Fig. 6.9 shows an example of student drawing with necessary dimensions and tolerances.

Example of Technical Drawing with Tolerances

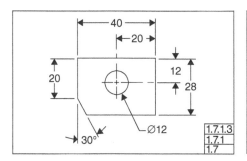
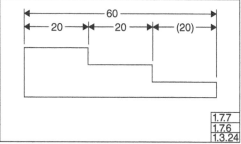

Figure 6.6 a) Dimensions and b) intermediate reference dimension [ASME Y14.5]. *Source:* ASME Y14.5.1M-1994, Mathematical Definition of Dimensioning and Tolerancing Principles, The American Society of Mechanical Engineers.

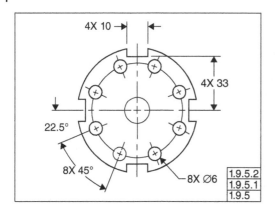

Figure 6.7 Repetitive features and dimensions [ASME Y14.5]. *Source:* ASME Y14.5.1M-1994, Mathematical Definition of Dimensioning and Tolerancing Principles, The American Society of Mechanical Engineers.

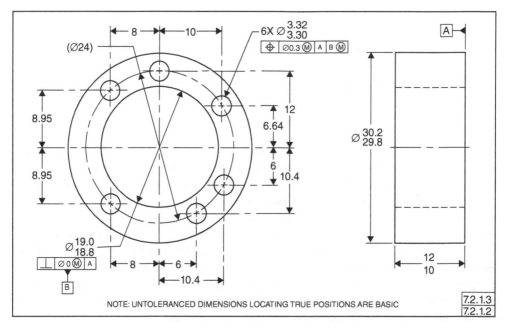

NOTE: UNTOLERANCED DIMENSIONS LOCATING TRUE POSITIONS ARE BASIC

Figure 6.8 Dimensions and positional tolerancing with datum references [ASME Y14.5]. *Source:* ASME Y14.5.1M-1994, Mathematical Definition of Dimensioning and Tolerancing Principles, The American Society of Mechanical Engineers.

6.2.3 Coordinate Dimensioning

The dimensions are shown using dimension lines, extension lines, from a dimension specified to a feature (Fig. 6.10–11). Common information will be introduced. The remaining information can be found in the ASME Y14.5-2018 standard.

The dimensioning defines the exact amount of material that should remain after successive machining. Measurements are taken to inspect the part with respect to its known datum specified in the drawing.

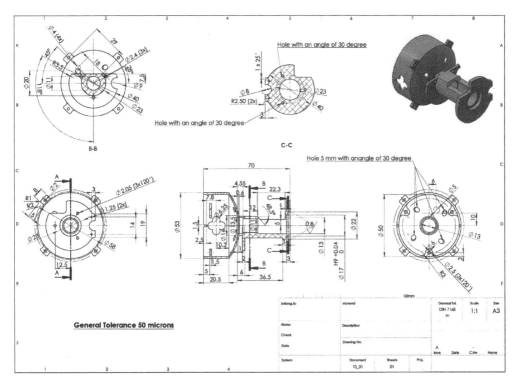

Figure 6.9 Technical drawing of a workpiece.

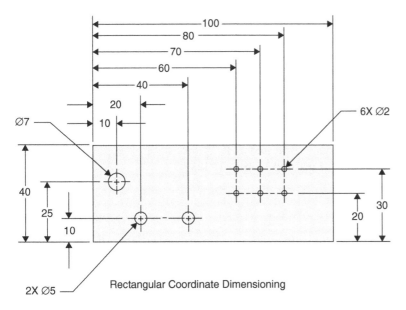

Rectangular Coordinate Dimensioning

Figure 6.10 Technical drawing of a workpiece [ASME Y14.5]. *Source:* ASME Y14.5.1M-1994, Mathematical Definition of Dimensioning and Tolerancing Principles, The American Society of Mechanical Engineers.

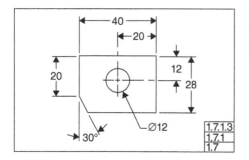

Figure 6.11 Dimensions of a part. [ASME Y14.5]. *Source:* ASME Y14.5.1M-1994, Mathematical Definition of Dimensioning and Tolerancing Principles, The American Society of Mechanical Engineers.

We consider eight basic rules in dimensioning:

1) Dimensions should not be duplicated, and each information should be given in one way only.
2) Dimensions should be attached to the view that best shows the contour of the feature to be dimensioned.
3) Avoid dimensioning hidden lines whenever possible.
4) Avoid dimensioning over or through the object.
5) Whenever possible, locate dimensions between adjacent views.
6) A circle is dimensioned by its diameter and an arc by its radius, e.g., R0.80.
7) Holes are located by their centerlines, which may be extended and used as an extension line.
8) Holes should be located and dimensioned in the view showing them as circles.

6.2.4 Tolerance Types

There are four common types of tolerances in measurements:

1) Limit dimensions;
2) Plus and minus tolerances:
   a) Unilateral tolerances;
   b) Bilateral tolerances;
3) Compound tolerances;
4) Block tolerance.

a-Limit Dimensions: In this case, the dimension shows the maximum and minimum size that a feature in a part can have, and hence, within this limit the part can pass inspection.

Example: The diameter of a shaft can be expressed as: $\emptyset_{1.016}^{1.020}$ *or* $\emptyset_{12.980}^{13.123}$
If placed on one line, then: $\emptyset$ 1.016 – 1.020 or $\emptyset$ 12.980 – 13.123

b-Plus and Minus Tolerances

b-1 Unilateral Tolerances

If the two limit dimensions of a feature are both above the nominal size or below the nominal size, then the tolerance is said to be unilateral (Fig. 6.12).

Figure 6.12 Unilateral tolerance

b-2 Bilateral Tolerances

If the two limit dimensions of a feature are above and below the nominal size, then the tolerance is said to be bilateral (Fig. 6.13).

Figure 6.13   Bilateral tolerance

c-Compound Tolerances

In this case, the tolerance can be composed and calculated from established tolerances around the feature (Fig. 6.14.).

Figure 6.14   Compound tolerance

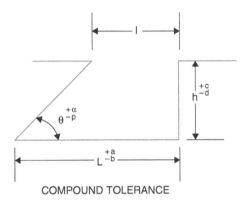

COMPOUND TOLERANCE

d-Block Tolerance

When several dimensions in a part to be manufactured have the same tolerance in a technical drawing, there is no need to mention this in every single dimension but to add in the left corner a block as in the following example:

A block tolerance (Fig. 6.15) is actually a general note that applies to all dimensions not covered by some other tolerancing type. Block tolerances are placed in the tolerance block to the left of the title block and above the projection block.

UNLESS OTHERWISE SPECIFIED
DIM ARE IN MM
LIN TOL ±0.2
ANG TOL ± 0.5°
INTERPRET DIM AND TOL PER
ASME Y14.5 – 2016

Figure 6.15   Page or block tolerances

6.2.5 Characteristics of Features and Their Tolerances

The variable geometric characteristics of part features and the associated types of tolerances are introduced. All features on a part are subject to variation, and the features must be completely defined with tolerances.

This defines also the geometric characteristics of the feature itself, such as its size and its form and the relationship of the feature to the rest of the part, such as where it lies or how much it tilts relative to another feature or a datum reference frame. It is important to understand the variation that will be assigned to each geometric characteristic of every feature must be fully defined, as well as the sources of the variation. The variation of the relationship between parts also need to be understood and mentioned in the technical drawing.

There are six geometric characteristics that describe the feature geometry and the interrelationship of part features. The dimensioning and tolerancing covers in practice beside the datum various aspects:

Part of the role of a tolerance is also to indicate six aspects to include:

- **Tolerance of size**
- **Tolerance of form**
- **Tolerance of location**
- **Tolerance of orientation**
- **Tolerance of profile**
- **Tolerance of runout**

These are also taken into consideration by the standard ASME Y14.5 2018, and four of them are materialized in the following drawing in gray boxes in Figures 6.16 and 6.17.

The feature control frame states the requirements or instructions of the feature (Table 6.3). There is one message in one feature control frame (Fig. 6.18). An example of a drawing with GD&T is shown in Figure 6.19.

A list of all ISO and ASME Y14.5 symbols is given in the appendix of this chapter.

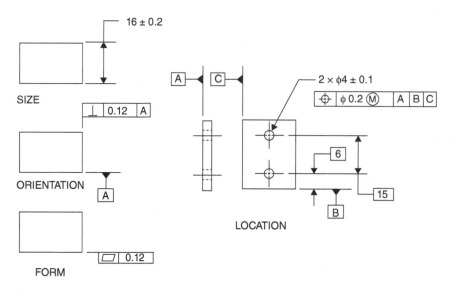

Figure 6.16 Moving from dimensions with SLOF to GD&T.

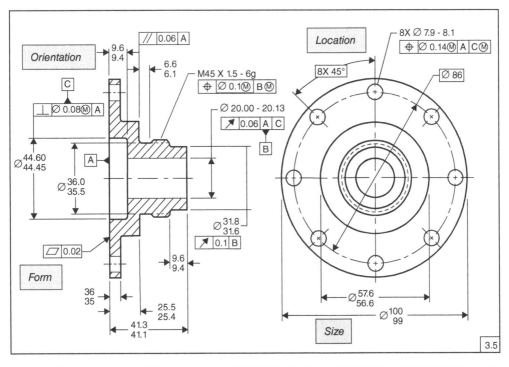

Figure 6.17 Feature control frame placement [ASME Y14.5]. *Source:* ASME Y14.5.1M-1994, Mathematical Definition of Dimensioning and Tolerancing Principles, The American Society of Mechanical Engineers.

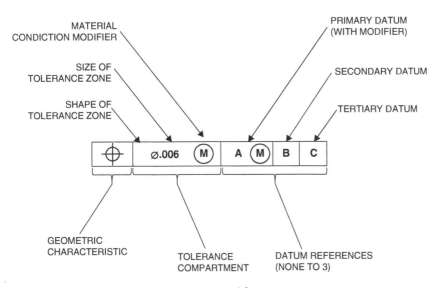

Figure 6.18 Components of the feature control frame.

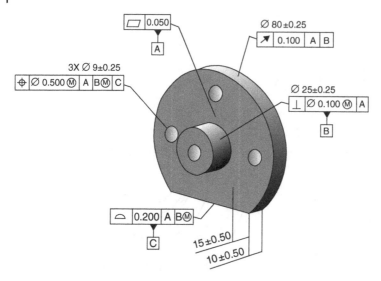

Figure 6.19 Component with GD&T control.

6.3 Tolerance Format and Decimal Places

Four standard formats are practiced for linear and angular dimensions and tolerances. Formats are included for US inch and metric dimensioning and tolerancing. According to the standard ASME Y14.5-2009, the rules for angular dimensions and tolerances are the same for drawings prepared using US inch and metric units.

- A nominal value is not specified by the limit dimensions. A high (maximum) value and a low (minimum) value are specified to define the range of variation. When a limit dimension is stated in a horizontal format, the smaller value precedes the larger value, with the values separated by a dash.
- When a limit dimension is stated in a vertical format, the larger value (upper limit) is placed above the smaller value (lower limit). It makes no difference whether the tolerance feature is an internal feature or an external feature.
- Equal-bilaterally tolerance dimensions specify a nominal value and the amount a dimension may deviate from nominal. The tolerance values are equal in each direction.
- Unequal-bilaterally tolerance dimensions specify a nominal value and the amount a dimension may deviate from nominal.
- The tolerance values are not equal in each direction, and neither value is zero.
- Unilaterally toleranced dimensions specify a nominal value and the amount a dimension may deviate from nominal in one direction only. The tolerance is in one direction only, either larger or smaller. The other tolerance value is zero.

The technical terms are defined hereafter:

- **Bilateral Tolerance (Limit Dimension)**
  A tolerance that allows the dimension to vary in both the plus and minus directions.
- **Equal Bilateral Tolerance**
  Variation from the nominal is the same in both directions.

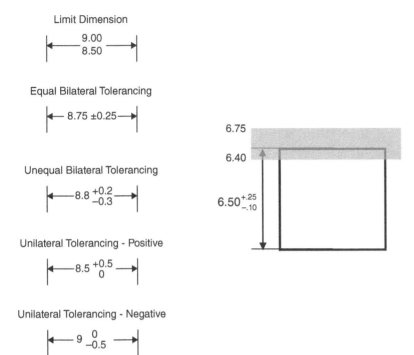

Figure 6.20 Tolerance range specified plus/minus tolerances

- **Unequal Bilateral Tolerance**
  The allowable variation is from the target value and the variation is not the same in both directions.
- **Unilateral Tolerance**
  The allowable variation is only in one direction and zero in the other. It can be positive or negative.

The nominal dimension value must be part of the tolerance range independently from which of the above methods is selected. The upper and lower limits must include the nominal dimension, even if it is at one extreme of the range, but obviously, the tolerance range cannot ever be "off the part." (Fig. 6.20).

From this figure, it is clear that the max and min dimension of the feature throughout the four ways of showing the tolerance is [8.50 – 9.00].

6.4 Converting Plus/Minus Dimensions and Tolerances into Equal-Bilaterally Toleranced Dimensions

The method of tolerance analysis is the process of considering the tolerances and analyzing their combination at an assembly level.

We will define the process for analyzing tolerance stacks. It will show how to set up a loop diagram to determine a nominal performance/assembly value and four techniques to calculate variation from nominal.

The most important goal for the reader is to understand the assumptions and risks that go along with each tolerance analysis method. This method of tolerance stack-up requires all dimensions and tolerances to be converted into equal-bilateral format whether they are plus/minus or GD&T. Techniques of conversion will be shown

Converting Limit Dimension to Equal Bilateral Format If a limit dimension is given, it is required to convert it to equal bilateral format as shown in Figure 6.18

- Determine the upper limit 10.00 (metric) of the dimension;
- Determine the lower limit 9.55 of the dimension;
- Calculate the total tolerance = $10 - 9.55 = 0.45$;
- Define the equal bilateral value = $0.45/2 = 0.225$;
- Add equal bilateral value to lower limit: adjusted nominal value to obtain $9.55 + 0.225 = 9.775$;
- Equal bilateral equivalent = 9.775 ± 0.225.

Converting Unequal Bilateral Format to Equal Bilateral Format
- Suppose an unequal-bilateral tolerance dimension format $8.50^{+.25}_{-.10}$;
- Its upper limit: $8.50 + 0.25 = 8.75$;
- Its lower limit: $8.50 - 0.10 = 8.40$;
- Total tolerance: $8.75 - 8.40 = 0.35$ or $0.25 + 0.10 = 0.35$;
- It is divided by 2 to obtain the equal bilateral tolerance $0.35/2 = 0.175$;
- Adjusted nominal value: $8.40 + 0.175 = 8.575$;
- Hence, the bilateral format becomes 8.575 ± 0.175.

Converting Unilaterally Positive Format to Equal Bilateral Format
- Suppose a unilateral positive tolerance dimension $8.50^{+.25}_{-.00}$;
- Its upper limit is $8.50 + 0.25 = 8.75$;
- Its lower limit is then, $8.50 - 0.00 = 8.50$;
- The total tolerance range is therefore $8.75 - 8.50 = 0.25$ or $0.25 + 0.00 = 0.25$;
- This is divided by 2 to obtain the equal bilateral tolerance $0.25/2 = 0.125$;
- Adjusted nominal value: $8.50 + 0.175 = 8.625$;
- Hence, the bilateral format will be 8.625 ± 0.125.

Converting Unilaterally Negative Format to Equal Bilateral Format
- Suppose a unilateral negative tolerance dimension $8.50^{+.00}_{-.25}$;
- Its upper limit: $8.50 + 0.00 = 8.50$;
- While its lower limit: $8.50 - 0.25 = 8.25$;
- Total tolerance: $8.50 - 8.25 = 0.25$, or $0.25 + 0.00 = 0.25$;
- Equal bilateral tolerance $0.25/2 = 0.125$;
- Adjusted nominal value: $8.50 - 0.125 = 8.375$;
- Bilateral format: 8.375 ± 0.125.

Mean Shift The nominal dimension is not always given in the middle of the tolerance range. These conversions can shift the nominal dimension to be in the center. Examples are given below for better understanding. The mean refers to the nominal dimension (Fig. 6.21).

Mean shift calculation for limit dimension converted into equal bilateral format:

- Initial dimension is $8.50^{+.25}_{-.10}$ converted into 8.575 ± 0.175;
- Mean shift: $8.575 - 8.500 = 0.075$.

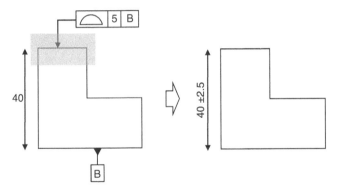

Figure 6.21 GD&T tolerance and equivalent normal tolerance.

Mean shift calculation for unilateral positive format converted into equal bilateral format:

- Initial dimension is $8.50^{+.25}_{-.00}$ converted into 8.625 ± 0.125 with mean shift of 0.125.

Mean shift calculation for unilateral negative format converted into equal bilateral format:

- Initial dimension is $8.50^{+.00}_{-.25}$ converted into 8.375 ± 0.125 with mean shift of -0.125.

6.5 Tolerance Stack Analysis

Tolerance stack-up is the study of dimensional relationship of one part or between parts in an assembly to avoid conflicts [12–14]. This is a very important step in the design process as it is a decision-making tool. The parts we receive are never to their assigned size but within a tolerance range. This due mainly to:

- The manufacturing process errors that are sometimes dynamic;
- The tool wear;
- The parts fixtures on the machine;
- The errors introduced by the operator;
- Variation in the materials;
- Environmental conditions affecting both machine and instruments;
- Inadequate maintenance;
- Assembly process;
- Inspection procedures not reliable.

The common question asked frequently is, why do we need tolerance analysis? Basically, the reasons are various and will be discussed hereafter:

1) How to ensure that parts fit together at assembly?
2) How imperfect part fit together at assembly?
3) How much imperfection or variation is allowable?
4) Does it matter if a part is manufactured larger than nominal part and the mating part is smaller than nominal?
5) What if both parts are manufactured on the small side and mating holes in each part are slightly tilted or out of position?
6) What happens to a feature on one part if a surface on the mating part is tilted?

Hence, the answer to all these questions is the tolerance stack-up and analysis. In addition, the benefits are large, as follows:

1) Optimization of the part tolerance;
2) Better understanding of the part and assembly function;
3) Allow to make intelligent design decisions [15, 16];
4) After analysis, this allows to discover and solve design problems in virtual condition before manufacturing;
5) Overall evaluation of the design under consideration;
6) Exploring design alternatives using modified parts due to tolerance assignment;
7) Balancing the accuracy and cost with manufacturing capabilities.

The tolerance analysis is a study of individual tolerances and their meaning, and it is the study of the cumulative variation between part features. With respect to this and as an engineer, you can:

- Tighten up the tolerances on each component so the sum of the tolerances is lower;
- Include a spacer that comes in different sizes to take up any slack resulting from the tolerance addition (like shimming);
- Design so that the tolerance stacks are not relevant to function;
- Consider that the variation in each part is likely to be statistically distributed.

In the following part of this chapter, the basics of tolerance stack-up analysis will be discussed. The following questions summarize the expected knowledge to acquire.

- Where to begin a stack?
- Designating positive and negative routes.
- Which geometric tolerances are factors?
- Finding the mean.
- Calculating boundaries for GD&T, MMC.
- LMC and RFS material condition modifiers (to be discussed later).
- Mean boundaries with equal bilateral.

Tolerance Build-Up The following example is a basic one showing a simple method to calculate the overall dimension and its related tolerance. The four parts are stack and can be held in a holder where the dimension and its tolerance T ± t are determined.

The stack-up dimension is shown in Figure 6.22; the total dimension and the tolerance are as follows:

$$T = A + B + C + D \tag{6.1}$$

$$t = \pm(a + b + c + d) \tag{6.2}$$

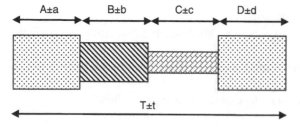

Figure 6.22 Accumulation of tolerances on a drawing.

6.5.1 Worst-Case Tolerance Analysis

The worst-case tolerance analysis (WC) is a particular case of tolerance analysis, and it determines the absolute possible maximum variation for a given distance or gap that is not dimensioned or toleranced and is usually functional, hence its importance. All parts participating in the determination of the requested dimension and tolerance through stack-up have their dimensions and their tolerances algebraically added initially and see their respective tolerances modified after analysis.

It is assumed in the method that all dimensions in the tolerance stack-up may be at their worst-case maximum or minimum, regardless of the improbability. Tolerance stack-ups is defined through a chain of dimensions and tolerances that are linked from start to end.

Procedure: The procedure is given below. It is simple and easy to implement immediately.

1) Identify the start point and finish point;
2) Determine the positive and the negative directions;
3) Convert all dimensions and tolerances to equal-bilateral format;
4) Place each positive dimension in one column and negative dimensions in another column;
5) Place the tolerance value for each dimension in the tolerance column adjacent to each dimension;
6) Add entries in each column.

Analysis: Each dimension in WC will be represented by a minimum and maximum value showing the range of variation. Hence, part of the analysis is to check if the maximum range is considered for each input, how would be the maximum in the output considered feature or stack-up. Therefore limits of acceptance can be defined. This is not probability.

Assembly Shift Assembly shift is the amount that part can move during assembly due to the clearance between a hole and a fastener, a hole and a shaft, a width and a slot, or between any external features within an internal feature. Assembly shift accounts for the freedom parts have to move from their nominal locations due to the clearance between mating internal and external features at assembly (Fig. 6.23).

6.5.2 Rules for Assembly Shift

- Assembly shift amount is by how much parts can move in the assembly due to the clearance between an internal feature including a hole and an external feature such as a fastener.
- In floating fastener cases, assembly shift is added to the tolerance stack-up twice, each line representing the amount the clearance holes in each part can shift about the fastener. The amount

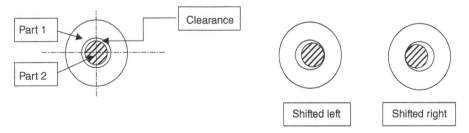

Figure 6.23 Shift about the fastener in part 2.

each part may shift about the fastener is independent of the mating parts and must be calculated separately.

- In fixed fastener cases assembly shift is added to the tolerance stack-up once, representing the amount the clearance holes can shift about the fastener.
- Assembly shift is typically not calculated for fasteners within a threaded hole because fasteners are commonly assumed to self-center within the threaded holes.
- In cases where the results of the tolerance stack-up are very critical and the tolerances are tight, it may be necessary to calculate or estimate the amount that a threaded fastener may move within a threaded hole.
- In cases where oversized holes or slots are used to allow for adjustment at assembly, the assembly shift may be eliminated or even subtracted from the total tolerance. This must be done with utmost caution, as the tolerance analyst must be absolutely certain that the assembly process will allow time for adjustment, the assemblers understand the purpose of this extra adjustment, and the parts can be adjusted at assembly, i.e., they are not too heavy or awkward to properly be adjusted to an optimal position.

Example: Worst-case maximum/minimum dimensions (Fig. 6.24).

$$2.5 + 3.2 + 2.5 = 8.2$$

$$2.4 + 3.0 + 2.4 = 7.8$$

Example: Assembly of individual parts (Fig. 6.25).
Suppose we have two blocks A and B that are to fit into a slot in C, all with tolerances shown. Let's look at the largest and smallest gap that we could have.

- Largest Gap $= (C + c) - (A - a) - (B - b) = C - (A + B) + (a + b + c)$
- Smallest Gap $= (C - c) - (A + a) - (B + b) = C - (A + B) - (a + b + c)$

Consequently, the dimension and tolerance of the gap is effectively:

$$\{C - (A + B)\} \pm (a + b + c)$$

If we have 10 parts (say a clutch pack) that all have to fit into a housing, the addition of all those tolerances can be significant.

If controlling the spacing of the clutches is important to function, then there is a problem because the gap determined tolerance becomes variable, unless these spaces between clutches are specifically known. If wear is introduced then the gap becomes variable in time.

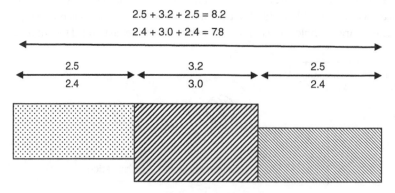

Figure 6.24 Accumulation of tolerances on a drawing.

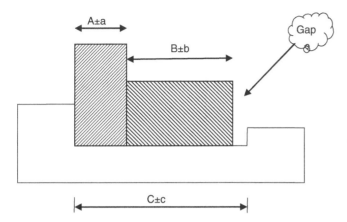

Figure 6.25 Accumulation of tolerances on a drawing.

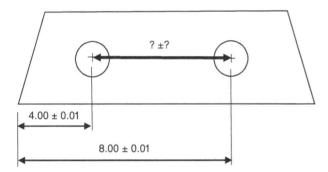

Figure 6.26 Accumulation of tolerances on a drawing.

Example: What is the effective dimension and tolerance between the two holes?

In this case, as shown in Figure 6.26, the tolerances add directly. Considering the extremum of the dimensions to find the largest and smallest distance between the two holes.

Hence, for the largest:

$$8.01 - 3.99 = 4.02 \text{ mm.}$$

The closest will be:

$$7.99 - 4.01 = 3.98 \text{ mm.}$$

In summary, the dimension and tolerance is

$$4.00 \pm 0.02 \text{ mm.}$$

This addition of tolerances may make it hard to join with a mating part that has two pins that fit in those holes.

6.5.3   Worst-Case Tolerance Stack-Up in Symmetric Dimensional Tolerance

Considering the assembly below, we would like to determine the range of variation of the gap AB, bearing in mind that the dimensions and tolerances given are symmetrical (Fig. 6.27).

Using Figure 6.28, the following table (Table 6.1) can be built.

The part with gap AB has now the dimension and tolerance: **2.0 $\pm$ 1.7 mm.**

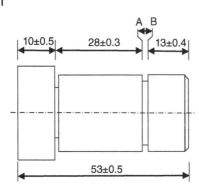

Figure 6.27 Missing dimension and tolerance in a part.

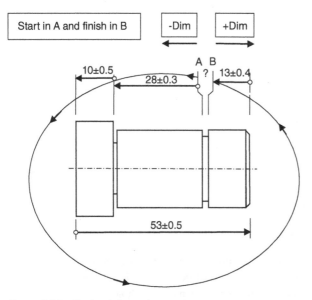

Figure 6.28 Chain of dimensions and tolerances.

Table 6.1

Dimensions with signs	+	−	Tolerance
		28	± 0.3
		10	± 0.5
	53		± 0.5
		13	± 0.4
Total	53	51	± 1.7
AB dimension	53 − 51 = **2**		± **1.7**
	Maximum distance **3.7**		
	Minimum distance **0.3**		

Figure 6.29 Chain of dimensions and tolerances.

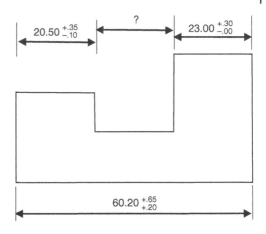

Table 6.2

Dimensions with signs	Old Dim and Tol		New Dim and Tol		-	+	Tolerance
	20.50	+0.35, -0.10	20.625	± 0.225	20.625		± 0.225
	23.00	+0.30, -0.00	23.15	± 0.15	23.15		± 0.15
	60.20	+0.65, +0.20	60.625	± 0.225		60.625	± 0.225
Total					43.775	60.625	
AB dimension					60.625 − 43.775 = **16.85**		± **0.60**
					Maximum distance **17.55**		
					Minimum distance **16.25**		

6.5.4 Worst-Case Tolerance Stack-Up in Asymmetric Dimensional Tolerance

We consider Figure 6.29 with given tolerances. Determine the missing dimension with symmetrical tolerance in worst case.

The corresponding table (Table 6.2) using Figure 6.30 showing the dimensions under the rules discussed before show that the part has now the following dimension and tolerance: **16.85 ± 0.60 mm.**

This dimension of the length AB is within the range of [16.25 – 17.55] mm. A nominal value with symmetrical tolerances can be selected.

6.6 Statistical Tolerance Analysis

6.6.1 Definition of Statistical Tolerance Analysis

A statistical tolerance analysis serves the determination of the most probable variation of an output of interest tolerance, for example, a dimension and its related tolerance considering the variation of a set of inputs. It will be in the form of Gaussian or normal distribution, as shown in Figure 6.31.

A mechanical part is composed of multiple features, each one having tolerance values that control the variable aspects. Statistical tolerance analysis helps to understand how these tolerances

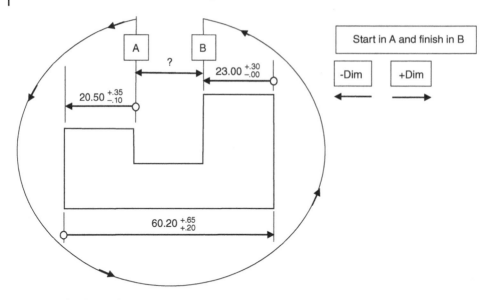

Figure 6.30 Chain of dimensions and tolerances.

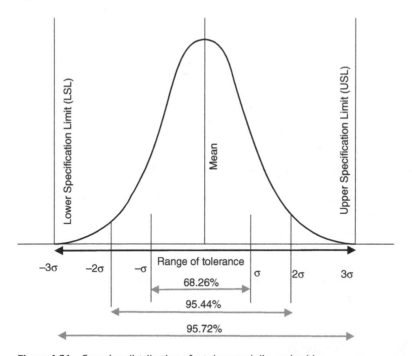

Figure 6.31 Gaussian distribution of a toleranced dimension(s).

contribute to the various performance characteristics of the design and hence contribute to the understanding of the design functionality and possibly coming up with design alternatives.

The simplest form of tolerance analysis is the 1D tolerance stack-up in one direction. The variation in each part with features contributes to the overall output.

Several statistical methods exist to treat tolerance analysis, among them root-sum squared for spreadsheet analysis and Monte Carlo simulations.

Table 6.3

Symbol	Modifier(s) applicable to tolerance	Modifier(s) applicable to datum feature
⌖	Ⓜ Ⓛ	Ⓜ Ⓛ
⌒ (concentric arc)	CANNOT BE MODIFIED	Ⓜ
⌒ (arc)	CANNOT BE MODIFIED	Ⓜ
○	CANNOT BE MODIFIED	NO DATUM FEATURE
⌭	CANNOT BE MODIFIED	NO DATUM FEATURE
▱	CANNOT BE MODIFIED	NO DATUM FEATURE
— LINE	CANNOT BE MODIFIED	NO DATUM FEATURE
— AXIS	Ⓜ	NO DATUM FEATURE
— C'PLINE	Ⓜ	NO DATUM FEATURE
⊥	Ⓜ	Ⓜ
//	Ⓜ	Ⓜ
∠	Ⓜ	Ⓜ
↗	CANNOT BE MODIFIED	CANNOT BE MODIFIED
↗↗	CANNOT BE MODIFIED	CANNOT BE MODIFIED
◎	CANNOT BE MODIFIED	CANNOT BE MODIFIED
≡	CANNOT BE MODIFIED	CANNOT BE MODIFIED

From Figure 6.31, we understand that the dimension will have an RSS tolerance stack-up as a standard deviation known as σ and there will be a distribution $\pm 1\sigma$, $\pm 2\sigma$, or $\pm 3\sigma$ depending how the obtained tolerances will be.

6.6.2 Worst-Case Analysis vs RSS (Root-Sum Squared) Statistical Analysis

Each dimension in WC will be represented by a minimum and maximum value showing the range of variation. Hence, part of the analysis is to check if the maximum range is considered for each input, how would be the maximum in the output considered feature or stack-up. Therefore, limits of acceptance can be defined. This is not probability.

While in RSS (root-sum squared), the statistical analysis does not focus on the extrema but focuses on the distribution of the variation for each dimension. Each dimension will be defined by a unique distribution (curve) of values based on the manufacturing process and its variations, for instance, change in materials, tool wear, and temperature.

The probability for the total will result from the combination of all these probabilities. Hence, the analysis is, when given the distribution of variation on each dimension, what is the probability that the performance characteristic will fall within defined acceptable limits.

RSS is limited by the fact that all inputs are assumed normally distributed and all performance characteristics have a linear relationship with the dimension. These assumptions do not account for the breadth of conditions that exist in typical scenarios found in manufacturing.

6.6.3 Second-Order Tolerance Analysis

Different types of parts are made with various manufacturing methods and hence followed by different distribution moments or parameters. RSS only uses standard deviation and does not include the higher moments of skewness and kurtosis that better characterize the effects of tool wear, form aging and other typical manufacturing scenarios.

However, second-order tolerance analysis (known as SOTA) incorporates all distribution moments: it can determine what and how the behavior of the output is when the assembly function is not linear.

Kinematic adjustments and other assembly behaviors result in nonlinear assembly functions usually in typical mechanical engineering scenarios. The method is discussed by Glancy et al. [17]

6.6.4 Cases Discussions

CASE #1

Worst-case maximum/minimum dimensions of assembly (Fig. 6.32).

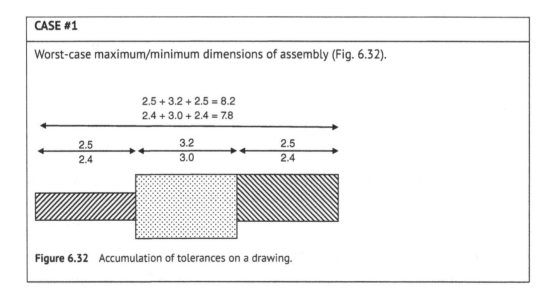

$$2.5 + 3.2 + 2.5 = 8.2$$
$$2.4 + 3.0 + 2.4 = 7.8$$

Figure 6.32 Accumulation of tolerances on a drawing.

CASE #2

Follow up on Figure 6.33:

1) Equate the tolerance to an interval in the probability distribution, e.g., 6σ.
2) Add up the variances and determine the interval, e.g., 6σ.

$$\sigma_{total}^2 = \sum_{i=1}^{n} \sigma_i^2$$

$6\sigma1 = 2.5 - 2.4 = 0.1$, then $\sigma1 = 0.0167$

$6\sigma2 = 5.7 - 5.5 = 0.2$, then $\sigma2 = 0.0333$

$\sigma^2 total = 0.0167^2 + 0.0333^2 = 0014$

$3\sigma_{total} = 3\sqrt{0.0014} = 0.114$

$\sigma^2 total = 0.0167^2 + 0.0333^2 + 0.0167^2 = 0.0017$

$3\sigma_{total} = 3\sqrt{0.0017} = 0.125$.

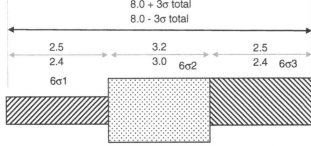

Figure 6.33 Accumulation of tolerances on a drawing.

CASE #3

Follow up on Figure 6.34:

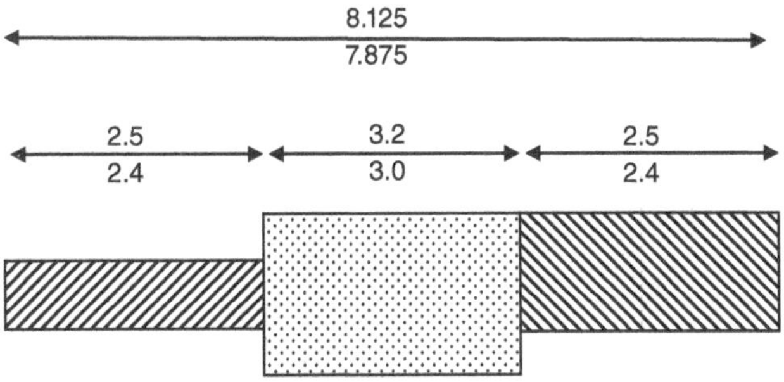

8.125
7.875

2.5
2.4

3.2
3.0

2.5
2.4

Figure 6.34 Accumulation of tolerances on a drawing.

This case is much tighter than worst-case scenario! Limits to references.

CASE #4

Using reference surface (Fig. 6.35): baseline dimensioning, the worst-case dimensions and tolerance.

$$5.7 - 2.4 = 3.3$$

$$5.5 - 2.5 = 3.0$$

$$\sigma^2_{total} = \sum_{i=1}^{n} \sigma_i^2$$

$$6\sigma1 = 2.5 - 2.4 = 0.1, \text{then}\, \sigma1 = 0.0167$$

$$6\sigma2 = 5.7 - 5.5 = 0.2, \text{then}\, \sigma2 = 0.0333$$

$$\sigma^2 total = 0.0167^2 + 0.0333^2 = 0014$$

$$3\sigma_{total} = 3\sqrt{0.0014} = 0.114$$

The mid-point can be located as:
$$(5.6 - 2.45) + 3\sigma_{total} = 3.264$$
$$(5.6 - 2.45) - 3\sigma_{total} = 3.036.$$

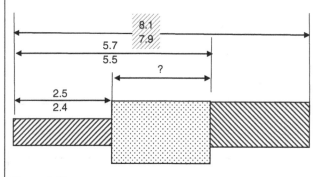

Figure 6.35

6.6.5 Understanding Material Condition Modifiers

The limits of the tolerances dimension are called material conditions. Material condition modifiers (MCM) materialized by the minimum or least material conditions help the designer to better describe what is dimensionally acceptable by providing insight into additional tolerances caused by the presence or absence of material on a part or feature. Three types of MCM exist:

The maximum material condition (MMC) describes the condition of a feature or part where the maximum amount of material in volume or size exists within its dimensional tolerance. It is used to indicate tolerance for mating parts such as a shaft and its housing.

- Symbol in GD&T is (ASME Y14.5M-1994, 2.7) (Fig. 6.36) Ⓜ
- If it is a hole or internal feature MMC = smallest hole size
- If it is a pin or external feature MMC = largest size of the pin.

Figure 6.36 MMC for tolerance allowed at a) MMC and b) MMC to a datum.

Least material condition (LMC) is used to indicate the strength of holes near edges as well as the thickness of pipes. LMC is a feature of size symbol that describes a dimensional or size condition where the least amount of material (volume/size) exists within its dimensional tolerance. The callout also overrides GD&T Rule#2 or the Regardless of Feature Size (RFS) rule shown by symbol S.

- Symbol in GD&T is (ASME Y14.5M-1994, 2.7) Ⓛ
- If it is a hole or internal feature LMC = largest hole size
- If it is a pin or external feature LMC smallest size.

Regardless of Feature Size (RFS) is the default condition of all geometric tolerances by rule #2 of GD&T and requires no callout. Regardless of feature size simply means that whatever GD&T callout you make, is controlled independently of the size dimension of the part.

This rule can be overridden by MMC or LMC, which specify the GD&T conditions at the max or min size of the part. LMC or MMC must be called out on the drawing specifically though to eliminate the regardless of feature size (RFS) default (Fig. 6.37).

Under MMC, the gaging is as follows: The two parts (Fig. 6.38) will always fit, that is, go-gauge. Hence, if you made sure that the MMC of the shaft was always smaller than the MMC of the hole, there is a guarantee that there will be always clearance between the parts. This is important for any tolerance stack to ensure that when the tolerances are at their least desirable condition, the part still functions properly. Currently, the pin large diameter is 10.1 mm MMC, and minimum hole is 9.9 mm MMC, so it's not a fit. The interference is 0.2 mm.

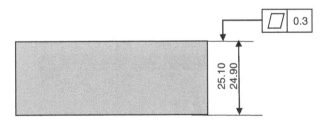

Figure 6.37 RFS example of flatness remaining constant irrespective of the dimension variation within tolerances.

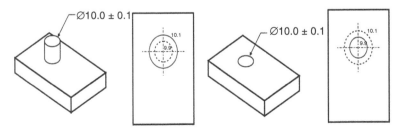

Figure 6.38 Sample of tolerance for fit parts.

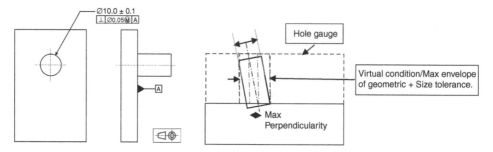

Figure 6.39 Gauging the pin.

Gauging a Pin with Perpendicular Callout MMC becomes important if a gauge is functional. To limit the size, use specified MMC with callout using GD&T. If we add a perpendicularity condition together with dimension, then the pin has to satisfy both conditions together.

The gauge diameter, known as hole gauge, is equal to the maximum diameter of the pin (MMC) added to GD&T symbol tolerance. So the gauge hole becomes 10.15 mm (Figure 6.39).

Any difference obtained with actual size from MMC is considered as bonus tolerance. This means that in the worst case the perpendicularity and size of the hole and the pin will fit together.

All Applicable Geometric Tolerances (Rule #2). RFS applies, with respect to the individual tolerance, datum reference, or both, where no modifying symbol is specified. MMC or LMC must be specified on the drawing where it is required.

Bonus Tolerance As the size of the pin departs from MMC toward LMC, a bonus tolerance is added equal to the amount of change. Bonus tolerance equals the difference between the actual feature size and the MMC of the feature.

Bonus Tolerance = MMC − LMC.

Example 6.1 *Stack-up analysis to secure functional gap*

The contribution of the components' tolerances to a functional gap in a 2D wheel assembly in worst case of several parts (Fig. 6.40) can be analyzed in CAD software such as Creo. Adjustment

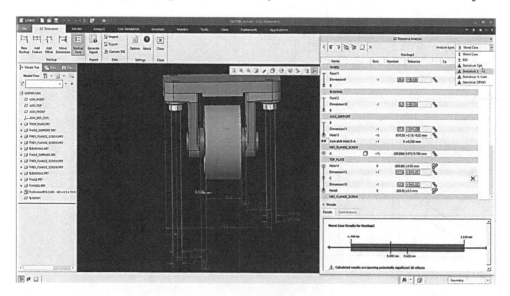

Figure 6.40 Snapshot of the wheel assembly and functional gap. *Source:* Courtesy of Sigmetrix

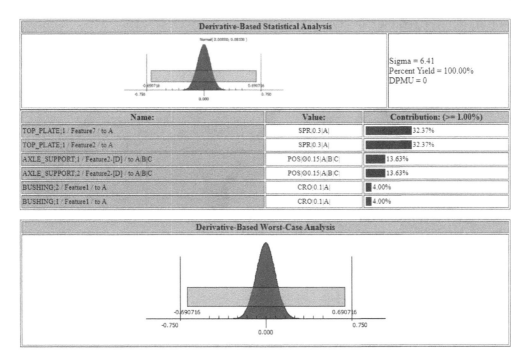

Figure 6.41 Snapshot of stack-up analysis improved gap in CREO software (Sigmetrix courtesy).

of tolerance for each part is possible with the observation of sensitivity. Fig. 6.41 shows improved functional gap compared to Fig. 6.40. GD&T data helps showing the effect of each tolerance visually on screen.

Example 6.2 *Validating design and assembly*

Consider a swing weight governor similar to the one shown (Fig. 6.42). It has two balls that will swing as the rotational speed increases. Hence, the collar will move up and down depending on the speed. This will control the fuel rack level. This is a precision mechanical system where the bars and balls are precisely located and assembled. What is the tolerance range of the height if the angle α moves between 180 and 60 degrees? The following are the practical assumptions:

- The bars length is given with tolerance (Table 6.4);
- All holes have high-precision location without tolerance;
- The six pins and holes have zero clearance and move freely.

The height, H, can be defined according to Figure 6.42 with respect to the triangle as:

$$H^2 = L_1{}^2 + L_2{}^2 - 2L_1L_2 \cos \alpha$$

where α is the angle between the bars.

The extremum heights H_{max} and H_{min} can be defined from the previous relationship shown next and the results are graphically shown in Figure 6.43 and in Table 6.5.

$$H_{max}{}^2 = (L_1 + \delta l_1)^2 + (L_2 + \delta l_2)^2 - 2(L_1 + \delta l_1)(L_2 + \delta l_2) \cos \alpha$$

$$H_{min}{}^2 = (L_1 - \delta l_1)^2 + (L_2 - \delta l_2)^2 - 2(L_1 - \delta l_1)(L_2 - \delta l_2) \cos \alpha$$

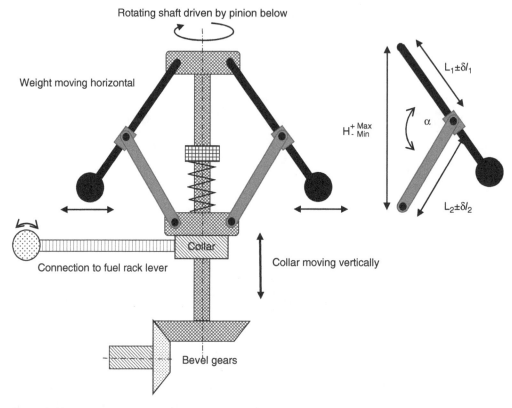

Figure 6.42

Table 6.4

L_1	4	L_{1max}	4.02
l_1	0.02	L_{1min}	3.98
L_2	3	L_{2max}	3.02
l_2	0.02	L_{2min}	2.98

Example 6.3   *Tolerance in assembly*

Calculate the minimum gap of the assembly to make sure the two parts will never touch in the assembly using the two pins (Fig. 6.44). A method as described before is shown in Table 6.6.

Result: the minimum gap will be 4 − 3.4 = 0.6 mm.

Example 6.4   *Tolerances stack-up analysis*

Figure 6.45 shows an assembly composed of crankshaft, a coupling, the flywheel, and 4 bolts to hold all parts together. The purpose is to define the tolerances that are nonfunctional from those that are functional. The calculation process will show those tolerances that are important and influencing in this assembly, for instance, what is the importance of the alignment and perpendicularity to the assembly?

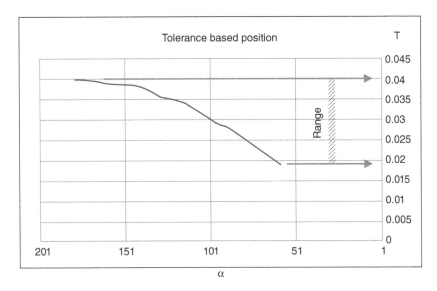

Figure 6.43 Results of the simulation.

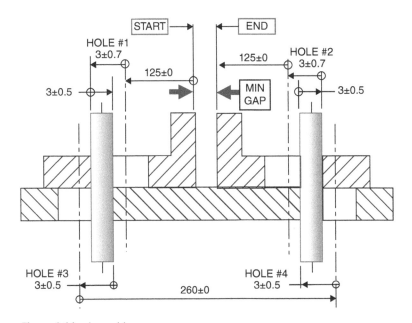

Figure 6.44 Assembly

As initial observation when looking at the drawing with GD&T in Figure 6.45, there is an LMC on features D and B showing a difference of 50.1 mm and 49.97 mm, giving 0.13 mm that is the largest clearance they can have.

The LMC of part D as 50.1 mm and LMC of part B as 49.97 mm shows a difference of 0.13 mm as the maximum possible clearance.

On the coupling part bolt hole location, the LMC is 8.90 mm with an 8 mm diameter giving a clearance of 0.9 mm that is a lot higher than the allowed fit calculated as 0.13 mm of the fit B and D. Hence, B and D are controlling the alignment of the two parts. Consequently, the clearances of

Table 6.5 Tolerance stack of swing weight governor.

Angle α	H_{max}	H_{min}	H_{nom}	T(+/-)
180	7.040	6.960	7.000	0.040
170	7.014	6.934	6.974	0.040
160	6.935	6.856	6.896	0.039
150	6.805	6.728	6.766	0.039
140	6.624	6.549	6.587	0.038
130	6.394	6.322	6.358	0.036
120	6.117	6.048	6.083	0.035
110	5.795	5.730	5.763	0.033
100	5.431	5.370	5.401	0.030
90	5.028	4.972	5.000	0.028
80	4.590	4.539	4.564	0.025
70	4.120	4.075	4.098	0.022
60	3.625	3.586	3.606	0.019

Table 6.6 Tolerance stack.

#		−	+	Tolerance ±
1		125		0.0
2		3		0.7
3			3	0.5
4		3		0.5
5			260	0.0
6		3		0.5
7			3	0.5
8		3		0.7
9		125		0.0
10		−262	+266	± 3.4
11	**Dimension**	−262 + 266 = 4		

the threaded holes and OD as factors are filtered out. The outer diameters of B and D are factors, but do they have any effects? We consider two different scenarios ignoring the perpendicularity and then considering it.

Case (a)

Extreme case by pushing the coupling part up and calculate the maximum gap in stationary state and in rotation state to see any variation. The minimum gap from the GD&T drawing is only 229.8 mm, while the maximum is as follows (Fig. 6.46).

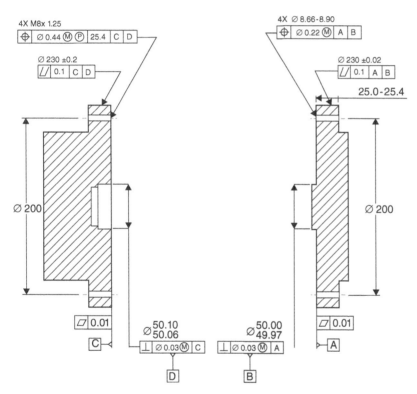

Figure 6.45   a) Crankshaft assembly. b) Drawing.

MMC

On the left part in the drawing, the GD&T indicates that Ø 230.2 mm is Maximum Material Condition (MMC) for outer diameter (OD).

There is a total runout tolerance ⟋⟋ ; Runout Tol = +0.1; this makes the OD out of boundary to Ø 230.3 where the radius is then 115.15 mm.

The perpendicularity ⊥ is the same and will not allow us the maximum gap, hence, not to consider.

The side diameter 50.10 has Least Material Condition (LMC) with radius Ø 25.05 mm.

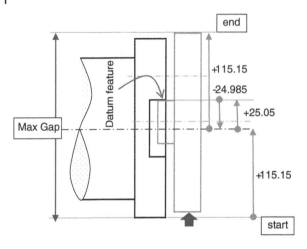

Figure 6.46 Crankshaft assembly.

The opposite fitting diameter of the coupling is LMC of B with diameter Ø 49.97 with radius 24.985. The budget can be calculated in the next table without considering perpendicularity.

Direction	−	+
		+115.15
		+25.05
	−24.985	
		+115.15
Results	−24.985	+255.35

Hence, 255.35 − 24.985 = Ø 230.365 mm as max gap in stationary state.

But if rotating 180⁰, the difference between surface B and D will be: 25.050 − 24.985 = 0.065 mm (shows the offset of the axes). This is to be added to Ø 230.365 + 0.065 = Ø 230.43 mm, which is maximum gap when rotating.

Case (b)
Perpendicularity Control between A and C
If the perpendicularity is perfect between B and D, only gaps are active in B and D and on the outer diameter OD of the assembly (Fig. 6.47).

If perpendicularity is considered, then we have the following scenario in assembly where on top of the radial gap, we must add the perpendicularity tolerance.

If out of perpendicularity, then estimate its effect between features B to A; see case in Figure 6.48.

If perpendicularity is considered, it is clear that the edge of feature B may result in touching the edge of D inside prematurely; hence, we are not able to give the maximum gap.

Perpendicularity Will Not Reduce the Max Gap under Consideration
230.365 mm is the max gap in stationary state calculated with another method. The corresponding maximum rotating state is 230.365 + 0.065 = 230.43 mm as maximum in rotating state (Fig. 6.49). The second max is for when rotating of spinning application showing the dynamic variation of the gap.

Figure 6.47 Crankshaft assembly.

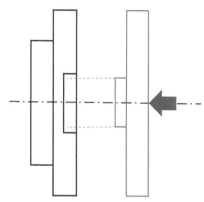

Figure 6.48 Crankshaft assembly.

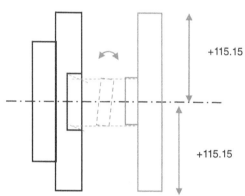

+115.15

+115.15

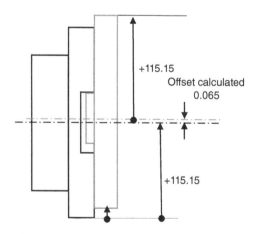

+115.15

Offset calculated
0.065

+115.15

Figure 6.49 Crankshaft assembly.

Direction	-	+
	0	+115.150
	0	+0.065
	0	+115.150
Results	0	Ø+230.365

Appendix A from ISO and ASME Y14 Symbols

SYMBOL FOR:	ASME Y14.5	ISO
STRAIGHTNESS	—	—
FLATNESS	▱	▱
CIRCULARITY	○	○
CYLINDRICITY	//	//
PROFILE OF A LINE	⌒	⌒
PROFILE OF A SURFACE	⌒	⌒
ALL AROUND	↗⊖	↗⊖
ALL OVER	↗⊜	↗⊜ (proposed)
ANGULARITY	∠	∠
PERPENDICULARITY	⊥	⊥
PARALLELISM	//	//
POSITION	⊕	⊕
CONCENTRICITY (Concentricity and Coaxiality in ISO)	◎	◎
SYMMETRY	≡	≡
CIRCULAR RUNOUT	*↗	*↗
TOTAL RUNOUT	*↗↗	*↗↗
AT MAXIMUM MATERIAL CONDITION	Ⓜ	Ⓜ
AT MAXIMUM MATERIAL BOUNDARY	Ⓜ	NONE
AT LEAST MATERIAL CONDITION	Ⓛ	Ⓛ
AT LEAST MATERIAL BOUNDARY	Ⓛ	NONE
PROJECTED TOLERANCE ZONE	Ⓟ	Ⓟ
TANGENT PLANE	Ⓣ	NONE
FREE STATE	Ⓕ	Ⓕ
UNEQUALLY DISPOSED PROFILE	Ⓤ	UZ (proposed)
TRANSLATION	▷	NONE
DIAMETER	Ø	Ø
BASIC DIMENSION (Theoretically Exact Dimension in ISO)	50	50
REFERENCE DIMENSION (Auxiliary Dimension in ISO)	(50)	(50)
DATUM FEATURE	*⊔̸ Ⓐ	*⊔̸ or *⊔̸ Ⓐ

* May be filled or not filled

Multiple Choice Questions of this Chapter

Multiple Choice Questions are given for each chapter with solutions in an online extension of this book. Please use link: www.wiley.com\go\mekid\metrologyandinstrumentation\

References

1 Gramenz, K., 1925, *"Die Dinpassungen und ihre Anwendungen,"* Dinbuch 4.

2 Wade, O.R., 1967, *Tolerance Control in Design and Manufacturing*, Industrial Press Inc., New York.

3 Harry, M.J., and Stewart, R., 1988, "Six sigma mechanical design tolerancing," Motorola Government Electronics Group, Scottdale, Arizona.

4 Griess, K.H., 1990, "Tolerancing—design guide," (proprietary), The Boeing Company, D6-25382-90.

5 ASME Y14.5.1M-1994, "Mathematical definition of dimensioning and tolerancing principles," The American Society of Mechanical Engineers.

6 Mekid, S., 2013, "High speed desktop ultra precision CNC micro/meso-machine," *Advanced Materials Research*, 739 (2013), pp. 640–646.

7 Ogedengbe, T., and Mekid, S., 2011, "Application of finite element analysis and Taguchi method to the design of a micro milling machine structure," *Int. J. of Design Engineering*, 4(3), 2011, pp. 197–219.

8 Mekid, S., 2008, *Introduction to Precision Machine Design and Error Assessment*, (ISBN13: 9780849378867), CRC Press.

9 Mekid, S., 2000, "High precision linear slide. Part_1: design and construction," *Int. Journal of Machine Tools and Manufacture*, 40(7), pp. 1039–1050.

10 Mekid, S., and Olejniczak, O., 2000, "High precision linear slide. Part_2: control and measurements," *Int. Journal of Machine Tools and Manufacture*, 40(7), pp. 1051–1064.

11 Ogedengbe, T., and Mekid, S., 2011, "An investigation of influence of machining conditions on machining error," *Int J. Computer Aided Engineering and Technology*, (3)3/4, pp. 230–239.

12 Mekid S., and Ogedengbe, T., 2010, "A review of machine tool accuracy enhancement through error compensation in serial and parallel kinematic machines," *Int. J. Precision Technology*, (1)3/4, pp. 251–286.

13 Mekid, S., "Design strategy for precision engineering: second order phenomena," *J. Engineering Design*, (16)1, pp. 63–74.

14 Mekid S., and Vaja, D., 2008, "Propagation of uncertainty: expressions of second and third order uncertainty with third and fourth moments," *Measurement*, (41)6, pp. 600–609.

15 Mekid, S., Pruschek, P., and Hernandez, J., 2009, "Beyond intelligent manufacturing: a new generation of flexible intelligent NC machines," *Mechanism and Machine Theory*, (44)2, February 2009, pp. 466–476.

16 Mekid S., and Bonis, M., 1997, "Conceptual design and study of high precision translational stages: application to an optical delay line," *J. American Society for Precision Engineering*, (21)1, July 1997, pp. 29–35.

17 Glancy, C.G., and Chase, K.W., 1999, "A second-order method for assembly tolerance analysis," *Proceedings of the ASME 1999 Design Engineering Technical Conferences. Volume 1: 25th Design Automation Conference*, Las Vegas, Nevada, USA. September 12–16, 1999, ASME, pp. 977–984.

7

Instrument Calibration Methods

"Modern calibration always improves power plant performance and reduces side costs."

—Literature.

7.1 Introduction

Humans have used parts of the body as tools of measurements, for instance, the forearm, hand, and finger. The word "foot" is still in use. The word "calibration" joined the English language within the last century, and it was used mainly in defense. But throughout history, calibration existed already in the ancient civilizations of Egypt, Mesopotamia, and the Indus Valley. The excavations carried out in these regions have revealed the use of angular graduation for civil engineering.

The calibration was used for division of linear distance and angles using specific instruments, such as a dividing engine (Fig. 7.1), and on the other side the measurement of gravitational mass to cover most commerce and technology development since early civilization about AD 1800.

7.2 Definition of Calibration

According to the *International Vocabulary of Basic and General Terms in Metrology*, a calibration is the assessment of the uncertainties in the results (measurand) of the measurement task. Note: Here "uncertainty" is used as a synonym for error. The uncertainty may contain uncorrected known and unknown systematic effects as well as random effects.

According to *The Automation Systems and Instrumentation Dictionary*, the word "calibration" is defined as "a test during which known values of measurand are applied to the transducer to observe the corresponding output reading recorded under specified conditions."

The definition includes the capability to adjust the instrument to zero, which is the lowest value in the scale, and to set the desired span—or instrument range—needed for measurement.

A reference such as a standard instrument with higher accuracy is required for the sake of comparison (Fig. 7.2). This comparison includes detection of error, correlation, adjustment, and rectification. The activity of calibration is expected to be documented in a standard way to be discussed later.

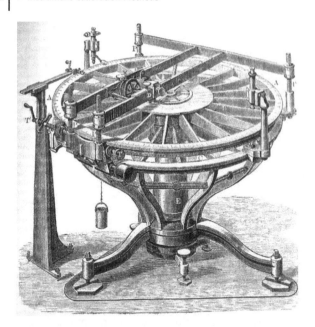

Figure 7.1 Initial circular dividing engine.

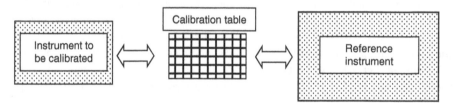

Figure 7.2 Calibration components.

Example: An instrument is set to measure in the range of 0 to 400 psig, so the low value is zero and the high value in the range is 400 psig. The span is therefore: 400 − 0 = 400 psig. The corresponding output voltage is 5 − 25 mA.

7.3 Need for Calibration

The calibration is always required for a new instrument. Since a calibration is performed by comparing or applying a known reference signal to the instrument under test, errors are detected by performing a calibration. An error is the algebraic difference between the indication and the actual value of the measured variable.

Instrument error can occur due to a variety of factors including thermal drift, environment conditions, electrical supply, addition of components to the output loop, or sometimes other changes in the process [1–5]. To detect errors in instruments and correct them, periodic calibrations should be planned even if the instrument has not shown any sign of error.

Instruments can observe drifts during their life time or because of a sudden physical issue, such as heat or mechanical shock, which may degrade the accuracy of the measurement. The calibration of instruments is needed to minimize any measurement uncertainty. It serves to maintain the accuracy of the equipment in use.

Calibration compares a known measurement (the standard) to the measurement (unknown) using a given instrument. Usually, the accuracy of the standard should be around ten times the accuracy of the measuring device being tested. However, an accuracy ratio of 3:1 is acceptable by most standards organizations. Calibration is carried out two times before and after using the measuring instrument. The interval in between must be defined and used systematically. This is the calibration interval.

Calibration assures a common reference, that is, traceability to national and international standards. It ensures that equipment has been satisfactorily working since the last calibration and helps it to continue working satisfactorily until next calibration.

Overall a calibration can be static or dynamic. All those common types we have discussed are known as static calibration. A known value is input to the system under calibration, and the system output is recorded. Static means that the points of measurement remain constant; they do not vary with time and space.

In dynamic calibration, the variables of interest are time or space dependent, and such varying information is requested. The dynamic variables are time and space varying for both magnitude and frequency.

As a summary, the calibration of an instrument is essential to maintain instrument uncertainty within known limits that are traceable with respect to a reference. It is a process of configuring an instrument with respect to another more accurate to provide a result for a sample within an acceptable range [6, 7].

7.4 Characteristics of Calibration

Like any measurement and verification process, calibration is characterized by several parameters on top of the procedure of calibration to secure a proper and reliable calibration. The following are commonly used parameters in calibration

Calibration Range
This is the region between the limits of measurement capabilities of the instrument.

Span
This is the true range defined algebraically from the range as maximum minus minimum.

Static Sensitivity
The slope of static calibration curve (Fig. 7.3) provides the static sensitivity of the measurement defined as

$$K = K(xi) = \left(\frac{dy}{dx}\right)_{x=xi}. \tag{7.1}$$

Calibration Tolerance
A tolerance is defined for the calibration in progress.

- *Accuracy*
  The ratio of error of the full output or the ratio of the error to the output expressed in percent span.
- *Tolerance*
  The permissible deviation from a specified value. It is expressed in measurement units or percent of span.

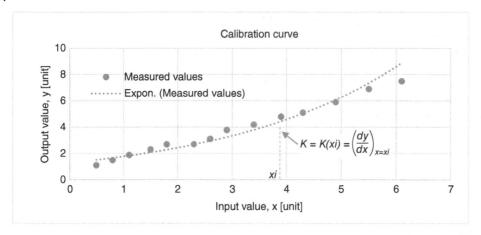

Figure 7.3 Static calibration curve.

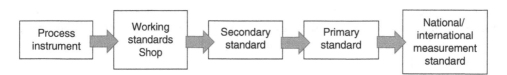

Figure 7.4 Traceability route.

If the specified calibration tolerance of the previous instrument is ±3 psig, the corresponding output 1 psig will be 0.05 mA. Hence, the tolerance on each side of the range ±3 psig is 3 × 0.05 = 0.15 mA . This tolerance should appear in the records as ±3 psig and ±0.3 mA.

Accuracy Ratio
It is known as the relationship between the accuracy of the test standard and the accuracy of instrument under calibration. The usual rule is to make sure the test standard instrument is highly more accurate than the instrument under test, and the usual ratio is 4:1 over the whole range of measurement.

Traceability
It is important that all calibration should be performed traceable to a national or international standard. It is defined as the property of a result of a measurement whereby it can be related to appropriate standards, generally national or international standards, through an unbroken chain of comparison.

Hence, when any instrument is calibrated by another highly accurate instrument, the latter must also be calibrated with respect to a reference. The relationship to a national or international reference is conserved through this chain (Figure 7.4).

Uncertainty
This is associated with the result of a measurement that characterizes the dispersion of the values that could reasonably be attributed to the measurand (see Chapter 8). Uncertainty analysis for calibration labs should conform to ISO17025 requirements.

Instrument Repeatability

This is the ability for the measurement system to indicate the same value on repeated but independent activity of the same input to provide a measure of the instrument repeatability. In other words, this is the standard deviation measuring the variation of the output for a given input.

Reproducibility

This is the closeness of agreement in results obtained from duplicate tests carried out under similar conditions of measurements.

7.5 Calibration Overall Requirements and Procedures

7.5.1 Calibration Methods/Procedures

The calibration methods and calibration procedures of an accredited calibration laboratory are verified and tested in terms of accuracy as part of the accreditation process. The method is not a goal in itself. There is a difference between calibration and validation. It is important to get updated on a regular basis for example here[1]. The common calibration procedures can be using a curve of data calibration, or standard method, as explained next .

Data Calibration

The method is associated with collected measured data with a regression plot. It is related to accredited calibrated but not accredited to ISI standard.

Standard Calibration

This calibration is for instruments that are not critical to quality and may not need to be accredited, but it is necessary to document the process.

Calibration Procedure

The calibration method can either be linear or nonlinear depending on the instrument and delivered measurements.

Example: Calibrating a thermometer needs to be in a laboratory that has a calibrator, for instance, a liquid bath calibrator. This is a new reference temperature measurement for now. It has been calibrated to a known accuracy. The thermometer (device under test) is placed inside the calibrator, and the measurements are recorded over the allowed range while observing the difference between your thermometer and the calibrator. If there is a difference, then it needs to be adjusted. Two ways exist to adjust the thermometer; it is either through adjusting the display or using the calibration results to determine new offsets. This operation can be repeated to check the thermometer at different times and must be recorded, as explained.

In the method the following can be encountered. The first adjustment to start with in calibrating an instrument is usually the zero.

---

1 https://www.nist.gov/pml/weights-and-measures/laboratory-metrology/calibration-procedures.

First type of instrument calibration error(s) can be of zero type as shown in the graph below (Fig. 7.5):

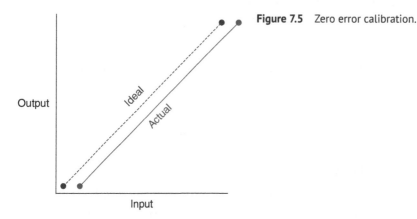

Figure 7.5 Zero error calibration.

The second type of instrument calibration error(s) can be linearity error, as shown in the graph below (Fig. 7.6):

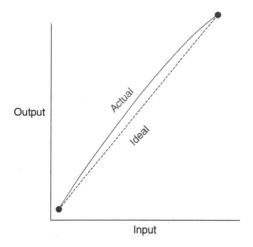

Figure 7.6 Linearity error calibration.

The calibration procedure should include the following items:

- **Purpose,** stating the reason for the procedure of calibration.
- **Scope,** stating to what and to whom this procedure is applied.
- **Definitions,** describing all key terms for clarity.
- **References and attachments,** mentioning all documents and guides used for this process.
- **Test equipment/materials required,** defining the test equipment and materials required to perform the procedure. As part of data recording, it is important to list the specific test equipment. This will help to meet uncertainty requirements.
- **Safety measures,** mentioning any potential human health, facility, equipment or process safety issues. All needed safety certificates must be provided.
- **Prerequisites and initial conditions,** mentioning any conditions that should be met prior to staring the calibration. Otherwise, they can be included in the test procedure or notes.

- **Test procedure,** providing a concise, step-by-step description of the procedure or method; this is part is crucial.
- **Acceptance criteria,** describing the passing criteria included in the test procedure.
- **Approvals,** specifying the operator and the approvals signatures authority including the calibration procedure, should appear in the final calibration sheet.

ISO 17025 Calibration

This ISO 17025 is the most rigid form of calibration. Under ISO, the companies are maintaining high international quality by maintaining rules and regulations.

ISO 17025, "General Requirements for the Competence of Testing and Calibration Laboratories," is the main ISO standard used by testing and calibration laboratories. This is the standard according to which most labs must hold accreditation to be deemed technically competent. To achieve the level of quality required, there is a need to do the following:

a) Maintain records of calibrated instruments with all details throughout the calibration;
b) Keep an accurate list of instruments according to ISO to avoid rejection of certification if the instrument is not physically there;
c) Have well-framed quality modules;
d) Inspect the documentation and records of the calibration process. Changes can be easily detected if the calibration process is audited every single time. Following the previous conditions is necessary.

7.6 Calibration Laboratory Requirements

ISO/IEC 17025:2005 specifies the general requirements for the competence to carry out tests and/or calibrations, including sampling. It covers testing and calibration performed using standard methods, nonstandard methods, and laboratory-developed methods.

It is applicable to all organizations performing tests and/or calibrations. These include, for example, first-, second-, and third-party laboratories and laboratories where testing and/or calibration forms part of inspection and product certification.

ISO/IEC 17025:2005 is applicable to all laboratories regardless of the number of personnel or the extent of the scope of testing and/or calibration activities. When a laboratory does not undertake one or more of the activities covered by ISO/IEC 17025:2005, such as sampling and the design/development of new methods, the requirements of those clauses do not apply.

ISO/IEC 17025:2005 is for use by laboratories in developing their management system for quality, administrative, and technical operations. Laboratory customers, regulatory authorities, and accreditation bodies may also use it in confirming or recognizing the competence of laboratories. ISO/IEC 17025:2005 is not intended to be used as the basis for certification of laboratories.

Compliance with regulatory and safety requirements on the operation of laboratories is not covered by ISO/IEC 17025:2005.

A couple of requirements itemized for each to check for any laboratory carrying out calibration are as follows:

a) **Location**
 - Permanent
 - On-site

b) **Environment**
- Temperature
- Humidity (50 ± 10%)
- Air flow (min to standards)
- Filtration (high degree)
- Electromagnetic screening
- Noise level (less than 60 dB)
- Vibration (0.25 μm max displacement amplitude from 0.1 Hz to 30 Hz and 0.001 g max 30 Hz to 200 Hz.)
- EMI/EMC (use of shielding/filtering to be less than 100 μV/m)
- Cleanliness (good housekeeping with particles control)
- Lighting (recommended 450–700 lux)
- Power supply (to check stability)
- Accessibility (firefighting devices, quality system document, and instrument manuals)
- Receiving and calibrating area

c) **Equipment**
- Reference standards
- Transfer standards
- Working standards
- Connecting leads
- Adaptors/accessories
- Subsidiary equipment
- Computers and automation
- Software

d) **Staff**
- Training
- Authority
- Responsibility
- Technical
- Competence

e) **Management**
- Documents
- Records
- Calibration

7.7 Industry Practices and Regulations

The calibration in industry follows the same international standards and procedures. The purpose of instrument calibration is to maintain adherence to industry standards and government regulations. It also helps to maintain quality assurance.

Control of the Calibration Environment

In calibration laboratories, most test equipment and reference standards are affected by environmental changes of several parameters:
- temperature;
- relative humidity;

- barometric pressure;
- electromagnetic interference;
- vibration;
- air cleanliness and airflow; and
- several other environmental conditions.

There is a strong need by the laboratory to monitor, control, and record those environmental conditions that influence the measurement results (Table 7.1). A continuous monitoring system should be in place to record those environmental conditions at specified time intervals. The monitoring and control must also extend to encompass the storage and handling of equipment within the laboratory.

It is important to understand that the control of those environmental conditions in calibration laboratories constitute a trade-off between the dimensional and electronic equipment calibrations and the level of calibration accuracy required.

For any reliable measurement, several parameters can affect the measurement depending on its nature. For example, a laser beam is affected by the pressure, temperature, and humidity. A function exists to take the current values and input them into measurement for compensation. The most usual parameter is temperature. The room temperature must be 68°F or 20°C as defined by ISO. Temperature is also important since all instruments may undergo thermal expansion if temperature has a drift change [2, 8–11].

To ensure that the affecting parameters are under control, the calibration process has to be carried out in a professional way.

Calibrating procedures that are related to any measurements used for quality purpose are subject to the ISO9000 standard and require appropriate training.

7.8 Calibration and Limitations of a Digital System

Currently, all measurements are becoming digitized and hence all connected instruments and sensors are important to be correctly calibrated, along with all downstream signal processing and results outputs.

Example: A dynamometer measuring motor torque has a transducer with a 50 N range, when the expected signal level is only 15 N. It is usually advisable that the full-scale output of the transducer matches the expected signal levels, rather than the whole range of the transducer to avoid using too coarse digitization at the lower end of the measured range. This will not help much if the resolution or repeatability of the signal itself has a limitation.

Converting from a digital to analog output, for example when controlling motor speed, consists of changing from a series of steps to a linear signal by smoothing. Rounding errors can be a problem. When converting to a digital signal from an analog input, the linear signal has to become a series of steps so it can be seen that if the steps are too coarse or the information sampled at too low a rate, then information can be lost.

Nyquist proposed for a limited bandwidth signal being f_{max}, the equally spaced sampling frequency f_s must be greater than twice the maximum frequency f_{max}, that is,

$$f_s > 2 \times f_{max}, \tag{7.2}$$

so that the signal is uniquely reconstructed without aliasing.

Table 7.1 Recommended environmental conditions.

| Laboratory type | Field | Temperature (degrees celcius) | | Relative humidity (%) | Airborne particle count[b] |
		Set point and limits	Maximum rate of Change (K/hour)[a]		
I	MECHANICAL/DIMENSIONAL	20 ± 1	0.5	30–55	50 000
	ELECTRIC/ELECTRONIC	23 ± 1	1.0	30–55	100 000
	OTHER	The environmental conditions for other fields of measurements should be developed and assessed with respect to appropriate applicable influencing factors.			
II	MECHANICAL/DIMENSIONAL	20 ± 2	1.0	30–55	150 000
	ELECTRIC/ELECTRONIC	23 ± 2	1.5	30–55	250 000[c]
	OTHER	The environmental conditions for other fields of measurements should be developed and assessed with respect to appropriate applicable influencing factors.			
III	GENERAL	Range of 18–28 with preferred setpoint of 23	1.5 (1.0–for Mechanical/Dimensional)	10–60	[d]

a) Certain measurement or comparison equipment may require more stringent control.
b) Measured in accordance with U.S. Federal Standard No. 209, "Clean Room and Work Station Requirements. Controlled Environment".
c) There should be no accumulation of particles on or under benches, cabinets, equipment, instrumentation, etc.
d) Careful housekeeping with no accumulation of particles on or under benches, cabinets, equipment, instrumentation, etc.

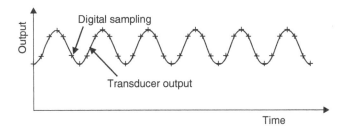

Figure 7.7   Digital sampling @ 30° intervals.

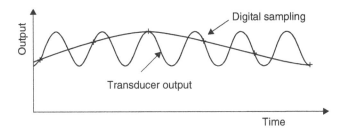

Figure 7.8   Aliasing digital sampling @ 390° intervals giving a long sine wave.

The frequency $2 \times f_{max}$ is called the Nyquist sampling frequency.

Occasionally sampling at too low a frequency can give completely misleading results, with the apparent output signal looking completely different from the input. Aliasing is a problem that can occur when digitally sampling data at too low a rate. Figure 7.7 shows a sine wave being sampled at intervals of 10°; the output then shows up roughly as a sine wave of the correct frequency.

Figure 7.8 shows the same sine wave being sampled at intervals of 390°. This also gives a sine wave, but a very long wavelength one in this case, with a frequency completely different from the original signal. Aliasing produces output frequencies that are not real and that are lower than the actual frequency being sampled. The output frequency due to aliasing is the difference between the original frequency and the sampling frequency.

7.9   Verification and Calibration of CNC Machine Tool

The maintenance of a machine tool does not cover only the operations and system failures but includes also maintenance of the precision to be within an acceptable range, especially if the machine is used for production.

This maintenance can be either a verification or repair of every component contributing to the degradation of the machine performance as part of the preventive maintenance, including:

- CNC machine diagnostics;
- CNC machine tool condition monitoring; and
- Data analysis and corrective action creation.

This predictive maintenance is incredibly effective while new tools are becoming more comprehensive of error characterization in machine tools in its various aspects. Intelligent machines can in the near future self-check since the calibration and compensation software will be embedded in the controller and correct most errors through compensation or possibly with in-process measurement support CAM altering for compensation of errors.

The tools used today for verification and calibration are:

- Grid plate encoder;
- Double-ball bar system;
- Capacitor gauge;
- Rotary encoder for rotary motion;
- Laser interferometer with double laser beams;
- Laser Doppler calibration system.

A typical 3-axis machine tool is subject to 21 degrees of freedom including linear positioning, pitch, yaw, straightness, roll, and squareness to the other axes, as discussed in Chapter 5.

Each of these degrees of freedom can have a detrimental effect on the machine's overall positioning accuracy and hence contributes directly to the degradation of the accuracy reflected on machined parts.

Furthermore, the dynamic of moving axes may degrade the accuracy because some gains on the controller are not precise anymore.

Usually, to verify a CNC machine tool two techniques are used according to ISO 230:

a) **Laser interferometry**

Laser interferometry was discussed in Chapter 5

b) **The double-ballbar**

The double-ballbar is the tool that can detect this type of failure when combining two axes together to build a circle. Potential errors causing the machine to deviate from the programmed circle path are:

Potential errors in one axis:
- Backlash, reversal spikes, lateral play, cyclic error, straightness, scale error,

Potential errors between axes:
- Servo mismatch and squareness.

The ballbar plot shapes are obtained under specific data including size of the bar and feed rate in the *XY* plane. The graphs are obtained for both CW and CCW showing differences; the CW data is shown in blue and CCW data in red. A plot scale of 5 μm/division is used (Fig. 7.9). The ballbar has a clear traceability as shown in Figure 7.10.

7.10 Inspection of the Positioning Accuracy of CNC Machine Tools

Calibration of a computer numerical controlled (CNC) machine, also known as digital controlled machine tool, works together with compensation techniques of errors. If the machine is within tolerance, it can go with feed rates and still maintain tolerance. The calibration serves also for:

1) Time saving in online part inspection and reducing back and forth moves between machine tool and CMM [12].
2) Good prediction of when the machine will go out of tolerance.
3) Good prediction of tool wear through tolerance of parts.

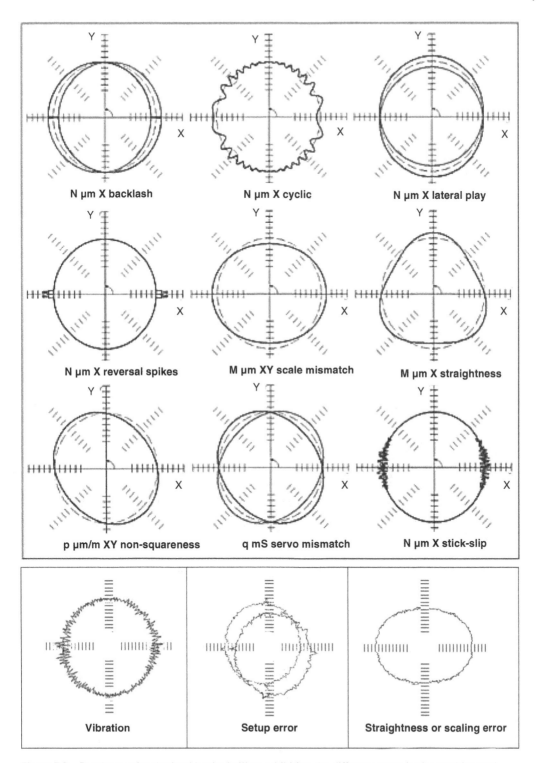

Figure 7.9 Results graphs obtained by the ballbar exhibiting the different errors in the machine tool.

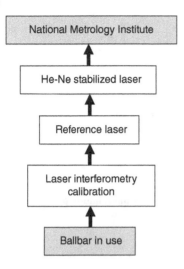

Figure 7.10 Ballbar traceability chart.

Once the errors are quantified and known, a 3-D positioning errors can be tabulated as lookup tables or compensation tables, allowing the used software to correct measured probe positions. With 3-D error correction, inherent errors in machine tool geometry and positioning can be eliminated to provide accurate dimensional measurement.

Therefore, by satisfying the 4:1 gauge accuracy ratio, discussed earlier, with volumetric error compensation, a CNC machine tool can provide the same high accuracy as a CMM.

The positioning accuracy of CNC machine tools refers to the position accuracy achieved through the movement of each coordinate axis of the machine tool under the control of the numerical control device. The positioning accuracy of CNC machine tools is known as the motion accuracy of the machine tool.

Ordinary machine tools are manually fed. The positioning accuracy is mainly determined by the reading error. The movement of the CNC machine tool is realized by digital program instructions, so the positioning accuracy is determined by the numerical control system and mechanical transmission error. A second external sensor is always needed to compare the machine position read by position sensors through its controller [13–16].

All axes of this machine have their motion completed and controlled by the controller. The precision that each moving part can achieve under the controller reflects directly the precision that the machined part can have. The positioning accuracy is an important test that must be calibrated regularly.

Standards such as ISO 230-2, ASME B 89, VDI 2617, GB-10931-89, ASME B5.54, VDI 3441, and JSI-B6330 are used.

A- Linear Motion Positioning Accuracy Detection

Equipment: According to the national standards and the provisions of the International Organization for Standardization (ISO standards, e.g., ISO 230), the detection of CNC machine tools should be based on laser measurements (laser interferometry). If not available, it is also possible to use a standard scale with an optical reading microscope for comparative measurements [17–20]. However, the accuracy of the measuring instrument must be one to two levels higher than the accuracy of the measurement, as discussed earlier.

Method: Linear motion positioning accuracy is performed under no-load conditions on machine tools and benches.

With multiple positioning, all the errors must be reflected; hence, the ISO standard stipulates that each positioning point calculates the average value and the dispersion difference based on five measurement data and the dispersion difference band formed by the dispersion band.

B- Repeatability of Positioning Accuracy in Linear Motion

Equipment: Keeping the same instrument and configuration, the measurement is carried out at any three positions near the midpoint and the two ends of each coordinate stroke.

Method: Each position is quickly moved, and the positioning is repeated 7 times under the same conditions. The stop position measurement is recorded with the maximum reading difference. Subtracting one-half of the most significant difference among the three positions, the positive and negative signs are attached as the repeated positioning accuracy of the coordinates, which is the most basic index reflecting the stability of the axis motion accuracy.

C- Origin Return Accuracy in Linear Motion and Repeatability

This is to inspect and correct the return to origin accuracy as a specific point in the coordinate system. Repeatability of this move is also checked.

D- Inspection of Reverse Error in Linear Motion

In reverse motion of a linear move, a loss in the amount of motion occurs. It includes the reverse dead zone of the drive position, for instance, servo motor and stepping motor. On the coordinate axis feed chain, each mechanical motion transmission pair, for example, gears, causes errors such as backlash and elastic deformation. The larger the error, the lower the positioning accuracy and the lower the repeated positioning accuracy.

Method: The inspection method of the reverse error consists of moving to a position forward or backward in the stroke of the measured coordinate axis and using the stop position as a reference and then giving a specific movement command value in the same direction to move a distance. Then run the same distance in the opposite direction and measure the difference between the stop position and the reference position. The measurement has performed a plurality of times (generally 7 times) at three points near the midpoint and both ends of the stroke, the average value at each position is obtained, and the maximum value among the obtained average values is the reverse error value.

E- Inspection of the Positioning Accuracy Detection of Rotary Tables

Equipment: Measuring tools include standard turret, angle polyhedron, circular grating, and collimator (collimator), and so forth, which can be selected according to the specific conditions.

Method: The measurement method is to rotate the table forward (or reverse) to an angle and stop, lock, and position. Use this position as a reference, then quickly turn the table in the same direction and measure every 30 locks. Each of the forward rotation and the reverse rotation is measured for one week, and the maximum value of the difference between the actual rotation angle of each positioning position and the theoretical value (command value) is the division error.

A CNC rotary table needs targeted positions for every 30 degrees. Each target position needs to quickly locate 7 times from the positive and negative directions, the difference between the area and the target position is actually reached, and then according to a standard such as GB10931- 89, the "Method for Evaluating the Position Accuracy of Digital Control Machines" requires to calculate the average position deviation and standard deviation, the difference between the maximum value of all the average position deviations, the standard deviation, and the sum of all the average position deviations and the standard deviation. It is the positioning accuracy error of the CNC rotary table.

It is generally needed to measure several equal-angle points in degrees such as 0, 90, 180, 270, and so forth, and the accuracy of these points is required to be improved by one level compared with other angular positions.

F- Inspection of the Repeated Indexing Accuracy of Rotary Tables

Equipment: same as previous.

Method: The measurement method is repeated three times in three places of the rotary table, and the inspection is performed in the forward and reverses directions. The maximum value of the difference between the values of all readings and the theoretical value of the corresponding position is considered as the indexing precision.

In a CNC rotary table, one measurement point is recorded every 30 as the target position, and five fast positionings are carried out for each target position from positive and negative directions, respectively, and then the difference between the actual arrival position and the target position is measured. That is, the position deviation is measured, and then the standard deviation is calculated according to the method specified in GB10931-89, which is six times of the maximum value of the standard deviation of each measuring point. This is the repeating indexing precision of the numerical control rotary table.

G- Inspection of the Accuracy of the Return to Origin in Rotary Tables

Equipment: similar equipment as previous.

Method: it is required to perform the machine origin return from 7 arbitrary positions, measure the stop position, and use the maximum difference read as the origin return accuracy.

It should be pointed out that the inspection of the current positioning accuracy is measured under the condition of fast motion and positioning. When feeding system is not adequate, several positioning accuracy values will be obtained when positioning with varying speeds of feed. Note that all measurements are related to the ambient temperature and the working state of the coordinate axis. At present, most of the numerical control machine tools adopt a semi-closed loop system, and the position inspecting components are mostly mounted on the driving motor generating an error of 0.01 to 0.02 mm over a stroke of 1 m. This is not uncommon. It is an error caused by thermal elongation, and some machines use a pre-stretch (pre-tightening) method to reduce the impact.

The repeatability in positioning accuracy of each coordinate axis shows the most basic accuracy index of the shaft reflecting the stability of the motion accuracy of the axle, and it cannot be assumed that the machine tool with poor precision can be stable in production. At present, due to the increasing number of functions of the NC system, system errors such as pitch accumulation error and backlash error can be compensated for by the motion accuracy of each axis. Only the random error cannot be attempted to be compensated, and the repeatability positioning accuracy is secured. It reflects the absolute random error of the feed drive mechanism. It can't be corrected by the CNC system compensation. When it is found to be out of tolerance, only the fine adjustment of the feed drive chain is performed. When selecting a CNC machine tool it is better to check that it has a high repeatability.

7.11 CNC Machine Error Assessment and Calibration

The calibration of a CNC machine tool includes diagnosis and errors corrections that may deviate over time according to the manufacturer's original specifications. The reasons for calibration may include:

- Newly set machine;
- Machine has been moved from its original place;
- Machine has crashed or one of the axes has crashed for various reasons;
- Out of tolerance observation on the machine parts.

The verification of the machine includes positioning, straightness, and squareness. Flatness is added on request, and if the machine has a rotary axis, then angularity is added.

It is aimed to inspect a CNC machine performance by characterizing its axes in positioning, repeatability, systematic errors, accuracy, dead motion, and the effects of Abbe-offset during calibration and compensation using laser interferometry (Fig. 7.11).

To attain the desired level of accuracy, good calibration of equipment is required. Before taking the measurement, the following aspects are considered:

a) The behavior of the instrument must be considered;
b) The precision of the measuring instrument must be 5–10 times more than the expected precision of the measured object;

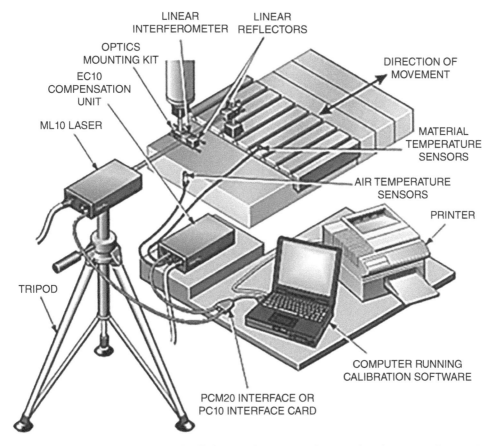

Figure 7.11 Positioning accuracy of a CNC machining center using laser interferometer with wavelength compensation.

c) When taking the measurement of the object, one must know the atmospheric conditions, for instance, temperature, coefficient of expansion, radiation from light, and sunlight, making sure the machine is not exposed to the sunlight directly.

The objectives of the test are as follows:

a) To determine the positioning accuracy performance of an axis of a CNC machine tool;
b) To identify possible sources of the accuracy degradation.

A-Experimental Procedure:

a) A laser interferometer was set to take readings along the x-axis of the CNC Machining Centre having MDSI open CNC controller. Figure 7.10 shows the arrangement of the inspection equipment.
b) Target positions: 30 mm equally spaced target positions were selected over a stroke of 480 mm with five runs in each direction along the x-axis.
c) Positions at previous targets over the stroke were recorded for several runs without any compensation values in the CNC controller.
d) Wavelength compensation: Since the wavelength of light is dependent on the reflective index of air, it is necessary to compensate for environmental conditions. Air temperature, pressure, and humidity were recorded to obtain the wavelength compensation using Edlen's equation. The compensation for machine temperature deviation from 20 °C reference temperature was considered.
e) Results of the systematic and random errors data along the x-axis based on ISO 230-1 to 230-4 specifications were obtained.
f) Using the Talyvel electronic level, the pitch motion of the x-axis at 30 mm intervals was measured to evaluate any possible Abbe error corresponding to the x-axis.

Figure 7.12b shows the setup and the required equipment for the test:

- CNC Machining Centre with Fanuc 6™ CNC controller;
- HP laser interferometer head and display;
- Laptop with inspection software;
- Printer to print results and graphs.

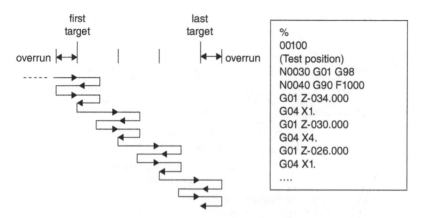

Figure 7.12 The bidirectional pendulum target sequence.

B-Nomenclature of the Variables Used in the Measurement Analysis

P_i	Target position.
i	The particular position among other selected target positions along the axis.
P_{ij}	Actual position.
j	Number of approaches.
x_{ij}	Deviation of actual position from target position ($P_{ij} - P_i$).
↑	Refers to data collected from a measurement in forward (positive) move to target.
↓	Refers to data collected from a measurement after a reverse (negative) move to target.
$\bar{x}_i \downarrow$	Mean unidirectional positional deviation, either reverse or forward depending on specified symbol, ↑ or ↓, respectively.
$\bar{x}_i$	Mean bidirectional positional deviation, with no direction symbol specified.
B_i	Reversal value at a position.
B	Reversal value of an axis.
$\bar{B}$	Mean reversal value of an axis.
$S_i \downarrow$	The standard deviation of positioning at a position, i. In this case, it is in the reverse direction indicated by the symbol ↓, but it could also be in the forward direction indicated by ↑.
P	Air pressure [mm/Hg].
T	Air temperature [°C].
H	Relative humidity [% relative].
N	Refraction index.
C	Correct wavelength of light compensation factor for the measurement conditions.
M	Range of the mean bidirectional positional deviation of an axis.
$R_i \downarrow$	Unidirectional repeatability of positioning at position i. The approach direction is indicated by either ↑ or ↓.
R_i	Bidirectional repeatability of positioning at position i.
R↓	Unidirectional repeatability of positioning in the reverse or forward direction depending on the symbol, ↑ or ↓.
R	Bidirectional repeatability of positioning of an axis.
E↓	Unidirectional systematic positional deviation of an axis. Approach direction is indicated by the arrow: forward, ↑, or reverse, ↓.
E	Bidirectional systematic positional deviation of an axis.
A↓	Unidirectional accuracy of positioning of an axis. Approach direction is indicated by the arrow; forward, ↑, or reverse, ↓.
A	Bidirectional accuracy positioning of an axis.

The bidirectional pendulum target sequence requires the moving part to be progressively moved through the target positions gathering data from all runs at each individual target. If two runs are required, a target is visited twice before the moving part is moved to the next target. The NC program for traveling the machine axis to test points is written and transferred to the machine controller via an RS 232 port connection. A sample is given in Figure 7.12. The test is run according to standards ISO 230-1 to 4. The test cycle carries out five runs stopping at 17 different positions, and each run will consist of a forward and reverse element, as requested by the standard shown in Figure 7.13.

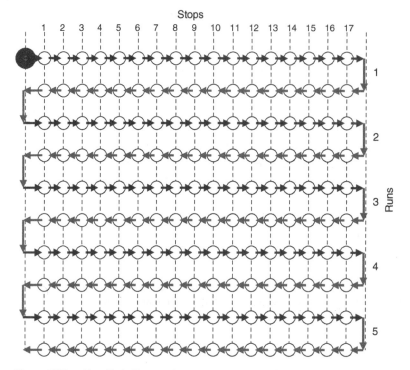

Figure 7.13 - Test Cycle Executed.

C-Performance Criteria According to ISO Standards

1- Mean Unidirectional Positional Deviation at a Specified Position

$$\bar{x}_i \uparrow = \frac{1}{n}\sum_{j=1}^{n} x_{ij} \uparrow \tag{7.3}$$

Equation (7.3) has been used to calculate the mean unidirectional positional deviation at a position. The values used to determine the mean unidirectional positional deviation at a specified position are the raw values recorded from measurements. An example of this calculation is detailed for run 1:

$$\bar{x}_i \uparrow = \frac{1}{n}(x_{11} \uparrow + x_{12} \uparrow + x_{13} \uparrow + x_{14} \uparrow + x_{15} \uparrow)$$

$$\bar{x}_i \downarrow = \frac{1}{n}\sum_{j=1}^{n} x_{ij} \downarrow \tag{7.4}$$

2- Mean Bidirectional Positional Deviation at a Specified Position

$$\bar{x}_i = \frac{\bar{x}_i \uparrow + \bar{x}_i \downarrow}{2} \tag{7.5}$$

Equation (7.3) has been used to find the mean of the values found from Equations (7.4) and (7.5) to provide a bidirectional mean. Below is an example of this calculation for run 1.

3- Reversal Value at a Specified Position

$$B_i = \bar{x}_i \uparrow - \bar{x}_i \downarrow \tag{7.6}$$

Equation (7.6) has been used to calculate the difference between the mean unidirectional positional deviations obtained from both the forwards and reverse approach for each run.

4- Reversal Value of an Axis

$$B = \max[|B_i|] \tag{7.7}$$

Equation (7.7) has been used to determine the maximum of the absolute reversal values at all target positions along the *x*-axis.

5- Mean Reversal Value of an Axis

$$\bar{B} = \frac{1}{m} \sum_{i=1}^{m} B_i \tag{7.8}$$

Equation (7.8) simply determines the mean of the values calculated by Equation (7.7) to give the arithmetic mean of the reversal values B_i at all the target positions on the *x*-axis. For the current exercise, the average is written as follows:

$$\bar{B} = \frac{1}{17} \left(\begin{array}{c} B_1 + B_2 + B_3 + B_4 + B_5 + B_6 + B_7 + B_8 + B_9 \\ + B_{10} + B_{11} + B_{12} + B_{13} + B_{14} + B_{15} + B_{16} + B_{17} \end{array} \right).$$

6- Unidirectional Standard Uncertainty Estimator of Position at a Specified Position

$$S_i \uparrow = \sqrt{\frac{1}{n-1} \sum_{j=1}^{n} (x_{ij} \uparrow - \bar{x}_i \uparrow)^2} \tag{7.9}$$

Equation (7.9) determines the standard uncertainty of the positional deviation. This equation has been used as demonstrated below for run 1:
Equation (7.10) is like Equation 7.9

$$S_1 \downarrow = \sqrt{\frac{1}{5-1} \left[\begin{array}{c} (x_{11} \downarrow - \bar{x}_1 \downarrow)^2 + (x_{12} \downarrow - \bar{x}_1 \downarrow)^2 + (x_{13} \downarrow - \bar{x}_1 \downarrow)^2 \\ + (x_{14} \downarrow - \bar{x}_1 \downarrow)^2 + (x_{15} \downarrow - \bar{x}_1 \downarrow)^2 \end{array} \right]}$$

$$S_1 \downarrow = \sqrt{\frac{1}{n-1} \sum_{j=1}^{n} (x_{ij} \downarrow - \bar{x}_1 \downarrow)^2} \tag{7.10}$$

7- Unidirectional Repeatability of Positioning at a Specified Position

$$R_i \uparrow = 6S_i \uparrow \tag{7.11}$$

Equation (7.11) is used to determine the unidirectional repeatability of positioning at a specified position. A coverage factor of 2 has been applied. The opposite approach direction is presented by the following equation $R_i \downarrow = 6S_i \downarrow$.

8- Bidirectional Repeatability of Positioning at a Specified Position

$$R_i = \max[3S_i \uparrow + 3S_i \downarrow + |B_i|; R_i \uparrow; R_i \downarrow]. \tag{7.12}$$

9- Unidirectional Repeatability of Positioning

$$R \uparrow = \max [R_i \uparrow]. \tag{7.13}$$

The reverse direction is written as $R \downarrow = \max [R_i \downarrow]$.

10- Bidirectional Repeatability of Positioning of an Axis is defined as

$$R = \max [R_i]. \tag{7.14}$$

11- Unidirectional Systematic Positional Deviation of an Axis is defined as

$$E \uparrow = \max [\bar{x}_i \uparrow] - \min [\bar{x}_i \uparrow]. \tag{7.15}$$

Equation (7.15) determines the difference between the algebraic maximum and minimum of the mean unidirectional positional deviations for one approach direction at any specified position. The following expression is used for reverse motion $E \downarrow = \max [\bar{x}_i \downarrow] - \min [\bar{x}_i \downarrow]$.

12- Bidirectional Systematic Positional Deviation of an Axis Is Defined as

$$E = \max [\bar{x}_i \uparrow; \bar{x}_i \downarrow] - \min [\bar{x}_i \uparrow; \bar{x}_i \downarrow] \tag{7.16}$$

Equation (7.16) is the difference between the algebraic maximum and minimum of the mean unidirectional positional deviations for both the forward and reverse approaches.

13- Mean Bidirectional Positional Deviation of an Axis Is Defined as

$$M = \max [\bar{x}_i] - \min [\bar{x}_i] \tag{7.17}$$

Equation (7.17) is used to determine the difference between the algebraic maximum and minimum of the mean bidirectional positional deviations at any specified position along the x-axis.

14- Unidirectional Accuracy of Positioning of an Axis

$$A \uparrow = \max [\bar{x}_i \uparrow + 3S_i \uparrow] - \min [\bar{x}_i \uparrow - 3S_i \uparrow] \tag{7.18}$$

The opposite direction is covered by the following expression $A \downarrow = \max [\bar{x}_i \downarrow + 3S_i \downarrow] - \min [\bar{x}_i \downarrow - 3S_i \downarrow]$.

15- Bidirectional Accuracy of Positioning of an Axis

$$A = \max [\bar{x}_i \uparrow + 3S_i \uparrow; \bar{x}_i \downarrow + 3S_i \downarrow] - \min [\bar{x}_i \uparrow - 3S_i \uparrow; \bar{x}_i \downarrow - 3S_i \downarrow]. \tag{7.19}$$

Equation (7.19) is the most important equation used in this exercise as it describes the range derived from combining the bidirectional systematic deviations and the estimator of the standard uncertainty of bidirectional positioning using a coverage factor of two.

16- Edlen's Equation for Wavelength of Light Compensation

The refractive index of air can be calculated given the environmental conditions estimated and measured at the start of the verification and monitored throughout the test.

$$N = 0.3836391P \left[\frac{1 + 10^{-6}P(0.817 - 0.0133T)}{1 + 0.0036610T} \right] - 3.033 \times 10^{-3} \times H \times e^{0.05762T} \tag{7.20}$$

For the values of T, P, and H shown in table 7.1, $N = 245.14154$.

From the refractive index above, the wavelength of light compensation factor can be calculated and used in the corrective calibration as follows:

$$C = \frac{10^6}{N + 10^6}, \text{hence } C = 0.999975486. \tag{7.21}$$

D- Tests and Data Acquisition

The environmental parameters have been measured and shown in Table 7.2. Position values were recorded and shown in Table 7.3. As 5 runs were performed for 17 stops, only one stop was considered to show examples of machine performance in one axis.

A Talyvel inclinometer was used to measure the axis inclination (deflection) for each position, and results are shown in Table 7.4. The final assessment of the *x*-axis is found in Table 7.5.

Position Error for Forward, Reverse, and Bidirectional

Figure 7.14a shows as anticipated that the *x*-axis positional error increases as the machine table moves further away. Fig.7.14b shows the same behavior as in Figure 7.14a with reversal error. The two graphs are very similar to one another. Figure 7.14c contains the forward error, reverse error, and systematic (bidirectional) error in the *x*-axis of the CNC Machining Centre. From Fig.7.14c, a direct comparison can be seen between the forward and reverse error.

E- Example of Calculations

1) x-axis positional results for target position @ 210.00 mm (Table 7.6).
2) Forward average error (Table 7.7).
3) Standard deviation 3σ, forward $= 3\sqrt{\frac{\sum(\overline{X_f} - X_i)^2}{n-1}} = 3\sqrt{\frac{1.97282}{4}} = 2.0168\,\mu m.$
4) Reverse average error (Table 7.8), $\overline{X_r} = \frac{(-4.51) + (-5.11) + (-5.28) + (-5.19) + (-5.32)}{5} = -5.028\,\mu m.$
5) Standard deviation 3σ, reverse $= 3\sqrt{\frac{\sum(\overline{X_r} - X_i)^2}{n-1}} = 3\sqrt{\frac{0.47436}{4}} = 1.0331\,\mu m.$

Table 7.2 Initial environmental conditions

Air Temperature	20.2 °C
Air Pressure	732 mm/Hg
Humidity	44% relative
Machine Temperature	20.0 °C
Machine expansion co-efficient	11.5 μm/m/°C
Wavelength Compensation	755.1 ppm
Total Compensation	707.0 ppm

Table 7.3 Raw data collected by laser interferometer measurements.

Target position	Run number 1		Run number 2		Run number 3		Run number 4		Run number 5	
	Forward	Reverse	Forward	Reverse	Forward	Reverse	Forward	Reverse	Forward	Reverse
	x_{ij}	x_{ij}	x_{ij}	x_{ij}	x_{ij}	x_{ij}	x_{ij}	x_{ij}	x_{ij}	x_{ij}
0	0.11	3.06	0.25	3.02	0.40	2.04	−0.27	2.02	−0.47	1.74
30	−0.33	1.59	−0.30	0.76	−0.71	1.42	−0.23	0.45	−0.91	0.74
60	−1.03	−0.18	−1.03	−1.05	−2.15	−1.20	−2.31	−0.68	−2.09	−1.38
90	−1.63	−0.59	−1.26	−0.83	−1.47	−0.96	−2.27	−1.41	−2.46	−1.44
120	−2.32	−1.53	−2.51	−1.35	−2.89	−1.39	−2.42	−1.88	−3.08	−1.98
150	−4.57	−2.99	−4.81	−3.81	−4.99	−3.78	−4.86	−3.71	−4.84	−3.93
180	−5.89	−3.89	−5.81	−4.06	−6.21	−5.47	−6.52	−5.37	−6.44	−4.93
210	−7.32	−6.37	−7.24	−6.29	−7.43	−6.49	−7.24	−6.70	−8.12	−6.94
240	−8.88	−7.94	−9.40	−7.64	−9.49	−8.29	−9.07	−7.87	−9.18	−8.08
270	−10.76	−9.19	−10.27	−8.66	−10.83	−9.46	−10.84	−9.33	−11.11	−9.43
300	−13.07	−11.71	−13.27	−11.46	−12.69	−10.77	−13.26	−11.75	−13.09	−11.39
330	−14.77	−12.97	−15.06	−13.38	−15.58	−13.27	−15.04	−12.87	−15.85	−13.41
360	−16.55	−13.94	−17.13	−13.98	−17.08	−14.31	−16.87	−14.27	−16.98	−13.72
390	−18.48	−14.90	−18.46	−14.84	−18.32	−14.96	−18.63	−15.16	−18.30	−15.05
420	−20.09	−15.97	−20.54	−16.48	−20.45	−15.80	−20.39	−15.52	−20.39	−16.05
450	−23.26	−17.46	−23.31	−18.13	−23.59	−18.30	−23.72	−18.04	−23.50	−17.51
480	−26.24	−18.91	−26.03	−18.69	−25.81	−18.79	−26.40	−19.24	−26.34	−19.25

Table 7.4 Pitch Motion Error (Raw Data).

Target Position (mm)	Angle (Sec-arc)	Angle (Degrees)	Error in Position (μm)
0	0	0	0
30	1.0	0.000277786	0.077160283
60	2.5	0.000694444	0.482253009
90	4.0	0.001111111	1.234567393
120	5.5	0.001527778	2.334103122
150	7.0	0.001944444	3.780859433
180	8.5	0.002361111	5.574835319
210	10.0	0.002777778	7.716029537
240	11.5	0.003194444	10.2044406
270	13.0	0.003611111	13.04006678
300	14.5	0.004027778	16.2229061
330	16.0	0.004444444	19.75295636
360	17.0	0.004722222	22.29921696
390	18.0	0.005000000	24.99979167
420	19.5	0.005416667	29.33999083
450	21.0	0.005833333	34.02739182
480	22.0	0.006111111	37.34521411

Table 7.5 Further Calculated Values from the Calibration.

Reversal Value of an axis, B	7.19 μm
Mean Reversal Value of an axis	−2.36 μm
Bidirectional Repeatability of Positioning at a Position, R_i	8.69 μm
Unidirectional Repeatability of Positioning R↑	2.53 μm
Unidirectional Repeatability of Positioning R↓	2.93 μm
Bidirectional Repeatability of Positioning R of an axis	8.69 μm
Unidirectional Systematic Positional Deviation of an axis, E↑	26.17 μm
Unidirectional Systematic Positional Deviation of an axis, E↓	21.35 μm
Bidirectional Systematic Positional Deviation of an axis, E	28.54 μm
Mean Bidirectional Positional Deviation of an axis, M	23.76 μm
Unidirectional Accuracy of Positioning of an axis, A↑	28.46 μm
Unidirectional Accuracy of Positioning of an axis, A↑	23.16 μm
Bidirectional Accuracy of Positioning of an axis, A	31.14 μm

6) The system error $= \frac{\text{Forward Average Error} + \text{Reverse Average Error}}{2}$.

Therefore, system error $= \frac{(-6.736) + (-5.082)}{2} = -5.909$ μm

Also the reverse error $=$ reverse average error $-$ forward average error.

Therefore, the reverse error $= (-5.082) - (-6.736) = -1.654$ μm.

Observations and Comments

On-Axis Accuracy

On-axis accuracy is the uncertainty of position after all sources of linear error are eliminated. Linear (or monotonically increasing) errors include displacement inaccuracy, planar inaccuracy, volumetric inaccuracy, inaccuracy of the lead screw pitch, the angular deviation effect at the measuring point (Abbe error), and thermal expansion effects. Graphically these errors are represented by the slope of a best-fit line on a plot of position versus error (Fig. 7.13a). Knowing the slope of this line (error/travel), we can approximate the *absolute accuracy* as:

Absolute Accuracy $=$ On-Axis Accuracy $\pm$ Abbé error.

Repeatability

Repeatability is the ability of a motion system to reliably achieve a commanded position over many attempts. Manufacturers often specify *unidirectional repeatability,* which is the ability to repeat a motion increment in one direction only. This specification sidesteps issues of backlash and hysteresis and therefore is less meaningful for many real-world applications where reversal of motion direction is common.

As revealed in the results, the *unidirectional repeatability* is 4.22 μm at target position of 210.000 mm in forward direction. A more significant specification is *bidirectional repeatability,* or the ability to achieve a commanded position over many attempts regardless of the direction from which the position is approached. In this case the bidirectional repeatability achieved at target position of 480.000 mm is 10.16 μm.

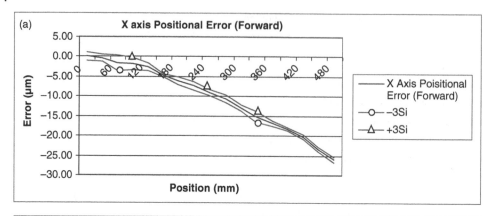

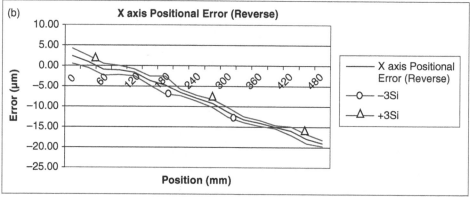

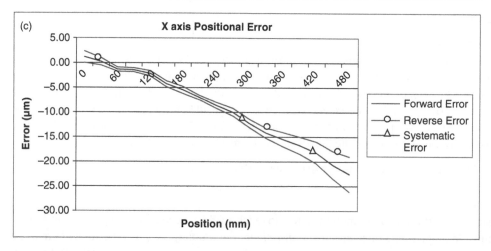

Figure 7.14 a) X-axis Positional Error (Forward), b) X-axis Positional Error (Reverse), c) X-axis Positional Error (Bidirectional).

Systematic Errors

The systematic errors do not lend themselves to averaging out over the time of measurement and may be constant. Accordingly, a systematic error is particularly dangerous in that no matter how many times a measurement is repeated, this error can remain undetected. The techniques to identify systematic errors include measuring the quantity repeatedly over such a time period that some of the systematic errors may have had time to become random.

Table 7.6

Target Position120 mm	Run No 1	Run No 2	Run No 3	Run No 4	Run No 5
Forward error [µm]	−5.62	−6.70	−7.01	−6.81	−7.54
Reverse error [µm]	−4.51	−5.11	−5.28	−5.19	−5.32

Table 7.7

	X_i	$\overline{X_f}$	$(\overline{X_f} - X_i)$	$(\overline{X_f} - X_i)^2$
Run 1	−5.62	−6.736	−1.116	1.244556
Run 2	−6.70	−6.736	−0.036	0.001296
Run 3	−7.01	−6.736	0.274	0.075076
Run 4	−6.81	−6.736	0.074	0.005476
Run 5	−7.54	−6.736	0.804	0.646416
				Total = 1.97282

Table 7.8

	X_i	$\overline{X_r}$	$(\overline{X_r} - X_i)$	$(\overline{X_r} - X_i)^2$
Run 1	−4.51	−5.082	−0.572	0.327184
Run 2	−5.11	−5.082	0.028	0.000784
Run 3	−5.28	−5.082	0.252	0.063504
Run 4	−5.19	−5.082	0.162	0.026244
Run 5	−5.32	−5.082	0.238	0.056644
				Total = 0.47436

In this test, the systematic errors (average) can be measured by taking the average of the forward error average and the reverse error average at each target position.

Correlation of the Pitch Motion with the Abbe Offset and the Measured Positioning Performance

To achieve a high-precision length measuring system, the measured length and the measurement scale must lie on a single line. Additional linear off-axis error is introduced through amplification of tilt and wobble with a long moment arm (offset). The distance from the source amplifies angular errors. Figure 7.15 between the pitch motion error with Abbe offset as constant and changing-pitch angle with the change in target positions and the target positions is linear. That is, with the increase in the displacement, there is increase in the pitch motion error.

Influence of Selection of Target Position, Weight of Workpiece, and Thermal Errors

Target positions should not be selected randomly. Equally spaced 30 mm target positions were selected so that any cyclic error present in the positioning system would be detected along with the progressive component of systematic error.

If the machine fails to return to the same target position, within certain tolerances, then any compensation based upon prediction, will of course, fail to operate with any reliability. The standard deviation results for each test were therefore of the utmost interest. The removal of systematic error

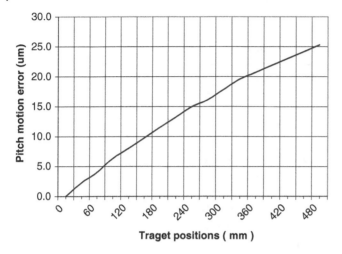

Figure 7.15 Pitch motion error.

is now quite commonplace, but the more demanding task of maintaining this accuracy still depends largely upon the initial design of the machine tool structure, guide ways, and so forth. If backlash, hysteresis, systematic errors, and so on, are compensated for the machine center, the structure of the machine has to be stiff enough to withstand any deformation induced by the machine or components own weight under various configurations.

The errors induced by the change in temperature during machining will have adverse effect on the positional accuracy. It is therefore mandatory to keep the temperature constant during the machining operations.

Closed-Loop Compensation

On the following figure (Fig. 7.16), an error compensation routine has been activated to compensate for Abbe error mainly[2].

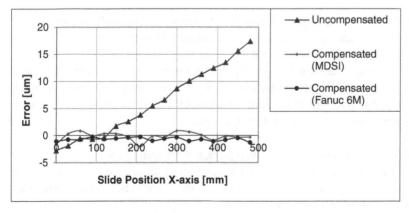

Figure 7.16 Compensated positioning.

---

2 The results used here were done a few days later following the test results presented previously.

Abbe error could be a significant source of error. The angular error could be caused by any of the following issues:

1) Curvature of ways;
2) Entry and exit of balls or rollers in recirculating ways;
3) Variation in preload along a way;
4) Insufficient preload or backlash in a way;
5) Contaminants between rollers and the way surface.

The CNC Machining Centre has a hysteresis error aspect. This is because once a position is reached, the machine can never return exactly back to that position again once it has moved away. Hysteresis is a problem when present in a CNC machine as it means the program carried out when machining may not achieve an acceptable accuracy and suggests that an optimal way to program a CNC machine for dimensional accuracy as each point should be visited only once. For example if a cut is made to the workpiece, then later in the same program a hole is to be drilled at the same point the error would have increased, and the machine would not be able to achieve the same position again.

The machine has also uncertainties in its calibration and initial setup. The data input to the software to determine the wavelength compensation may not have been accurate enough and is certainly a possible source of error. This is because some of the values input were just the team's "best estimates" of environmental conditions such as humidity, pressure, and temperature.

7.12 Assessment of the Contouring in the CNC Machine Using a Kinematic Ballbar System

The purpose of using a ballbar link (Fig. 7.17) is to characterize circular motion provided by either a milling machine with the combination of 2 axes or in a lathe machine. It is expected to verify the eccentricity and geometric errors such as squareness and deformation errors. These may have various sources in a CNC machine.

Several features will be discussed later to have a clear idea of what to expect from such a test.
The objectives of the test are:

a) To investigate the contouring characteristics of a CNC Machining Centre using a kinematic ball link system.
b) To investigate the salient features of the contouring/polar results with respect to error sources of the machine.

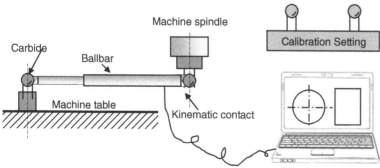

Figure 7.17   Experimental set-up.

Experimental Procedure:

a) Software[3] compensation from CNC Machining Centre with MDSI open controller was removed.

b) A 100 mm kinematic ballbar system complete with calibrated setting bar was used with the quick setting sleeve from the hardware. The center of the two reference spheres coincide with each other in the center of the *xy* reference plane of the machine. Figure 7.36 shows the experimental setup.

c) The machine was programmed to produce a 360 degrees clockwise circle with a 100 mm radius and a feed of 1000 mm/min with a tangential approach to the start point and a tangential exit. The start and exit point was at 202 degrees in the *XY*-plane.

d) Repeat step c) for counterclockwise movement.

e) The above test procedure was repeated for a feed rate of 4000 mm/min.

f) Measured results were displaced (polar plot) on the laptop and observed the lost motion (backlash) in the *x*- and *y*-axis using a software and laptop for online data acquisition and analysis of the results.

The results show clearly that there is a cyclic error in the machine. The plots also show a negative backlash, servo mismatch characteristic in the *XY*-plane, and radial deviation.

There are the two cases recorded in Figure 2.37 in which the feed rate is changed from 1000 to 4000. The diagrams show the influence of feed rate on parameters such as circularity and backlash.

The ball link tests show how the two axes work together to move the machine in a circular path. Two axes would make perfect circles in a perfect machine. The ball link measures any deviations the machine makes from a perfect circle and displays the data in a polar format.

Polar check simulator software automatically identifies the types and sources of errors.

1) **Backlash**

   The backlash error appears in two different ways, as positive and negative backlash. When moving to the defined test direction, positive backlash can be seen on the plot as a step outward starting on the axis (deformation or play in the machine structure). Negative backlash appears as a step inward starting on the axis (backlash overcompensation at the controller, i.e., the compensation value exceeds the value of the actual backlash error).

2) **Cyclic Error**

   Cyclic error appears as a cyclically changing motion error on the circular plot. It is important to know the machine tool and its driving mechanisms to understand the sources of error. The cyclic error may come from ball screws, encoders, transmission, and so forth.

3) **Radial Play**

   Lateral play usually occurs as a tangential deviation caused by looseness in the machine's guide ways. The graphs are usually symmetrical about both axes.

4) **Servo Mismatch**

   The magnitude of this error informs about the response time of the machine's position control loop to the given commands. An absolute difference between the nominal and actual values can be obtained. Otherwise, an unbalance will be noticed as a relative value.

   For an overshoot or lag of the response time, oval trace pattern will be seen at ±45° with the machine axes.

---

3 CONTISURE is a house software (UMIST, UK) initially developed by Dr. M. Burdekin as well as the very first ball bar link.

Al low feed rates, the gain of the axis that is lagging could be turned up, or the gain of the axis that is leading turned down.

5) **Squareness Error**

The squareness error is the deviation in the perpendicular axes at the test location. The test reveals whether the axes are bent locally or misaligned along their whole length with repeated tests.

6) **Directional Vibration**

If vibration is caused by a particular reason, the location and influencing direction of the error source will cause a directional bias in the vibration pattern.

7.13 Calibration of 3-axis CNC Machine Tool

The example (Fig. 7.18–20) presents the verification and calibration of the Spinner TC600 universal lathe. This is a 3-axis machine with real Y-axis and subspindles. The machine operates with a

Figure 7.18   a) Axes of TC 600 with turret disc BMT45. The *x*- and *z*-axis of this machine are inspected using Laser interferometry. b) Lathe machine TC 600. *Source:* Courtesy of Spinner.

Figure 7.19 X-Axis TC 600 with laser interferometry configured. *Source:* Courtesy of Spinner.

Figure 7.20 Z-Axis TC 600 with laser interferometry configured. *Source:* Courtesy of Spinner.

speed up to 6000 rpm and modern tool changer with tool holders according to VDI or BMT. The workpiece diameter can be 1–250 mm, workpiece length in chuck 1–250 mm, and with tailstock 1–600 mm.

Laser interferometry is installed for inspection as shown in the next two figures. Figures 7.21–22 show positioning error using ISO 232-2 for linear axis initially with a variation of 4 µm over the whole stroke, but in compensated mode the accuracy is within 1.3 µm over the same stroke. In the Z-axis (Fig. 7.23–24) the positioning error is 13.5 µm within the stroke, but after compensation it becomes 5.6 µm.

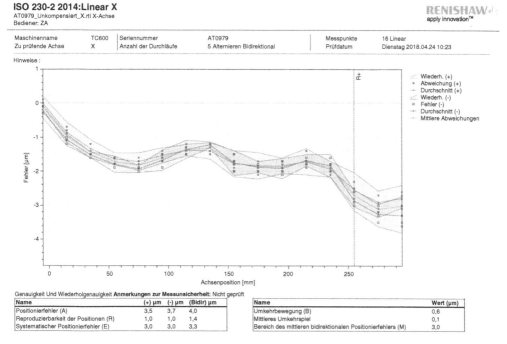

Figure 7.21 Non-compensated X-axis CNC TC 600 (courtesy Spinner).

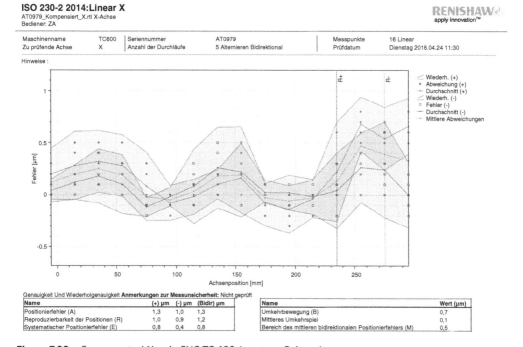

Figure 7.22 Compensated X-axis CNC TC 600 (courtesy Spinner).

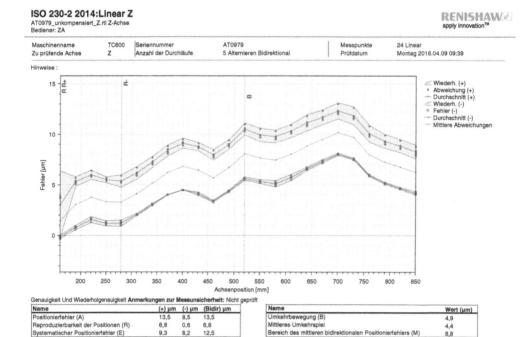

Figure 7.23 Non compensated Z-axis CNC TC 600 (courtesy Spinner).

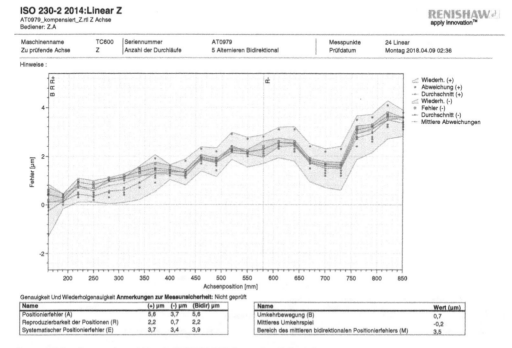

Figure 7.24 Compensated Z-axis CNC TC 600 (courtesy Spinner).

7.14 Calibration of a Coordinate Measuring Machine (CMM)

7.14.1 CMM Performance Verification

The international standard ISO 10360-1 defines a coordinate measuring machine (CMM) as a measuring system with the means to move a probing system and capability to determine spatial coordinates on a workpiece surface. Over the years, standards and guidelines have been developed to harmonize the performance specifications of a CMM to enable a user to make meaningful performance comparisons when purchasing a machine and, once purchased, to provide a well-defined way in which the specified performance can be verified.

There are various methods enabling a CMM user to specify and verify the performance of a CMM as well as acceptance testing. These have been discussed very actively within the CMM community since the mid-1970s. Currently, the verification and calibration of a CMM is guided with good practices using ISO 10360 series of specification standards. This standard component is described below.

The verification, also known as certification, is the process of confirming whether a CMM is performing within its stated specifications. The probability that a measurement error is smaller than the specified maximum permissible error is observed by a verification.

In a verification, the maximum permissible error can be specified as the largest error that the user can decide to accept. However, the process does not compensate for lack of trueness or applying a correction, which is where calibration comes in.

The CMM acceptance test, as set out in the appropriate specification standards, provides a mechanism by which a go/no-go decision on the ability of a CMM to perform a series of specific tests can be made. Similar tests are used to periodically check that this level of performance is maintained.

Many discussions have resulted in the publication of several national and international standards and guidelines describing the requirements for the performance verification, periodic reverification, and interim checking of CMMs. Furthermore, supplementary documents have been prepared to help the CMM user with the interpretation of the requirements of particular specification standards.

Most CMM performance verification standards and guidelines are based on sampling the length-measurement capability of a CMM, to decide whether its performance conforms to the specification. However, from such a sample, it is not possible to make an accurate statement about the overall length measurement capability of a performance-verified CMM. This is due mainly to the complicated ways in which errors combine within a CMM, which do not permit statements about the uncertainty of its measurements anywhere in its workspace to be directly derived from such a limited sample. Furthermore, there are no methods specified by which the uncertainty of other measurands can be calculated using only the length-measurement capability of a CMM.

Therefore, we do not consider the sampled length-measurement uncertainty alone as representative of all possible measurement tasks performed by a CMM. Hence, the performance verification does not guarantee traceability of measurements performed with a CMM for all measurement tasks.

However, we as a working group recognize that in an industrial environment, the currently practiced performance verifications and regular interim checks of CMMs are the state of the art to approximate traceability in case no calibration certificate of the measured components is required [21–25].

Structure of ISO 10360

International Standard ISO 10360 covers CMM verification. This standard currently has six parts:

- ISO 10360-1: 2000 Geometrical Product Specifications (GPS) –Acceptance and reverification tests for coordinate measuring machines (CMM) —Part 1: Vocabulary

- ISO 10360-2: 2009 Geometrical product specifications (GPS) –Acceptance and reverification tests for coordinate measuring machines (CMM) —Part 2: CMMs used for measuring linear dimensions
- ISO 10360-3: 2000 Geometrical Product Specifications (GPS) –Acceptance and reverification tests for coordinate measuring machines (CMM) —Part 3: CMMs with the axis of a rotary table as the fourth axis
- ISO 10360-4: 2000 Geometrical Product Specifications (GPS) –Acceptance and reverification tests for coordinate measuring machines (CMM) —Part 4: CMMs used in scanning measuring mode
- ISO 10360-5 :2010 Geometrical Product Specifications (GPS) –Acceptance and reverification tests for coordinate measuring machines (CMM) —Part 5: CMMs using multiple-stylus probing systems
- ISO 10360-6 :2001 Geometrical Product Specifications (GPS) –Acceptance and reverification tests for coordinate measuring machines (CMM) —Part 6: Estimation of errors in computing Gaussian associated features.

Part 2 and part 5 have been updated since the last revision of this guide. Part 6 has been added to this series of standards since this guide was last published.

ISO 10360-2 does not explicitly apply to non-Cartesian CMMs; however, it may be applied to non-Cartesian CMMs by mutual agreement.

This guide will concentrate on the tests listed in part 2 of the standard and will cover some aspects of parts 3 and 5. It is suggested that the reader regularly checks the catalog on the ISO website for further information and to see when new standards are published.

In addition, ISO/TS 23165: 2006 Geometrical product specifications (GPS) –Guidelines for the evaluation of coordinate measuring machine (CMM) test uncertainty provides guidance on how to calculate the uncertainty of measurement associated with the test [26].

7.14.2 Accreditation of Calibration Laboratories

The calibration of a mechanical component with a CMM (Fig. 7.25) having specs (Table 7.9) is a complex measurement problem that involves many factors. Each requirement for each geometrical feature, or relationship between features, of the component under test represents a separate measurement task for the CMM. Each measurement task and measurement strategy used to solve it involves a different combination of sources of uncertainties associated with the coordinate measuring system, that is, the CMM and the environment in which the CMM is sited. Hence, a general calibration of CMMs for all the measurement tasks it can handle is not practical. The concept of task-related calibration is introduced.

While a CMM is an independent machine and stand-alone to inspect parts, it can be also integrated inside CNC machine tools for in-process or post-process inspection, as shown in Figure 7.26.

Acceptance Test
Like a CNC machine, the test is carried out when the CMM is newly installed, moved, or overhauled.

Step 1 This includes CMM machine start-up, probe qualification, and probe configuration.

The single stylus-probing test that appeared in ISO 10360-2: 2001 does not appear in the current edition of ISO 10360-2. It has been moved to the new edition of ISO 10360-5 that will be replacing ISO 10360-5: 2000. ISO/PAS 12868 was prepared to allow the single stylus-probing test to be available until the publication of the new edition of ISO 10360-5. ISO/PAS 12868 has been withdrawn with the publication of ISO10360-5:2010. This test is advisable initially.

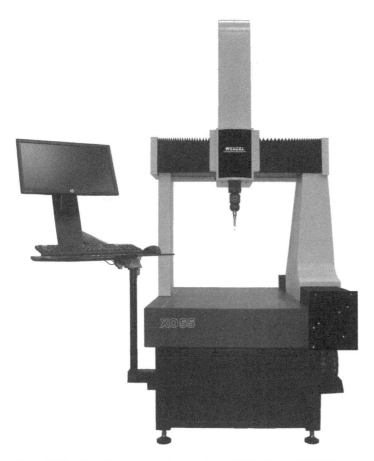

Figure 7.25 Coordinate measuring machine (CMM). *Source:* WENZEL Group.

Table 7.9 Specification of Laser Interferometer

Item	Specification
Maximum measurement range	10 m (using corner cube)
Length of optical fiber	Standard 3m / Optional 5m, 10m
Accuracy (MPE_E)	$+/- (L \times 10^{-7} + 0.01 \times 10^{-6})$ m
Sampling rate	1 MHz
Light source	Wavelength stabilized single mode He- Ne laser
Optical output approx.	0.3 mW
Wavelength stability	$\pm 1 \times 10^{-7}$
Power consumption	90 VA
Operating environment	Temperature: 10 to 40°C
	(Temperature change in operation within $\pm 10°$C)
	Humidity: 10 to 90%(Non-condensing)

Figure 7.26 Probing exist inside CNC machine tool. *Source:* Courtesy of Spinner.

The objective here is to verify that the machine meets its specification and then the manufacturer's specified conditions, such as length and type of stylus, probing speed, and reference sphere, should be used. The environmental conditions recommended by the manufacturer should be satisfied.

Step 2: Environmental Conditions The operating environment affects a CMM. Its position with respect to any heat source, vibration, and so forth, are fundamental possible issues at the time of installation. Limits for permissible environmental conditions, such as temperature conditions, air humidity, and vibration that influence the measurements are usually specified by the manufacturer. In the case of acceptance tests, the environment specified by the manufacturer applies. However, in the case of reverification tests the user can specify the environment. In acceptance or reverification, the user is free to choose the environmental conditions under which the ISO 10360-2 testing will be performed within the specified limits.

Step 3: Operating Conditions Under the acceptance test, the manufacturer's instructions are divided into two groups and should be followed.

a) Environment
 - location, type, number of thermal sensors;
 - machine start-up/warm-up cycles;
 - thermal stability of the probing system before calibration.
b) Stylus
 - stylus system configuration;
 - weight of stylus system and/or probing system;
 - probing system qualification; and
 - cleaning procedures for stylus tip.

Step 4: Workspace Loading Effects The length measurement performance of any CMM will be affected by its loading; hence specifications must be observed. The manufacturer must specify a limit on the maximum load per unit area (kg/m^2) on the CMM support, for instance, a table, and on any moving part, such as a gantry, in terms of individual point loads (kg/cm^2); for point loads, the load at any specific contact point shall be no greater than twice the load of any other contact point.

Figure 7.27 Step gauge with CMM calibration.

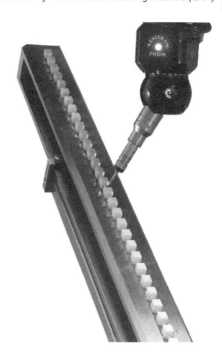

Step 5: Selection of Measuring Equipment and Checking the Probing System The basic single probing system is checked. A step gauge or series of gauge blocks (five different sizes) are selected based on ISO 3650 (Fig.7.27).

Example of Selection: CMM with 2040 mm by 1300 mm by 570 mm height has a diagonal calculated of 2485 mm; hence the shorted length of the material standard is 30 mm, and the longest length is minimum 66% and hence 1640 mm or higher.

- Length bar has an accuracy of 0.5 μm/m;
- Gauge blocks with accuracy of 0.5 μm/m;
- Step gauges with accuracy of 1.0 μm/m.

Other artifacts include laser interferometry with contact probing and ballbars, both used in bidirectional measurement.

Step 6: Length Measurement Error with Ram Axis Stylus Tip Offset of 0 mm and Results Qualification of the standard probe after all surfaces of the CMM are cleaned from any dust or particles is needed. It is important to check the alignment of the axis using the gauge blocks, for example.

For a minimal offset, the choice of location of the material standard in the CMM measuring volume will include some of the following positions. The four cross diagonals, the three in-plane diagonals (diagonals of *XY*-, *YZ*-, and *XZ*-planes at the mid position of the third axis), and measuring lines nominally parallel to an axis (the manufacturer may specify a separate maximum permissible error for these directions) as shown in Figure 7.28. Sample of outcome test is shown in Figure 7.29 over the whole line.

Step 7: Length Measurement Error with Ram Axis Stylus Tip Offset of 150 mm Similarly to the previous test, the default value of the ram axis stylus tip offset is 150 mm ± 15 mm.

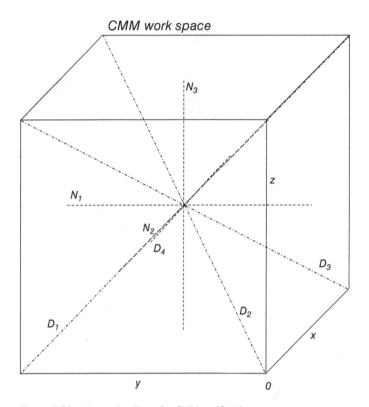

Figure 7.28 Measuring lines for CMM verification.

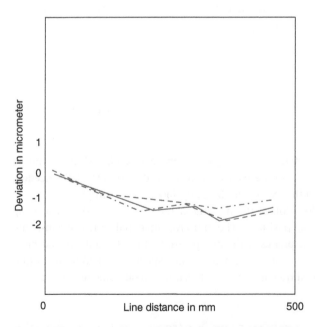

Figure 7.29 One line N_1 test (3 times) for CMM verification.

Step 8: Interpretation of Results/Acceptance This is to check that the error in both previous tests, that is 0 and 150 mm, are within the manufacturer's stated values. If this is the case, then the CMM is verified. Account should be made to ISO 14253-1 and ISO/TS 23165. It is also important to check the repeatability that should be with the permissible range.

The machine is considered to be reverified if the errors obtained are all less than the maximum permissible errors.

CMM Uncertainty

DD ISO/TS 23165: 2006 "Geometrical product specifications" (GPS) covers the uncertainties relating to ISO 10360 tests in "Guidelines for the evaluation of coordinate measuring machine" (CMM) test uncertainty.

The recommended equation for the standard uncertainty of the probing error, u(Pr), is shown by Equation (7.22) where *EF* is the form error reported on the calibration certificate of the test sphere and $u(EF)$ is the standard uncertainty of the form error stated on the certificate.

$$u(Pr) = \sqrt{\left(\frac{EF}{2}\right)^2 + u^2(EF)}. \tag{7.22}$$

The recommended equation for the standard uncertainty of the error of indication, u(E), is

$$u(E) = \sqrt{u^2(\varepsilon_{calib}) + u^2(\varepsilon_a) + u^2(\varepsilon_t) + u^2(\varepsilon_{align}) + u^2(\varepsilon_{fixture})}, \tag{7.23}$$

where

ε_{calib} is the calibration error of the material standard of size;
ε_a is the error due to input of the thermal expansion coefficient of the material standard of size;
ε_t is the error due to the input of the temperature of the material standard of size;
ε_{align} is the error due to the misalignment of the material standard of size; and
$\varepsilon_{fixture}$ is the error due to the fixturing of the material standard of size.

All related terminology is fully explained in ISO/TS 23165 with worked examples within the standard. The following tests are calibration lab procedures for various test instruments.

SECTION 1: SCOPE AND DESCRIPTION

1.1 SCOPE

    1.1.1 This document describes a calibration procedure for the measuring machine. The instrument being calibrated is referred to herein as the test instrument (TI).

    1.1.2 This procedure is intended only to provide a means of determining the accuracy as stated by the manufacturer.

Table 1. CALIBRATION DESCRIPTION

TI Characteristics	Performance Specifications	Test Method
Flatness and parallelism	Tolerance: Working surfaces flat and parallel within 1.27 μm (50 microinches)	Measured using an optical parallel flat and monochromatic light
Linear calibration	Range: 2000 mm Tolerance: ± 0.00254 mm	Measured using the complete set certified gauge blocks DoAII 8SM-142

SECTION 2: CALIBRATION REQUIREMENTS

2.1 CALIBRATION EQUIPMENT

Table 2. CALIBRATION EQUIPMENT

	Nomenclature	Manufacturer/Model
2.1.1	Gauge blocks	DoAll, 8-SM-142
2.1.2	Optical parallel flat	Starrett, OFP4
2.1.3	Monochromatic light	Van Keuren, LIA
2.1.4	25 mm nominal gauge block with calibrated value	Starrett, accessory

2.2 ENVIRONMENTAL CRITERIA

a. Temperature:	$20\,^{0}C \pm 0.5\,^{0}C$
b. Relative humidity:	less than 45%

SECTION 3: PRELIMINARY OPERATIONS

3.1 INSPECTION OF EQUIPMENT

    3.1.1 Ensure that the TI is clean and free from defects that would impair its operation. Check the carriage mechanism for smooth function operation. Inspect working surfaces for nicks and scratches.

3.2 PREPARATION OF LASER SYSTEM

    3.2.1 Ensure that the laser system automatic compensator has been calibrated in accordance with CP-DI-23-99.

    NOTE: For improved length measurements, the laser system may be operated without the compensator by separately measuring and monitoring material and laboratory conditions and applying calculated correction factors for the velocity of light and material linear expansion.

    3.2.2 Ensure that the laser system is operational and in optical alignment with the remote interferometer located beneath the fixed anvil.

    3.2.3 As applicable, set the automatic compensator DISPLAY-OPERATE-SCAN selector to DISPLAY and the AUTO-MANUAL switch to AUTO. Set the automatic compensator to TOTAL COMP. Ensure that the material compensation is operational as indicated by the TI switch being illuminated. Set the material compensation temperature coefficient thumbwheel switches to the correct setting for the standard gauge blocks ($+ 11.50\,^{0}C$ for DoAII 8-SM standards).

SECTION 4: CALIBRATION PROCESS

4.1 FLATNESS INSPECTION

NOTE: When using the monochromatic light, the distance from one fringe line to another will always be 0.295 μm (11.6 microinches). (The curvature of a light bank when a straight line drawn through the extreme ends of the bands cuts through four additional bands [4 × 0.295 = 1.18 or 4 × 11.6 = 46.4 microinches is equivalent to 1.27 μm (50 microinches])

4.1.1 Position the optical parallel flat on the TI spindle and set the monochromatic light source to observe the number of light bands on the spindle face. Apply a sufficient force to the optical parallel flat to obtain three to four light bands (fringes) across the spindle face.

4.1.2 Verify that the spindle is flat to within 1.27 μm (50 microinches).

4.1.3 Using the method described in steps 4.1.1 and 4.1.2 repeat the measurement for the anvil face.

4.2 PARALLELISM

4.2.1 Position on optical flat between the TI anvil and spindle. Adjust the monochromatic light source to observe the number of light bands on the spindle and the anvil face.

4.2.2 Gently slide the optical flat to remove the air film and apply a sufficient force from the anvil (adjustable) to obtain a minimum number of light bands on both anvil and spindle.

4.2.3 Count the total number of bands on the anvil and spindle and multiply by 0.295 μm (11.6 microinches).

4.2.4 Verify that the parallelism is within 1.27 μm (50 microinches).

NOTE: If the anvil has 1.5 bands and the spindle has 1.5 bands, then total number of bands is 3. (3 × 0.295 = 0.885 μm) or 3 × 11.6 = 34.8 microinches).

4.3 ZERO SETTING

4.3.1 Clean the TI measuring faces. Allow sufficient time for the TI to stabilize at the ambient temperature.

4.3.2 Ensure that both the fixed and null indicating anvils have been locked in position and that the proper contacts are in place. Set the TI zero refence position as follows:

NOTE: The "oz" symbol denotes "ounce" in the English System of units and is utilized only to be consistent with the manufacture's description of the part (1 oz. = 28.3495 grams).

4.3.2.1 Set the 16 oz weight to apply 16 oz. gauging force.

4.3.2.2 Bring the carriage into position with handwheel until the movable anvil touches the fixed anvil.

4.3.2.3 Lock the coarse and fine adjustment locks and turn the fine adjustment knob counterclockwise to move the carriage toward the fixed contact until the TI null indicator pointer approaches the null position. Final zero setting may be set with the indicator fine setting adjustment.

4.3.2.4 Reset the laser display to zero by pressing the "RESET" button of the laser display.

4.3.2.5 Make sure that the zero setting is repeatable by performing steps 4.3.2.2 to 4.3.2.4 two more times.

4.3.3 LINEAR MEASUREMENT TEST

NOTE: Use clean cotton gloves or insulated tongs to handle the gauge blocks to prevent raising the gauge block temperature above ambient.

4.3.3.1 Clean the gauge blocks and the TI measuring faces. Allow sufficient time for gauge blocks and TI to stabilize at the ambient temperature.

4.3.3.2 Select 125 mm nominal length gauge block to serve as the measurement standard for the first test point.

4.3.3.3 Move the carriage to a position slightly longer than the workpiece to be measured.

4.3.3.4 Place the V-roller rest, with the horizontal adjustment adjacent to the movable carriage contact and the other V-roller rest adjacent to the fixed contact. Position both rests in accordance with the airy support formula. (See Appendix).

4.3.3.5 Place the workpiece in the V-rest and attach the laser material temperature sensor to the body of the measuring machine, close to the workpiece. Move the workpiece against the fixed anvil contact.

4.3.3.6 Using the carriage handwheel, bring the null indicating contact up until the null indicator is activated.

4.3.3.7 Lock the fine adjustment locks and turn the fine adjust knob until a zero setting is achieved on the null indicator.

4.3.3.8 Adjust both the V-roller rests in the vertical position to establish the maximum reading of the null indicator. Now, adjust the horizontal position of the right-hand V-roller rest using the adjusting knob until a further maximum reading of the null indicator is obtained.

4.3.3.9 Repeat 4.3.3.7 observing the laser display until a minimum length reading close to the nominal value of the workpiece is obtained.

4.3.3.10 Record the minimum reading obtained on the laser display in the "Calibration Data Record Sheet."

4.3.3.11 Repeat steps 4.3.3.5 to 4.3.3.9 to obtain two further readings.

4.3.3.12 Repeat steps 4.3.3.1 to 4.3.3.10 for all nominal length gauge blocks given in the "Calibration Data Record Sheet."

SECTION 5: DATA ANALYSIS

5.1 Transfer the data recorded on the Calibration Data Record Sheet to a computer-based spreadsheet (Excel or equivalent) in the format indicated in the "Calibration Data Analysis" sheet attached to this document.

SECTION 6: CALIBRATION REPORT

6.1 Prepare a calibration report in the format indicated in the calibration procedure "CP-GN-03-99-Format of MSS Calibration Reports."

Multiple Choice Questions of this Chapter

Multiple Choice Questions are given for each chapter with solutions in an online extension of this book. Please use link: www.wiley.com\go\mekid\metrologyandinstrumentation\

Appendix

This chapter 7 continuous the calibration procedures in an online appendix.

References

1 Mekid, S., and De Luna, H., "Error Propagation in laser scanning for dimensional inspection," *Int. J. of Metrology*, 14(2), 2007, pp. 44–50.

2 Mekid, S., "Spatial thermal mapping using thermal stereo & wireless sensors for error compensation via OAC controllers," *Journal of Systems and Control Engineering*, 224(7), 2010, pp. 789–798.

3 Mekid, S., and Ogedengbe, T., "A review of machine tool accuracy enhancement through error compensation in serial and parallel kinematic machines," *Int. J. Precision Technology*, 1(3-4), 2010, pp. 251–286.

4 Guo, W.Z., Gao, F., and Mekid, S., "A new analysis of workspace performances and orientation capability for 3-DOF planar manipulators," *Int. J of Robotics and Automation*, 25(2), 2010.

5 Mekid, S., "Precision design aspects for friction actuation with error compensation," *Journal of Mechanical Science and Technology*, 23(11), 2009, pp. 2873–2884.

6 Mekid S., and Kwon, O., "Nervous materials: a new approach of using advanced materials for better control, reliability and safety of structures," *Science of Advanced Materials*, 1(3), 2009, pp. 276–285.

7 Mekid, S., "In-process atomic force microscopy (AFM) based inspection," *Sensors* (MDPI), 17(6), 2017 p. 1194.

8 Khalid, A., and Mekid, S., "Intelligent spherical joints based tri-actuated spatial parallel manipulator for precision applications," *Robotics and Computer-Integrated Manufacturing*, 54, 2018, pp. 173–184.

9 Mekid, S., *Introduction to Precision Machine Design and Error Assessment*, (ISBN13: 9780849378867), 2008, CRC Press.

10 Holmberg, K., Adgar, A., Arnaiz, A., Jantunen, E., Mascolo, J., and Mekid, S. (Eds.), *E-maintenance*, (ISBN 978-1-84996-204-9), 2010, Springer Verlag, New York.

11 Mekid, S., "Spatial thermal error compensation using thermal stereo via OAC controller in NC machines," *Proceedings of the 5th International Conference on Intelligent Production Machines and Systems*, 2009 Elsevier, Oxford.

12 Sartori, S., Pfeifer, T., Crossman, M., Brinkmann, K., Canuto, Ferraris, Gallorini, M., Moretti, L., Peisino, Ferrero, C., Balsamo, A. and Pogliano, U., "Coordinate Measuring Machine Calibration EAL," 2009, *European Cooperation for Accreditation of Laboratories Publication Reference*.

13 Sued, M.K., and Mekid, S., "Dimensional inspection of small and mesoscale components using laser scanner," *International Conference on Advances in Mechanical Engineering*, ICAME 2009, Malaysia.

14 Ogedengbe, T., Mekid, S., and Hinduja, S., "An investigation of influence of machining conditions on machining error," *Proceedings of the 5th Virtual International Conference on Intelligent Production Machines and Systems*, 2008, Elsevier, Oxford.

15 Ogedengbe, T., and Mekid, S., "Machine scaling optimization for on demand precision machining," *10th EUSPEN Conference*, 2008, Switzerland.

16 Mekid, S., Khalid, A., and Ogedengbe, T., "Common physical problems encountered in micro machining," *6PthP CIRP International Seminar on Intelligent Computation in Manufacturing Engineering - CIRP ICME '08*, (ISBN 978-88-900948-7-3), pp. 637–642, July 2008, Italy.

17 Vacharanukul, K., and Mekid, S., "In-process inspection of dimensional measurement and roundness," *International Conference on Manufacturing Automation*, National University of Singapore, May 2007, Singapore.

18 Vacharanukul, K., and Mekid, S., "Differential laser based probe for roundness measurement," *EUSPEN Conference*, 2007, Germany.

19 Khalid, A., and Mekid, S., "Analysis of Jacobian inversion in parallel kinematic systems," *35th International MATADOR Conference* 2007, Taiwan.

20 Mekid, S., and Vaja, D., "New expression for uncertainty propagation at higher order for ultra-high precision in calibration," *9th EUSPEN Conference*, 2007, Germany.

21 Khalid A., and Mekid, S., "Computation and analysis of dexterous workspace of PKMs," *Proceedings of the 3rd Virtual International Conference on Intelligent Production Machines and Systems*, 2007, Elsevier, Oxford.

22 Owodunni, O.O., Hinduja, S., Mekid, S., and Zia, A., "Investigation of optimum cutting conditions for STEP-NC turned features," *Proceedings of the 3rd Virtual International Conference on Intelligent Production Machines and Systems*, 2006, Elsevier, Oxford.

23 Khalid A., and Mekid, S., "Design of precision desktop machine tools for meso-machining," *Proceedings of the 2nd Virtual International Conference on Intelligent Production Machines and Systems*, 2006, Elsevier.

24 Khalid, A., and Mekid, S., "Design & optimization of a 3-axis micro milling machine," *6PthPInt. Conf. European Society for Precision Engineering and Nanotechnology*, 2006, Baden, Austria.

25 Mekid, S., and Khalid, A., "Robust design with error optimization analysis of a 3-axis CNC micro milling machine." *5PthP CIRP International Seminar on Intelligent Computation in Manufacturing Engineering - CIRP ICME '06*, July 25–28, 2006, Ischia, Naples, Italy.

26 www.Renishaw.com.

8

Uncertainty in Measurements

"As far as the laws of mathematics refer to reality, they are not certain; and as far as they are certain, they do not refer to reality."

—Albert Einstein

8.1 Introduction and Background

As it has been introduced in previous chapters, measurement is accompanied necessarily by an error of measurement due to various reasons including the instrument, the method of measurement, environmental conditions, and many more. The unreliability surrounding this measurement is the uncertainty of measurement. The background and fundamental definition of uncertainty and error will be discussed later based on international standards with all aspects in general practice.

The ISO 15189:2012 standard describes some enhanced expectations on the measurement of uncertainty and the ISO 17025 standard specifies requirements when reporting and evaluating uncertainty of measurement.

15189:2012 & 17025:2017 Requirements

ISO 15189: 2012 Section 5.5.1.4: "The laboratory shall determine measurement uncertainty for each measurement procedure, in the examination phases used to report measured quantity values on patients' samples. The laboratory shall define the performance requirements for the measurement uncertainty of each measurement procedure and regularly review estimates of measurement uncertainty."

Section 5.6.2: "Upon request, the laboratory should make its estimate of measurement of uncertainty available to laboratory users".

17025: 2017 Section 7.6.1: "Laboratories shall identify the contributions to measurement uncertainty. When evaluating measurement uncertainty, all contributions that are of significance, including those arising from sampling, shall be taken into account using appropriate methods of analysis."

Section 7.6.3: "A laboratory performing testing shall evaluate measurement uncertainty. Where the test method precludes rigorous evaluation of measurement uncertainty, an estimation shall be made based on an understanding of the theoretical principles or practical experience of the performance of the method."

<u>Section 7.8.3</u>: "Test reports: where applicable, the measurement uncertainty is presented in the same unit as that of the measurand or in a term relative to the measurand (e.g., percent) when:

- It is relevant to the validity or application of the test results;
- A customer's instruction so requires, or
- The measurement uncertainty affects conformity to a specification limit".

8.2 Uncertainty of Measurement

The result of any quantitative measurement has two essential components:

- A numerical value (expressed in SI units as required by ISO 15189) which gives the best estimate of the quantity being measured (the measurand). This estimate may well be a single measurement or the mean value of a series of measurements.
- A measure of the uncertainty associated with this estimated value. In any length measurement for example, it is the variability or dispersion of a series of similar measurements as a standard uncertainty (standard deviation) or combined standard uncertainty

Uncertainty has been previously defined and discussed under various aspects in this book. It is the unreliability that exists about the result of any measurement. High quality equipment and instruments, such as rulers, voltmeters, clocks, and thermometers, are not necessarily trustworthy and may not give the right answers. For every measurement—even the most careful—there is always a margin of doubt.

8.3 Measurement Error

The term "error" or "measurement error" is the difference between the true value and the measured value. The most likely or "true" value may then be considered as the measured value including a statement of uncertainty, which characterizes the dispersion of possible measured values. As the measured value and its uncertainty component are at best only estimates, it follows that the true value is not determined. Uncertainty is a consequence of the interchange of errors, which create dispersion around the estimated value of the measurand; the smaller the dispersion, the smaller the uncertainty. Even if the terms "error" and "uncertainty" are used somewhat interchangeably in everyday descriptions, they have different meanings. They should never be used as synonyms. There is a way to report the uncertainty as follows. The symbol ± (plus or minus) follows the reported value of a measurand and the numerical quantity that follows this symbol, indicates the uncertainty associated with the particular measurand and not the error. Since uncertainty is a property of a test result, the preferred reporting form is:

Measurement = (best estimate ± uncertainty) units

Example 8.1 The length is measured at 35 mm with an uncertainty of 1 mm at a confidence of 95%. In practice, it is 35 mm ± 1 mm. To show confidence that the estimated value is within 34 mm and 36 mm.

Whenever possible, there is a trial to correct for any known errors: for example, by applying corrections from calibration certificates. However, any error whose value is not known is a source of uncertainty.

Example of Reporting Experimental Results

Experimental uncertainties should be rounded to one significant figure. Experimental uncertainties are, by nature, inexact. Uncertainties are almost always quoted to one significant digit (example: ± 0.08 s). If the uncertainty starts with a one, some scientists quote the uncertainty to two significant digits (example: ± 0.017 kg).

It is wrong to write: 67.3 cm $\pm$ 3.2 cm; the correct form is: 67 cm $\pm$ 3 cm.

It is required to always round the experimental measurement or result to the same decimal place as the uncertainty. It would be confusing (and perhaps dishonest) to suggest that you knew the digit in the hundredths (or thousandths) place when you admit that you unsure of the tenths place.

It is wrong to write: 1.437 s $\pm$ 0.1 s; the correct form is: 1.4 s $\pm$ 0.1 s.

8.4 Why Is Uncertainty of Measurement Important?

Every engineer is interested in performing good quality measurement and identifying the uncertainty of measurement that will help her understanding the results. Several aspects exist already to look after measurement uncertainty such as:

a) **Tolerance** - where there is a need to know the uncertainty before you can decide whether the tolerance is met or a need to read and understand a calibration certificate or a written specification for a test or measurement.
b) **Calibration** - where the uncertainty of measurement must be reported on the certificate.
c) **Test** - where the uncertainty of measurement is needed to determine a pass or fail result or to meet a tolerance.

This is valid for one or several measurements but also valid in a system where there is a contribution of multiple factors to the final uncertainty, called *"budget of uncertainty."*

8.5 Components and Sources of Uncertainty

8.5.1 What Causes Uncertainty?

It is important to figure out the factors affecting the measurement uncertainty in order to reduce the uncertainty, and these are mainly channeled as follows.

- **Instrument**: This is a usual source of uncertainty, the instrument could be affected or damaged with age, wear, noise, and so forth. This can be worse if it is not calibrated.
- **Process**: The measurement process may not be well designed for the purpose, especially if the standard way of measurement is not defined allowing each operator to carry out the measurement the way they see fit and therefore contributing to the problem.
- **Operator**: Well-trained and qualified operators are needed for metrology. Sometimes, current skill and/or knowledge may be a cause of the problem, and this is caused by poor training and lack of adherence to standards.
- **Environment**: Factors such as changes in temperature, humidity, pressure, or other conditions that may affect the measurements of parts and hence the measurement results must be considered.
- **Sampling**: The sampling needs to be properly defined or adequate to allow the measurement errors to be adequately captured. This is very important in measurements.

Some of the above errors may be random or systematic. If the errors are random, there is not much we can do about it except to report it in the overall uncertainty numbers. If the errors are systematic, that is, constant, regular, then it is possible to offset the measurement results to account for this systematic error. For example, if there is always a bias of +0.5 mm in our measurements, then we can subtract this value and report the true measurement value such that there is no bias in the measurements [1, 2].

8.5.2 Uncertainty Budget Components

Example a   *Uncertainty Budget Components*
This an itemized table of all components that contribute to the overall uncertainty in measurement results. Hence, we understand which components are contributing more than others.
It is also called budget of errors in machine tool. The following example gives an overview.

Example b   *Budget of Errors of an Axis Motion*
Every axis of motion has 6 degrees of freedom (DOFs). In the design of a very accurate motion in one axis, the objective would be to precisely control one of these DOFs (e.g., noble motion) and to minimize the effect of the other five. The unwanted motions (parasitic) resulting from the other DOFs reduce the performance of the positioning system.
When we are dealing with XY or multi axis motion, the unwanted parasitic movements generated by the DOFs of each axis combine statistically to create the systems' error budget. These parasitic movements result from several sources, such as imperfections in the bearings, deflections due to loads, or thermal distortions. The goal in the error budgeting is to allocate allowable values for each error source and select components such that the ability of each component to meet its error allocation is not exceeded.

8.5.3 The Errors Affecting Accuracy

The contributions of the static and dynamic errors are estimated to be respectively about 65–70 % and 30–35%. The possible sources and types of errors that affect the accuracy are as follows:

- Failure to account for a factor;
- Environmental factors;
- Instrument resolution;
- Calibration (mostly systematic);
- Geometric errors of machine components and structure;
- Kinematic errors;
- Thermal distribution;
- Material instability;
- Machine assembly;
- Fixtures;
- Structural stiffness;
- Cutting forces;
- Tool wear;
- Instrumentation;
- Controllers;
- Lag time or hysteresis;

- Zero offset;
- Parallax (systematic or random);
- Actuators and sensors;
- Errors due to second-order phenomena;
- Personal errors (carelessness, poor training).

Note that the thermal errors are considered to be the most difficult errors to be compensated due to the variety of heating sources present in a machine and the selection of the appropriate temperature variables for a given model.

Gross personal errors, also called mistakes, should be avoided and corrected if discovered. As a rule, personal errors are excluded from the error analysis discussion as it is generally assumed that the experimental result was obtained by following correct procedures.

8.6 Static Errors and Dynamic Errors

The errors of movement are classified into two categories: errors linked to static sources and errors linked to the dynamic behavior of the machine. It is brought to the attention of the designer that often these errors are linked to each other. Actually, the static errors sometimes generate the dynamic errors. These two types of errors affect the precision and the repeatability of the movement.

Quasi-Static Errors

The quasi-static errors appear at very low speeds and are found in:

- The conceptual design that may induce the Abbe error, assembling of parts inducing errors from the tolerances, and excess in tighten torques;
- Machining errors (planarity, parallelism, straightness, surface waviness, etc.);
- Load errors (own weight, external charges, bearing preloads, cutting forces, etc.);
- Errors due to reversible influences (thermal dilatations) and nonreversible influences (material aging);
- Errors due to dimensional instability of materials;
- Metrology errors (sensor precision and Abbe errors).

Dynamic Errors

The dynamic errors are generated by the power supplied from external forces, which often are not controllable:

- Rolling on surface waviness [3];
- Machining processes (e.g., cutting force);
- Noise source induced by control system components (motors, actuators), pressure pulsations of hydraulic or pneumatic supply circuits, fluid turbulence, etc.

8.7 Types of Uncertainty

The uncertainty has two types commonly discussed in estimating measurement uncertainty. These are type A and type B. The GUM is the most referenced document explaining all related uncertainty calculations. Hence, there are two ways to evaluate the uncertainty.

The following definition is an exert from the "Guide to the Expression of Uncertainty in Measurement":

"3.3.4 The purpose of the Type A and Type B classification is to indicate the two different ways of evaluating uncertainty components and is for convenience of discussion only; the classification is not meant to indicate that there is any difference in the nature of the components resulting from the two types of evaluation. Both types of evaluation are based on probability distributions (C.2.3), and the uncertainty components resulting from either type are quantified by variances or standard deviations." – JCGM 100

Basically;

- **Type A uncertainty** is calculated from a series of observations. Essentially, Type A Uncertainty is data collected from a series of observations and evaluated using statistical methods associated with the analysis of variance (ANOVA).
- For most cases, the best way to evaluate Type A uncertainty data is by calculating the:
  - Arithmetic mean,
  - Standard deviation, and
  - Degrees of freedom.
- **Type B uncertainty** is evaluated using available information. It is data collected from anything other than an experiment you have carried out. Despite that these data can be analyzed statistically, it is not considered as Type A data if data was not collected from a series of observations. Most of the Type B data that you will use to estimate uncertainty will come from:
  - Calibration reports,
  - Manufacturer's manuals,
  - Proficiency testing reports,
  - Datasheets,
  - Standard methods,
  - Calibration procedures,
  - Journal articles,
  - Conference papers,
  - White papers,
  - Industry guides,
  - Textbooks, and
  - Other available information.

8.8 Uncertainty Evaluations and Analysis

Type A uncertainty is characterized through the calculation of the arithmetic mean, standard deviation, and degrees of freedom. Large number of repetitions is required and it applies also to random sources of errors.

a) **Arithmetic mean**

This is calculated through the following equation where x_i are all measurement made. Their number is n.

$$\bar{x} = \frac{1}{n} \sum_{i=1}^{n} x_i.$$ (8.1)

This is also uncertainty of repeated measurements.

b) **Standard deviation**

The average variance as explained previously for repeated measurement is calculated in type A as the standard deviation; hence

$$\sigma = \sqrt{\frac{1}{n-1}\sum_{i=1}^{n}(x_i - \overline{x})^2}, \tag{8.2}$$

where the parameters are those used in arithmetic mean.

c) **Degrees of freedom**

There is a need to calculate the degrees of freedom associated with the previous two calculations for the sample set in hands. It is the number of values in the final calculation in statistics that are free to vary.

$$v = n - 1, \tag{8.3}$$

where n is the number of values in the sample set under consideration.

Type B uncertainty is based on scientific estimate performed by the operator that has to use all information on the measurement and is the source of its uncertainty. It also applies when the rules of statistics cannot be used and for a systematic error or for a single result of measurement. We are interested in the variance in type B that is calculated differently from type A uncertainty.

The relevant information needed may include:

- Previous measurement data;
- Experience with, or general knowledge of, the behavior and property of relevant materials and instruments;
- Manufacturer's specifications;
- Data provided in calibration and other reports;
- Uncertainties assigned to reference data taken from handbooks.

Type A evaluations of uncertainty based on limited data are not necessarily more reliable than soundly based Type B evaluations.

Example 8.2 *Type B uncertainty on the length measurement of an arm.*

The measurement is 150 mm with uncertainty u(L) = 1 mm

$$Ur(L) = u(L)/L$$

$$= 1/150 = 0.0066$$

Percentage uncertainty is 0.7%.

8.9 Uncertainty Reporting

This is a common practice for ISO/IEC 17025 accredited laboratories to report measurement uncertainty in test and calibration certificates. It is required to include uncertainty in issued lab certificates. Hence, it is important to know how to report uncertainty correctly.

Some common deficiencies laboratories are:

- Not reporting uncertainty in certificates;
- Not reporting uncertainty correctly;
- Not reporting uncertainty to two significant figures;
- Not rounding uncertainty correctly;
- Reporting uncertainty smaller than their scope of accreditation.

Requirements for Reporting Uncertainty

To report uncertainty in certificates and tests [4], there is a need to cover the recommendations from the following:

a) JCGM 100:2008 (GUM);
b) ISO/IEC 17025:2017;
c) ILAC P14:01/2013.

These will be discussed next in a) and b) where excerpts from the related standards are shown on their related paragraphs.

a) JCGM 100:2008 (GUM)

1) **Reporting Expanded Measurement Uncertainty**

7.2.3 When reporting the result of a measurement, and when the measure of uncertainty is the expanded uncertainty $U=ku_c(y)$, one should:
a) Give a full description of how the measured y is defined;
b) State the result of the measurement as $Y = y \pm U$ and give the units of y and U.
c) Include relative expanded uncertainty $U|\,|y|$, $|y| \neq 0$, when appropriate;
d) Give the value of k used to obtain U (for the convenience of the user of the result, give both k and $u_c(y)$);
e) Give the approximate level of confidence associated with the interval $y \pm U$ and the state how it was determined;
f) Give the information outlined in 7.2.7. or refer to a published document that contains it.

2) **Reporting the Measurement Results and Measurement Uncertainty**

7.2.4 When the measure of uncertainty is U, it is preferable, for maximum clarity, to state the numerical result of the measurement as in the following example. (The words in parentheses may be omitted for brevity if U, u_c, and k are defined elsewhere in the document reporting the result.)

"$m_s = (100.021\ 47 \pm 0.000\ 79)$ g, where the number following symbol $\pm$ is the numerical value of (an expanded uncertainty) $U=ku_c$, with U determined from (a combined standard uncertainty) $u_c = 0.35$mg and (a coverage factor) k= 2.26 based on the t-distribution for v = 9 degrees of freedom, and defines an interval estimated to have a level of confidence of 95 percent."

3) **Round Uncertainty to Two Significant Digits**

7.2.6. The numerical values of the estimate y and its standard uncertainty $u_c(y)$ or expanded uncertainty U should not be given with an excessive number of digits. It usually suffices to quote $u_c(y)$ and U [as well as the standard uncertainties $u(x_i)$ of the input estimates x_i] to at most two significant digits, although in some cases may be necessary to retain additional digits to avoid round-off errors in subsequent calculations.

In reporting final results, it may sometimes be appropriate to round uncertainties up rather than to the nearest digit. For example, $u_c(y) = 10.47$ m might be rounded up to 11m. However, common sense should prevail and a value such as u(xi)-28,05 kH should be rounded down to 28 kH. Output and input estimates should be rounded to be consistent with their uncertainties; for example, if y = 10.057 62 with $u_c(y) = 27$ m, y should be rounded to 10.058. Correlation coefficients should be given with three-digit accuracy if their absolute values are near unity.

b) ISO/IEC 17025:2017 (E)

1) **Reporting Measurement Uncertainty in Test Reports**

   a) Where applicable, the measurement uncertainty presented in the same unit as the of the measurand or in a term relative to the measurand (e.g., percent) when:
      - It is relevant to the validity or application of the test results;
      - A customer's instruction so requires; or
      - The measurement uncertainty affects conformity to a specification limit.
   b) Where appropriate, opinions and interpretations are added (see 7.8.7).
   c) Additional information that may be required by specific methods, authorities, customers or groups of customers.

2) **Reporting Measurement Uncertainty in Calibration Reports**

   7.8.4 Specific Requirements for Calibration Certificates
   7.8.4.1 In addition to the requirements listed in 7.8.2, calibration certificates shall include the following:
   a) the measurement uncertainty of the measurement result presented in the same unit as that of the measurand or in a term relative to the measurand (e.g., percent);

   NOTE: According to ISO/IEC Guide 99, a measurement result is generally expressed as a single measured quantity value including unit measurement and measurement uncertainty.

   b) the condition (e.g., environmental) under which the calibrations were made that have an influence on the measurement results;

3) **Reporting Measurement Uncertainty When Reporting Sampling**

   f) information required to evaluate measurement uncertainty for subsequent testing or calibration.
   7.8.6 Reporting Statements of Conformity
   7.8.6.1 When a statement of conformity to a specification or standard is provided, the laboratory shall ...

8.10 How to Report Uncertainty

The following is the five-step process to follow when reporting uncertainty in measurement:

1) Record the measurement result;
2) Estimate the uncertainty in measurement;
3) Round uncertainty to two significant figures;
4) Round the measurement result to match the uncertainty;
5) Report the results.

1. Record the Measurement result
The first step to reporting uncertainty is to know the value of the measurement result. Therefore, you must start the process by performing a measurement and recording the result.

Using the example in the GUM, imagine that measure the resistance of a resistor and find its value to be 10.05762 ohms.

Record the resistance value as the measurement result:

$$y = 10.05762\,\Omega.$$

2. Estimate Uncertainty in Measurement

After recording your measurement result, you can estimate the uncertainty in measurement.

Following the recommendations of the ILAC P14, estimate your uncertainty in measurement including these three factors:

- CMC uncertainty;
- UUT resolution;
- UUT repeatability.

After combining these three factors and calculating the expanded uncertainty to 95% where k = 2, you should have a value for calibration uncertainty.

$$U = 0.0272\,\Omega.$$

3. Round Uncertainty to Two Significant Figures

Now that uncertainty has been estimated for the measurement result, it is time to round uncertainty to two significant figures.

To do this, find your first two significant figures. Then, use conventional rounding to round up or down to the nearest number.

$$U = 0.0272\,\Omega \Rightarrow 0.027\,\Omega.$$

4. Round the Result to Match the Uncertainty

Next, round the measurement result to be consistent with the measurement uncertainty. In fact, conventional rounding is the method recommended by the GUM (JCGM 100:2008) [5]. As shown previously in Chapter 2,

> *"In reporting final results, it may sometimes be appropriate to round uncertainties up rather than to the nearest digit... However, common sense should prevail..."*

Since the estimated uncertainty does not give you enough accuracy to justify the value of the measurement result, round the measurement result to match the accuracy of the uncertainty.

$$Y = 10.05762\,\Omega \pm 0.027\,\Omega$$

$$Y = 10.058\,\Omega \pm 0.027\,\Omega.$$

There is no point in reporting a measurement result beyond the number of digits given in the estimated uncertainty. Thus, just round your measurement result to match the number of digits given in your estimated uncertainty. More details in rounding are explained in Chapter 2.

5. Report the Results

The final step is to report the results in your test or calibration certificate. This can be done in the test measurements, a brief statement, or tabulated, especially if there are many to report. Table 8.1 shows a real example from FLUKE.

Measurements can be given with a normal distribution indication lower and upper limit values.

Table 8.1 Sample of uncertainty reporting (Courtesy of Fluke).

FLUKE®	Certificate Number: XXXX-XXXXXXXXXXXXX		Calibration Date: XX-XX-XXXX	
		Calibration Results		
				Manufacturer's Specifications
Function/Range	Nominal Value	Measured Value	Measurement Uncertainty	Lower Limit	Upper Limit
0.000000 V	0.000000	0.0000009	5.8e-007 V	−0.0000020	0.0000020
1.000000 V	1.000000	1.0000004	3.5e-006 V	0.9999890	1.0000110
−1.000000 V	−1.000000	−1.0000009	3.5e-006 V	−1.0000110	−0.9999890
3.290000 V	3.290000	3.2900007	9.3e-006 V	3.2899684	3.2900316
−3.290000 V	−3.290000	−3.2899991	9.3e-006 V	−3.2900316	−3.2899684
33 V Range					
0.00000 V	0.00000	−0.000005	4.4e-006 V	−0.000020	0.000020
10.00000 V	10.00000	10.000006	2.7e-005 V	9.999880	10.000120
−10.00000 V	−10.00000	−10.000032	2.8e-005 V	−10.000120	−9.999880
32.90000 V	32.90000	32.900031	1.0e-004 V	32.899651	32.900349
−32.90000 V	−32.90000	−32.900053	1.0e-004 V	−32.900349	−32.899651

8.11 Fractional Uncertainty Revisited

When a reported value is determined by taking the average of a set of independent readings, the fractional uncertainty is given by the ratio of the uncertainty divided by the average value. For this example, fractional uncertainty is the uncertainty/average:

$$(0.05/21.17) \text{ cm} = 0.0023 \approx 0.2\%$$

Example 8.3 A series of measurements (Fig. 8.1) have been recorded on a system showing the following results $x_1, x_2, ..., x_n$; n_k is a number of random experiments, in which the same result x_k has occurred. The frequency of the result is shown by n_k / n.

8.12 Propagation of Uncertainty

A quantity Q depends on a couple of other quantities a, b, c each one with its uncertainty δa, δb, δc. The uncertainty of Q will depend on the uncertainty of these components in various ways depending on the degree of interrelations between these components. These interrelations constitute the propagation of the uncertainty of Q throughout all components. The related rules are summarized hereafter.

a) Algebraic Addition
If Q has an algebraic addition of components, for example, a, b, and c as in Equation 8.5, then the uncertainty of Q is expressed in Equation 8.6

$$Q = a + b - c, \tag{8.4}$$

$$\delta Q = \sqrt{(\delta a)^2 + (\delta b)^2 + (\delta c)^2}. \tag{8.5}$$

x_k	n_k	n_k/n
5.2	1	0.011
5.3	1	0.011
5.4	2	0.021
5.5	4	0.043
5.6	7	0.075
5.7	10	0.106
5.8	14	0.149
5.9	16	0.170
6.0	13	0.138
6.1	12	0.128
6.2	6	0.064
6.3	4	0.043
6.4	4	0.032
6.5	1	0.011
Total	**95**	

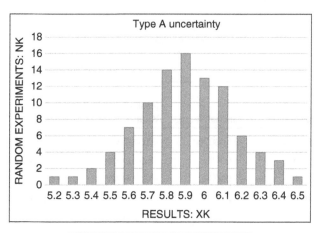

Arithmetic average is $x_m = 5.9$
Standard uncertainty: $u(x) = \sigma = 0.2$

Figure 8.1 Example of uncertainty A.

b) Product or Division

If Q is composed in product and division of components, for instance, a, b, c, and d as in Equation 8.6, then the uncertainty of Q is expressed in Equation 8.7.

$$= \frac{ab}{cd} \tag{8.6}$$

$$\frac{\delta Q}{|Q|} = \sqrt{\left(\frac{\delta a}{a}\right)^2 + \left(\frac{\delta b}{b}\right)^2 + \left(\frac{\delta c}{c}\right)^2 + \left(\frac{\delta d}{d}\right)^2}. \tag{8.7}$$

In the case of multiplication or division of independent measurements, the **relative uncertainty** of the product or quotient is the RSS of the individual **relative uncertainties**. When multiplying **correlated** measurements, the uncertainty in the result is just the sum of the relative uncertainties, which is always a larger uncertainty estimate than adding in quadrature (RSS). Multiplying or dividing by a constant does not change the relative uncertainty of the calculated value.

Example 8.4 Consider the volume of a box having the following sizes: L = 2 ± 0.01 m, h = 3 ± 0.01 m, and t = 5 ± 0.01 m.

$$V = K \times H \times t = 30 \text{ m},$$

$$\delta V = \sqrt{(0.01)^2 + (0.01)^2 + (0.01)^2} = 0.017 \text{ m}.$$

So the volume is expressed as: V = 30 ± 0.017 m or rounding up, V = 30 ± 0.02 m.

c) Power

If Q is expressed as in Equation 8.) with an exact number n, then the uncertainty is given in Equation 8.10.

$$Q = x^n, \tag{8.8}$$

$$\frac{\delta Q}{|Q|} = |n|\frac{\delta x}{|x|}. \tag{8.9}$$

Hence, if $Q = \sqrt{x}$,

$$\delta Q = Q \cdot \frac{1}{2} \frac{\delta x}{x}.$$

Example 8.5 -The period of oscillation is measured to be T = 0.5 ± 0.02 s; what is the uncertainty in the frequency?

The frequency is by definition $f = 1/T = (T)^{-1}$

$$n = -1,$$

$$f = 1/T = 2.0 \text{ Hz}.$$

Applying Equation 8.10, $\delta f = (1)(0.02/0.5)(2) = 0.08$ Hz.
Hence, the frequency can be written as: $f = 2.0 \pm 0.08$ Hz.

d) Generalized Technique: First-Order Average and Uncertainty of Measurement

The quantity to be measured and having the uncertainty assessed through its components can be carried out through the equation of the measurand as an indirect method. The expression of the measurand is treated as explained next considering the first-order truncation of the Taylor's expansion (Eq. 8.10) and applying the definition of the average (Eq. 8.11).

The measurand $g(X)$ is measured indirectly and determined from the quantity X. The Taylor's series expansion gives:

$$g(X) = \underbrace{\frac{(X-a)^0}{0!} \cdot g(a)}_{zero-order} + \underbrace{\frac{(X-a)^1}{1!} \cdot \frac{\partial g}{\partial X}(a)}_{first-order} + \underbrace{\frac{(X-a)^2}{2!} \cdot \frac{\partial^2 g}{\partial X^2}(a)}_{second-order} + \ldots + \underbrace{\frac{(X-a)^n}{n!} \cdot \frac{\partial^n g}{\partial X^n}(a)}_{n^{th}-order} + R_n$$

(8.10)

where a is a constant about which the expansion is carried out. R_n is a reminder term known as the Lagrange remainder, defined by [6] as:

$$R_n = \int \ldots \int_a^x g^{(n+1)}(X)(dX)^{n+1}$$

With respect to the current investigation, it seems obvious that the expansion should be carried out about the mean (i.e., central value). The average of measurement is obtained through the expectation value defined by the first statistical moment as:

$$x = E(X) = \int_{-\infty}^{\infty} x \cdot f_X(x) \cdot dx,$$

(8.11)

where $f_X(x)$ is the probability density function (PDF).

The standard deviation is known to be the square root of the variance, which in turn is written as the second statistical moment:

$$Var(X) = E(X - E(X))^2 = \int_{\infty}^{-\infty} (x - \mu_x)^2 \cdot f_X(x) \cdot dx$$

(8.12)

where $\mu_x \equiv E(X)$. It also expresses the uncertainty of measurement $u(x)$ defined in GUM [6], and the standard deviation σ_x, hence:

$$Var(X) = \sigma_x^2 = u^2(x)$$

(8.13)

Evaluating the expectation on either side of (Eq. 8.11) truncated to the first order will give

$$y = g(x)$$

(8.14)

where x is the best estimate of measurement (i.e., average of measurement).

The combined variance and standard uncertainty of $g(X)$ are calculated by subtracting $E(g(X))$ from either sides of Equation 8.11 truncated to the first order, squaring both sides and writing the expectation from either side to finally obtain in terms of uncertainty:

$$u_c^2(y) = \left(\frac{\partial g}{\partial x}\right)^2 \cdot u^2(x) \tag{8.15}$$

hence,

$$u_c(y) = \left(\frac{\partial g}{\partial x}\right) \cdot u(x) \tag{8.16}$$

or

$$\delta Q = \left|\frac{dQ}{dx}\right| \cdot \delta x. \tag{8.17}$$

This propagation of uncertainty in Equation 8.18 can be visualized graphically as shown in Figure 8.2 of the variable x within the function g. The uncertainty $u(x)$ induces an uncertainty $u(y)$ of the function.

For several measured $x_1, x_2..., x_n$, the maximum uncertainty of y can be calculated as:

$$\Delta y = \left|\frac{\partial y}{\partial x_1}\right||\Delta x_1| + \left|\frac{\partial y}{\partial x_2}\right||\Delta x_2| + ... + \left|\frac{\partial y}{\partial x_n}\right||\Delta x_n| \tag{8.18}$$

Example 8.6 If $Q = A \cdot x$, the uncertainty $\delta Q = |A| \cdot \delta x$ similar to Equation 8.5.

Example 8.7 If $Q = cos\,(\theta)$,

$$\delta Q = |\sin(\theta)| \cdot \delta\theta.$$

Example 8.8 Find the uncertainty of $\sin(\alpha)$ in radian if the measured angle in degrees with uncertainty is $\alpha = 15 \pm 1^o$

$$\sin(\alpha) = 0.258819$$

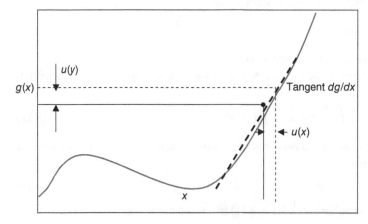

Figure 8.2 Propagation of error with respect to variable x.

The uncertainty of $\sin(\alpha)$ is:

$$\delta(\sin(\alpha)) = \left| \frac{d}{d\alpha} \sin(\alpha) \right| \cdot \delta\alpha$$

$$= |\cos(\alpha)| \cdot \delta\alpha$$

$$\delta\alpha = 1^\circ = 0.0174532 \, rad$$

$$\delta(\sin(\alpha)) = |\cos(15)| \times 0.0174532 = 0.9659258 \times 0.0174532 = 0.0168584$$

Hence, $\sin(\alpha) = 0.258819 \pm 0.0168584$

Example 8.9 The volume of a cube is determined by $V_c = x^3$, the length x has an uncertainty of δx. Establish the equation leading to calculate the uncertainty of the volume.

Based on Equation 8.19:

$$\delta V_c = 3x^2 \delta x \tag{8.19}$$

e) Function with Two or More Variables

Given a functional relationship between several measured variables (x, y), $Q = f(x, y)$, the uncertainty in Q if the uncertainties in x, y are known is as follows:

The variance in Q:

$$\sigma_Q^2 = \sigma_x^2 \left(\frac{\partial Q}{\partial x} \right)^2 + \sigma_y^2 \left(\frac{\partial Q}{\partial y} \right)^2 + 2\sigma_{xy} \left(\frac{\partial Q}{\partial x} \right) \left(\frac{\partial Q}{\partial y} \right). \tag{8.20}$$

If the variables x and y are uncorrelated, $\sigma_{xy} = 0$; therefore:

$$\sigma_Q^2 = \sigma_x^2 \left(\frac{\partial Q}{\partial x} \right)^2 + \sigma_y^2 \left(\frac{\partial Q}{\partial y} \right)^2 \text{ where } \sigma_Q = \sqrt{\sigma_x^2 \left(\frac{\partial Q}{\partial x} \right)^2 + \sigma_y^2 \left(\frac{\partial Q}{\partial y} \right)^2} \tag{8.21}$$

For more variables $(x, y, ..., z)$ if independent (uncorrelated):

$$\delta Q = \sqrt{\left(\frac{\partial Q}{\partial x} \cdot \right)^2 \delta x^2 + \left(\frac{\partial Q}{\partial y} \right)^2 \delta y^2 + \, ... \left(\frac{\partial Q}{\partial z} \right)^2 \delta z^2} \tag{8.22}$$

Q at the Average Values If x and y have several measurements: $x(x_1, x_2, x_3, ..., x_n)$ and $y(y_1, y_2, y_3, ..., y_n)$, the average of x and y will be:

$$\mu_x = \frac{1}{n} \sum_1^n x_i \text{ and } \mu_y = \frac{1}{n} \sum_1^n y_i. \tag{8.23}$$

The evaluation of Q at the average values: $Q \equiv f(\mu_x, \mu_y)$

The expansion of Q_i about the average using Taylor expansion (see Eq. 8.11):

$$Q_i = f(\mu_x, \mu_y) + (x_i - \mu_x) \left(\frac{\partial Q}{\partial x} \right) \bigg|_{\mu_x} + (y_i - \mu_y) \left(\frac{\partial Q}{\partial y} \right) \bigg|_{\mu_y} + higher\,order\,terms \tag{8.24}$$

If measurements are close to the average values:

$$Q_i - Q = (x_i - \mu_x) \left(\frac{\partial Q}{\partial x} \right) \bigg|_{\mu_x} + (y_i - \mu_y) \left(\frac{\partial Q}{\partial y} \right) \bigg|_{\mu_y}$$

and its variance is:

$$\sigma_Q^2 = \frac{1}{n} \sum_{i=1}^n (Q_i - Q)^2, \tag{8.25}$$

or

$$\sigma_Q^2 = \sigma_x^2 \left(\frac{\partial Q}{\partial x} \right)_{\mu_x}^2 + \sigma_y^2 \left(\frac{\partial Q}{\partial y} \right)_{\mu_y}^2 \tag{8.26}$$

for uncorrelated variables.

Higher-order expressions for average and uncertainty are developed for two variables in [7, 8].

Example 8.10 Calculate the variance in the power using error propagation:$P = I^2R$ with $I = 1.0 \pm 0.1$ A and $R = 10.0 \pm 1.0 \, \Omega$.

Ans. The power is calculated; $P = 10.0$ W.

$$\sigma_P^2 = \sigma_I^2 \left(\frac{\partial P}{\partial I} \right)_{I=1}^2 + \sigma_R^2 \left(\frac{\partial P}{\partial R} \right)_{R=10}^2$$

$$= \sigma_I^2 (2IR)^2 + \sigma_R^2 (I^2)^2$$

$$= (0.1)^2 (2 \cdot 1 \cdot 10)^2 + (1)^2 (1^2)^2 = 5W^2$$

Hence, $P = 10.0 \pm 2.2$ W.

If the true value of P is 10.0 W measured several times with an uncertainty of ± 2.2 W, from the previous statistics (e.g., normal) 68% of the measurements would lie in the range [7.8, 12.2] W.

Example 8.11 The variables x and y were measured for the quantity $Q = x^2y - xy^2$ as follows: $x = 3.0 \pm 0.1$ and $y = 2.0 \pm 0.1$. What is the value of Q and its uncertainty?

Ans. Q = 6.0 ± 0.9.

Example 8.12 An object is placed at a distance p from a lens and an image is formed at a distance q from the lens; the lens's focal length is expressed as follows:

$$f = \frac{pq}{p+q}$$

Using the general rule for error propagation, derive a formula for the uncertainty δf in terms of p, q and their uncertainties. Calculate the quantity f and its uncertainty if $p = 10.1 \pm 0.3$ and $q = 8.0 \pm 0.5$.

Multiple Choice Questions of this Chapter

Multiple Choice Questions are given for each chapter with solutions in an online extension of this book. Please use link: www.wiley.com\go\mekid\metrologyandinstrumentation\

References

1 Slocum, A.H., *Precision Machine Design*, 1992, Prentice Hall, Englewood Hill.

2 Mekid, S., *Introduction to Precision Machine Design and Error Assessment*, 2008, CRC Press, pp. 1–355.

3 Mekid, S., and Olejniczak, O., "High precision linear slide. Part 2: Control and measurements," *Int. Journal of Machine Tools and Manufacture* 40(7), 2000, pp.1051–1068.

4 "CIPM," *BIPM Proc.-Verb. Com. Int. Poids et Mesures* (49)8-9, 26, 1981 (in French); Giacomo, P. "News from the BIPM." *Metrologia* 18, 1982, pp. 41–44.

5 ISO, "Guide to the Expression of Uncertainty in Measurement," International Organization for Standardization, 1993, Geneva, Switzerland.

6 Blumenthal, L.M., "Concerning the remainder term in Taylor's formula," *Amer. Math. Monthly* 33, 1926, pp. 424–426.

7 Mekid, S., and Vaja, D., "Propagation of uncertainty: expressions of second and third order uncertainty with third and fourth moments," *Measurement*, Vol.41/6. 2008, pp. 600–609.

8 Mekid, S., "Design strategy for precision engineering: second order phenomena," *J. Engineering Design*, 16(1), 2005, pp. 63–78.

9

Dimensional Measurements and Calibration

9.1 Length Measurement

Measurement of length is very comprehensive and includes distance, displacement, position, dimensions, size, area, volume, and surface texture. This is one of the fundamental requirements in engineering and manufacturing industries and a critical part of setting up and operating many science experiments and facilities. The international expertise in dimensional measurement covers more than 12 orders of magnitude. This is ranging from the highest resolution measuring system based on x-ray interferometry (20 picometer accuracy) to a large volume, long range capability up to 100s of meters.

The chapter will cover some of instruments for displacement and length measurements. From length measurement calibration of instruments such as micrometers, calipers, gauges, or tape measures, to high-tech optics-based scales and comparators, the industry leading dimensional instrument calibration capabilities are available and well designed to reduce risk and inaccuracy in measurements. The following description will cover most common portable measuring devices and manual gauges.

9.2 Displacement Measurement

Rather than length, displacement is important to be measured in any mechanical system. Sensors that are used can be divided into contact and noncontact types. For any sensor to be used, a set of general characteristics are defined:

1) **Resolution**: the smallest change that can be detected by the sensor.
2) **Repeatability**: see whether the sensor gives the same output signal when it is subject to the same input.
3) **Accuracy**: sensor has to give for each input, an output signal, which measures the exact quantity variation.
4) **Frequency response**: effect on the output signal when the physical quantity varies in time.
5) **Linearity**: proportional constant variation between the output signal and the input.
6) **Step response**: the variation in time of the output signal.
7) **Delay time**: the time the sensor output needs to rise 10% of the nominal peak value.
8) **Rise time**: the time the sensor output needs to rise from 10% to 90% of the nominal peak value.

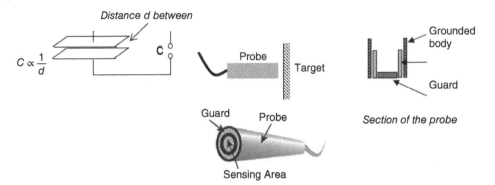

Figure 9.1 Capacitance sensor.

To measure displacement, the following techniques can be used:

A) Capacitance;
B) Linear/rotative variable differential transformer (LVDT, RVDT);
C) Laser interferometry.

A- Capacitance A capacitance sensor measures the distance between the probe and the target surface. The varying distance between plates induces variable capacitance. A wide variety of materials could be used as target surface after calibration of the sensor (Fig. 9.1).

The capacitance sensor is used for small displacement and the used range of capacitance is 90–0.01 pF. Some characteristics are given in Table 9.1.

$$C \propto \frac{1}{d}$$

The capacitance between terminals is given by:

$$C = \frac{kvA}{d}, \tag{9.1}$$

where

k: dielectric (permittivity) of free space (vacuum) of magnitude 8.85 pF/m
v: dielectric constant of the medium
A: overlapping surface of the two plates [m^2].

For a very small displacement of one of the plates, the capacitance will change as:

$$C = \frac{kvA}{d} \text{ to } C_\Delta = \frac{kvA}{d + \Delta}. \tag{9.2}$$

The relationship between the displacement and change in capacitance is obviously nonlinear, but for small changes in separation, the capacitance can be approximated through a Taylor series expansion. In general, any function, $f(d)$ can be approximated in the neighborhood of some nominal value as follows:

$$f(do + \Delta) = fo(d) + \Delta \frac{\partial f(do)}{\partial d} + \frac{\Delta^2}{2} \frac{\partial^2 f(do)}{\partial d^2} + \dots \tag{9.3}$$

The capacitance will be then:

$$C = \frac{kvA}{d} \left(1 - \frac{\Delta}{d} + \frac{\Delta}{d^2} \right), \tag{9.4}$$

where $\Delta = d - do$. For $\Delta \ll d$, the capacitance change is linear with respect to displacement.

Table 9.1 Capacitance characteristics.

	Capacitance
Resolution	Order of 1 angstrom
Measuring range	±0.15 mm
Frequency domain	30–40 kHz

Sensitivity

$$\frac{\partial C}{\partial d} = -\frac{k\upsilon A}{d^2}.$$ (9.5)

With the variation of d and fixed A, the sensitivity is defined as $S \equiv \frac{\Delta C}{\Delta d}$ Therefore, $S = -\frac{k\upsilon A}{d^2}$

Example: For a sensor having a width of 10 mm and d = 0.2 mm, the typical sensitivity is around 0.4425 pF/mm.

The characteristics of such sensors are briefly shown in Table 9.1

B. Linear/Rotative Variable Differential Transformer (LVDT, RVDT) This is a popular transducer based on a variable-inductance principle to sense linear displacement (10–20 cm) and rotary motion (one revolution). It consists of a magnetic core sliding with 3 coils: an AC-driven primary coil and two secondary coils connected in opposition. The voltage is proportional to displacement for AC and needs a phase sensitive displacement (PSD). The linearity error is around 0.5–1% (Fig. 9.2).

The position of the magnetic core controls the mutual inductance between the center of the primary coil and the two outer, or secondary, coils. The imbalance is developed in mutual inductance between the center location and an output voltage. The frequency applied to the primary coil can range from 50 to 25000 Hz. If the LVDT is used to measure dynamic displacements, the carrier frequency should be 10 times greater than the highest frequency component in the dynamic signal. In general, highest sensitivities are attained at frequencies of 1 to 5 kHz. The input voltages range from 5 to 15 V. Sensitivities usually vary from 0.02 to 0.2 V/mm of displacement per volt of excitation applied to the primary coil. The actual sensitivity depends on the design of each LVDT (Table 9.2).

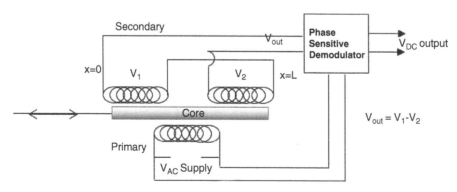

Figure 9.2 LVDT sensor

Table 9.2 LVDT, RVDT characteristics.

	LVDT, RVDT
Resolution	0.1 μm, 16 arcsec
Measuring range	0.1 m→1 m
Frequency domain	400 Hz – 10 kHz

C. Laser Measurements As part of the optical measuring systems, lasers are becoming part of many length and levels measurement systems. They are the improvement of the optical techniques especially when the lasers are becoming more stable and easily manipulated by optic lens combinations associated with measurement methods that are very crucial. These are, for example, triangulation, telemetry, and interferometry techniques. The latter was largely discussed in the previous chapters. We will discuss Michelson interferometry as a fundamental technique.

C.1 Michelson Interferometry The most accurate measurements in science are made possible by exploiting the interference properties of waves. Since waves depend on both time and spatial coordinates, both time and distance are measured this way. For example, the distance between the Earth and the Moon is determined to an accuracy of about one inch out of almost a quarter of a million miles by measuring the time taken for laser pulses to travel to the moon and back, after being reflected by a mirror on the Moon's surface.

The Michelson interferometer (Fig. 9.3) was first used to discover that the speed of light is independent of the observer's velocity relative to the light source. The interferometer can detect very small movements (on the order of nanometers). This asset will be used to our advantage; we will use the interferometer to measure wavelengths of light and to determine the index of refraction of air.

C.2 Theory The light of wavelength λ from the source is incident on a beam splitter, which reflects half of the light to mirror 1. The other half of the light is transmitted to mirror 2. The light beams reflected from mirror 1 and mirror 2 recombine at the half-silvered mirror.

If the difference in lengths of the two paths is an integral number of wavelengths, then constructive interference results. Thus, whenever mirror 2 moves by one-half wavelength ($\lambda/2$), the distance

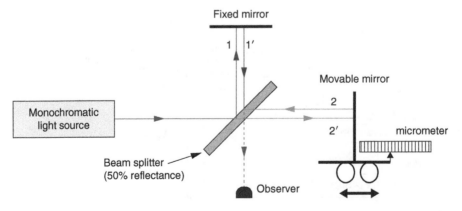

Figure 9.3 Michelson interferometer.

traveled by the light in that arm of the interferometer changes by λ, and therefore the interference pattern changes from a maximum to a minimum and back to a maximum as the mirror is moved: "one fringe has gone by." In other words,

$$2\Delta\ell_2 = \Delta n\,\lambda, \tag{9.6}$$

where:

$\Delta\ell_2$ is the distance the mirror 2 moves, and
Δn is the number of fringes that pass by.

To determine the wavelength of light, we will count a given number of fringes passing by (Δn), measure the distance the mirror moved ($\Delta\ell_2$), and calculate λ.

When air is evacuated from a length L of one arm of the interferometer, the optical path length traveled by the light in that arm changes by $2L(n_{air} - n_{vacuum})/\lambda = 2L(n_{air} - 1)/\lambda$. There is no mirror movement involved in this change of optical path length since only the material (air) in that arm of the interferometer is changing. Using Equation 9.6, the corresponding number of fringes passing by is given by $\Delta n\lambda = 2L(n_{air} - 1)$. By knowing the wavelength of light used and the length of the evacuable cell and by counting the number of fringes passing by as the air is slowly removed, we can calculate the index of refraction of air.

When two distinct wavelengths are present in the light beam, each one can interfere with itself as discussed above (i.e., due to path length differences in the two arms), or they can interfere with each other within a single arm. The different wavelengths will interfere constructively when the optical path of the shorter wavelength light is longer than the optical path of the longer wavelength light by an integer, m. That is, if $\lambda_2 > \lambda_1$, then the two different waves interfere when $\ell_2/\lambda_1 - \ell_2/\lambda_2 = m$. This results in the fringes appearing sharp.

Destructive interference will occur (the fringes will become fuzzy) if the optical path difference is an odd multiple of 1/2. If moving the mirror by $\Delta\ell_2$ changes the pattern from sharp to fuzzy and back to sharp, then

$$2\Delta\ell_2 = \lambda_1\lambda_2/(\lambda_2 - \lambda_1). \tag{9.7}$$

When the two wavelengths are nearly identical, a good approximation to Equation 9.7 is $2\Delta\ell_2 = \lambda_{av}^2/2\Delta\lambda$, where λ_{av} is the average of the two wavelengths, and $\Delta\lambda$ is their difference.

C-3- Calibration Mirror 2 is moved by a lever arrangement. A micrometer is attached that indicates the distance the lever moves. The mirror actually moves a much smaller distance, which is directly proportional to the distance the lever moves. In other words, once you take a distance measurement from the micrometer, you need to multiply by some constant to find the actual distance moved by the mirror. This constant is called K, and for this interferometer, K is about 0.2.

To determine K we must have a source of light for which the wavelength is known. We will use the He-Ne laser. When using the laser with the interferometer, never look directly into the instrument, but use the white screen provided to view the interference pattern. Shine the laser into the interferometer and place the white screen where you would normally look into the apparatus. You should see two images, one from each of the reflecting mirrors. By adjusting the screws on the back of mirror 1 you should be able to adjust the position of one of the images. Adjust the mirror until the images converge. Now insert a lens in front of the laser to provide a diverging light source. You should now see an interference pattern of bright and dark fringes on the screen. By adjusting mirror 1 you should be able to find the center of the interference pattern with alternating bright and dark circular fringes, as shown in Figure 9.4.

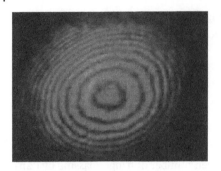

Figure 9.4 Interference pattern.

Table 9.3 Laser interferometry characteristics.

	Laser Interferometry
Resolution	≈ 12 Å (optics + env.?)
Measuring range	Up to 30 m
Frequency domain	Up to 1.8 m/s

Take an initial reading of the micrometer (the micrometer is marked off in millimeters). Then, slowly turning the micrometer, watch as several fringes pass by. Count 50 fringes pass by and again record the micrometer reading. Do this twice. Now, given $\lambda_{laser} = 632.8$ nm, the distance difference from the micrometer measurements, and the number of fringes, we can calculate K from Equation 9.6 by substituting $\Delta\ell_2 = K\,\Delta m$, where Δm is the difference in micrometer readings.

Now that we know K, we can use other light sources and, by counting fringes and taking micrometer readings, calculate wavelengths. General characteristics are shown in Table 9.3.

9.3 Manual Instruments

9.3.1 Caliper

The caliper in Figure 9.5 is a device used to measure the distance between two opposite sides of an object.

 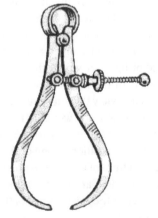

Figure 9.5 A caliper

9.3.2 Vernier Caliper

The Vernier caliper (Fig. 9.6) is a measuring instrument made in an L-shaped frame with a linear scale along its longer arm and an L-shaped sliding attachment with a Vernier, used to read directly the dimension of an object represented by the separation between the inner or outer edges of the two shorter arms. Analog and digital Vernier caliper exist in the market. The error of measurement is around 20 μm depending on the length.

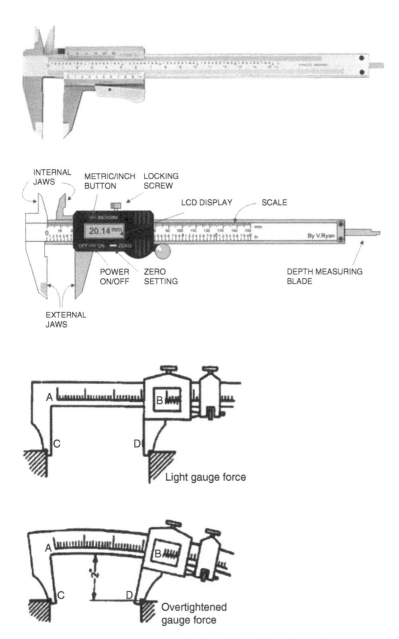

Figure 9.6   a) Vernier caliper and digital caliper. b) Associated Abbe error.

Figure 9.7 A micrometer (courtesy of Mahr).

The possible known associated error is the Abbe error (Fig. 9.2b) defined as follows. It is better to have the scale directly on the measurement line where the effects of pitch errors in the instrument are greatly reduced or eliminated. The micrometer has its scale (lead screw) aligned with the center of the contacts. This is applied to all instruments having the measuring probe away from the measurement scale.

9.3.3 Micrometer

To obtain very fine length measurements, a micrometer as a precision measuring instrument is used and is available in both metric and imperial versions (Fig. 9.7). Metric micrometers typically measure in 0.01 mm increments and imperial versions in 0.001 inches. The measurements are more accurate than those given by previous devices.

9.3.4 Feeler Gauge

A feeler gauge is a handheld measuring tool (Fig. 9.8). It is composed of a number of folding metal strips called blades that are machined to specified thickness levels. Measurements are given in millimeters or inches and can be found in a central location on each individual strip. Several thicknesses of strips made of steel are given with specific values ranging from 0.02 mm (0.0008") to as large as 5.08 mm (0.200"). The value of the size of each strip is engraved or printed with the corresponding fixed value.

Figure 9.8 Feeler gauge with multiple stripes.

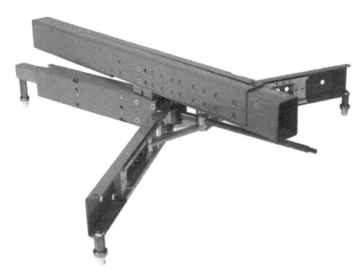

Figure 9.9 Liner Measurement Tool.

9.3.5 Liner Measurement Tool

The liner diameter measuring instrument, known as LDM (Fig. 9.9), is used to determine liner diameters in a fast and easy way in a fraction of time compared to traditional methods. In practice, a team of 3 people can measure two-cylinder liners in one long working day. If this system is used, a single operator can do the same operation in three hours. Wear and deformation can also be checked.

9.3.6 American Wire Gauge

The American wire gage (Fig. 9.10) is an index that shows, indirectly (inversely and logarithmically), the cross-sectional area of a round wire. The measurement of this area is straightforward in case of solid conductors.

$$A = \pi.r^2. \tag{9.8}$$

This area A is called circular MIL area and is often used instead; one circular mil is the area of a circle with a diameter of one mil (1/1000 inch), and the circular mil area of a solid wire, consequently, is always the diameter (2r) of the wire, in mils, squared.

9.3.7 Bore Gauge

Figure 9.11 shows a dial bore gauge defined as a comparative instrument having a digital or analog readout and is similar to a telescoping gauge. The first step is to set the dial bore gauge to the nominal value of the bore, and it will measure the variation and direction of the bore from nominal. Several methods exist to set these gauges to the nominal value. The most common method is using an outside micrometer that is set to the nominal value.

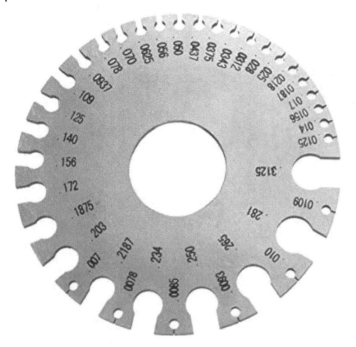

Figure 9.10 Wire gage.

Figure 9.11 Dial and digital bore gauge.

9.3.8 Telescopic Feeler Gauge

A telescopic gauge (Fig. 9.12) is a handheld measuring instrument that has retractable rods to secure a precise measurement. It is used by mechanics to measure the size of a bore in an engine. A bore refers to the diameter measurement of the cylinder where the pistons are positioned and need precision measurement of the diameters.

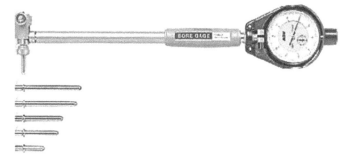

Figure 9.12 Telescopic gauge.

9.3.9 Depth Gauge

A depth gauge (Fig. 9.13) is like a pressure gauge displaying the equivalent depth in water. Most modern diving depth gauges have an electronic mechanism and digital display. The older types used mechanical mechanisms and analog displays.

9.3.10 Angle Plate or Tool

An angle plate (Fig. 9.14) is a work holding device that is used as a fixture in manufacturing. The angle plate is made from high quality material that has been stabilized to prevent further distortion. Slotted holes or "T" bolt slots are machined into the surfaces to enable a secure clamping of

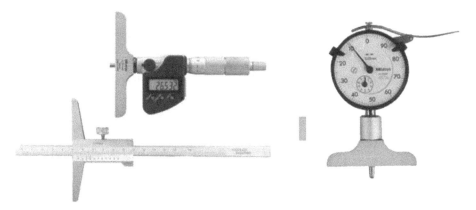

Figure 9.13 Digital and dial depth gauge.

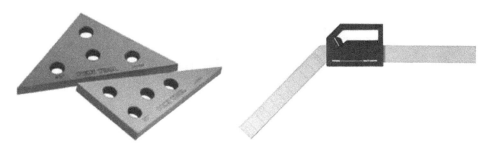

Figure 9.14 An angle plate.

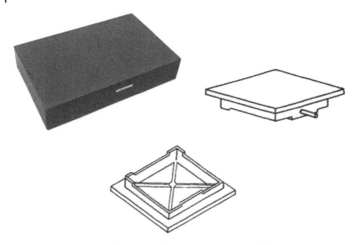

Figure 9.15 Flat plate. Cast iron surface plate, the underside.

workpieces to the plate and the plate to the worktable. More details for angle measurements are given later.

9.3.11 Flat Plate

This is also called a reference granite table (Fig. 9.15). It is a flat plane used as the foundation of all geometric accuracy and indeed of all dimensional measurements. It is usually provided in the workshops and inspection laboratories by the surface plate.

9.3.12 Dial Gauge

A dial gauge (Fig. 9.16) is a circular instrument with a graduated dial and a pointer actuated by a member that contact with the part being calibrated. It can be analog or digital.

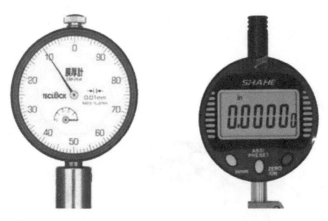

Figure 9.16 Angle plate dial and digital.

The basic parts of a dial indicator are:

a) The bezel used to rotate;
b) The zero to align with the needle markers used to provide reference points and rarely used by the mechanic point;
c) Circular graduated dial with pointer actuated by a member contacting the part being measured.

9.3.13 Oil Gauging Tapes

Oil gauging tapes are used to manually measure liquid levels in storage tanks. They are also known as oil gauging Derrick tapes (manufactured by US Tape) and are strong and durable with a corrosion-resistant frame (Fig. 9.17).

9.3.14 Thread Measurement

The thread pitch is known to be the distance between two threads in a bolt and is expressed in millimeters without putting the unit, for example, a thread pitch of 2.5 means the distance between the thread is 2.5 mm. there are two systems, the imperial and metric thread. The popular metric threads are M.5, M.75, M.9, M1.0, and M1.25, while the imperial threads are, for example, 50 TPI and 25 TPI (Fig. 9.18). Alternatively, an automatic inspection using CMM is also available (Fig. 9.18).

9.3.15 Planimeter

A planimeter or platometer (Fig. 9.19) is a measuring instrument used to determine the surface of any arbitrary two-dimensional (2D) shape.

Figure 9.17 Oil Gauging Tapes.

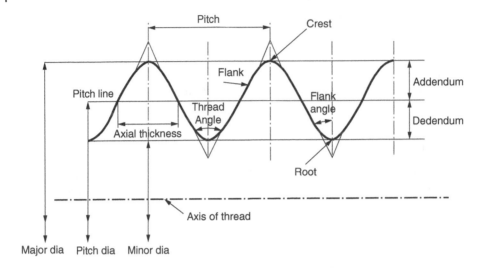

Figure 9.18 Thread definition, metric manual thread device and CMM thread inspection. *Source:* ssp48/Envato Elements Pty Ltd.

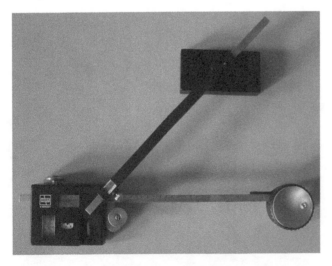

Figure 9.19 Planimeter. *Source:* Stefan Kühn/Wikimedia commons

9.4 Diameter and Roundness

It is obvious that diameters are important to be measured but there is a need to make sure a diameter is properly measured. Also, this measurement is tightly associated with out-of-roundness (OOR).

The measurement of roundness is of critical importance for many applications. There are several measurement methods from simple caliper measurements to highly accurate dedicated roundness measurement systems.

The simplest approach to determine the roundness of a component is to measure the consistency of its diameter at different orientations (Fig. 9.20). This is a practice carried out in-process for checking machine setup and can be adequate for assessing a component where the roundness is an optional, rather than functional, requirement. It can be functionally relevant, of course.

Roundness is defined in ISO 1101 as the separation of two concentric circles that just enclose the circular section of interest. It is clear that a measurement of a diameter as shown above will not yield the roundness of the component in accordance with this definition.

9.4.1 How to Measure a Diameter?

It is wrongly believed that it is sufficient to measure the diameter of a workpiece in several places meaning that the difference in readings will be giving an estimate of the out-of-roundness and a better value of the diameter. A good example is the measuring of a pentagon shape as in Figure 9.21.

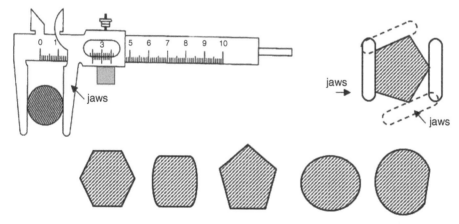

Figure 9.20   External diameter measurement.

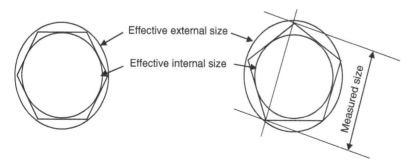

Figure 9.21   External diameter measurement.

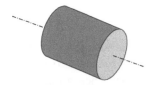 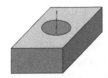

Figure 9.22   External and internal roundness.

The caliper reading when measuring the pentagon is identical irrespective of the shape's orientation, and yet it can be seen to be "out-of-round." Several shapes resembling a circle can be measured this way, as shown in this same figure.

Hence, it is not possible to rely on the measurement of out-of-roundness. It is important to understand what to measure, is it an inscribed or circumscribed circle? Figure 9.18 gives an example of measured size and effective size.

9.4.2   Roundness

Roundness is defined in ISO 1101 as the separation of two concentric circles that just enclose the circular section of interest. As previously explained, it is clear that a measurement of a diameter as shown above will not yield the roundness of the component in accordance with this definition. Manufactured components can always have elements that are round externally in cross section, for instance, a shaft and a hole (Fig. 9.22). Such simple geometries are widely used.

Among the most important fundamental form found in parts is the circular cross section. Circular forms exist in many applications, particularly in bearing surfaces such as rotating shafts and ball bearings.

Typical applications of a rotating shaft in a bearing housing forming a hydrodynamic bearing become then very sensitive to roundness and other factors since the clearance must be constant to some extent and within specific tolerances (Fig. 9.24). The measurement of out-of-roundness is an extremely important assessment. For example, a rotational bearing (Fig. 9.23) whose components have high roundness values will tend to be noisy or even nonrotating, leading to early failure. An accurate roundness measurement is therefore vital as functional measurement to ensure correct function of such parts.

Out-of-roundness can be measured using a V-block with three points or a coordinate measuring machine (CMM) or via a rotational datum method. Recently, noncontact measurement methods have been developed to measure diameters and out-of-roundness.

Out-of-roundness can be measured as follows:

a) The V-block measurement of roundness is based on a three-point method with several variations in the procedure to use it.
b) Large bores can be checked with an inverted arrangement, as shown in Figure 9.4, and can be moved around to collect data.

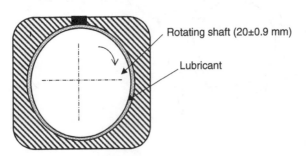

Rotating shaft (20±0.9 mm)

Lubricant

Figure 9.23   Hydrodynamic bearing with internal rotating shaft.

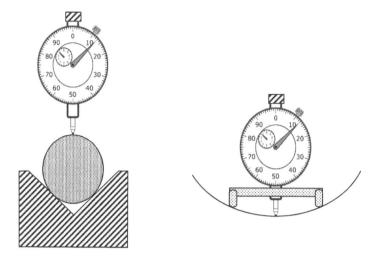

Figure 9.24 Out-of-roundness measured in (a) outer diameter and (b) inner diameter.

c) Large diameters can also be checked using the previous inverted arrangement.
d) Long shafts will need two similar V-blocks with the dial gauge positioned between them while the readings obtained will be further influenced by sag and lack of straightness of the shaft.

In general, the three-point method will always suffer from the limitation that the results may vary according to the V angle and the spacing of the irregularities. It may give a false impression when the undulations are regularly spaced, as each one influences the dial gauge three times as the part rotates; once when an undulation passes under the plunger and once when it contacts each arm of the V.

Coordinate Measuring Machine
The second possibility to measure out-of-roundness is to use a coordinate measuring machine (CMM), shown in Figure 9.25. This machine is usually very accurate from the fact that it is designed for dimensional measurements. The touch probe is used to collect points with coordinates around the shaft, a bore, and so forth. This helps to measure diameters, roundness, cylindricity, circularity, and so on.

Rotational Datum Method
To properly assess the roundness of good quality components, the best reference is a rotational datum. The part to be assessed is placed on a rotary table, and the movement of the surface toward or away from a gauge is monitored as the part is rotated (Fig. 9.26). To measure roundness, rotation is necessary coupled with the ability to measure change in radius. This is best achieved by comparing the profile of the component under test to an accurate circular datum.

The most precise way to calculate the roundness of a part is to use a scanning probe that stays in contact with the surface while collecting a high-density of data points for postprocessing to measure the variance of the radius from an accurate rotational datum. This approach involves fitting a circle to the data and calculating the roundness based on the component core.

There are a variety of roundness measurement instruments available. A device with a revolving table on which the part is placed is the most common configuration. A radial arm with a gauge mounted on it can be modified to bring the gauge into contact with the part. The arm is placed on

Figure 9.25 Hydrodynamic bearing with internal rotating shaft. *Source:* ssp48/Envato Elements Pty Ltd.

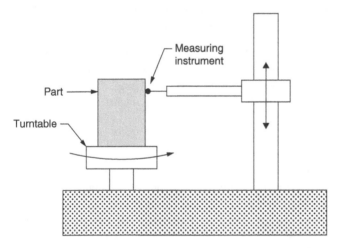

Figure 9.26 Hydrodynamic bearing with internal rotating shaft.

a column that allows the measurement plane's height to be changed. The advantages of such an instrument is that they can measure roundness extremely accurately in a short measurement time.

Case Study Measurement

On a precise lathe machine, a workpiece with several diameters was turned and tested for diameter and out-of-round measurements (Fig. 9.5). Taylor Hobson Limited uses a Talyrond 385 automatic centering and leveling spindle, as well as special styli (155/P56560) for workpiece measurement. The Talymin 5 Gauge is the gauge used on this stylus-based instrument, and it has a minimum resolution of 1.2 nm over a range of 100 nm.

The roundness results are compared to those of a high-quality ultra-precision air bearing spindle, and the radial limit of error (height above the table) is described in the following way ±(0.02 µm + 0.0003 µm/mm). The setup of a workpiece on the machine is shown in Figure 9.27.

Figure 9.27 Setup of the workpiece on the Talyrond machine.

The workpiece is positively secured on a three jaws chuck utilizing the workpiece hole (Fig. 9.23). It is also centered and leveled over the 1st and 3rd cylinders. Abnormal asperities (high peaks) or very low valleys on a raw profile are automatically removed, and the roundness of the workpiece is analyzed using a filter of 1–50 undulations per revolution (upr). Figure 9.24 shows an example of the results obtained from Taylor Hobson. Extensive roundness tests have been performed to validate the performance of the probe.

The results are shown in Table 9.4. The measurements made using the laser-based probe are averages of at least ten measurements (Figure 9.28). The out-of-round results are compared with the results obtained from Taylor Hobson. The probe performance gives a maximum error of 0.5 μm with an uncertainty of 1 μm for roundness measurement. The results show that the principle of roundness measurement of the developed probe seems applicable to measure the OOR of the rotating workpiece and has errors within 0.5 μm. The cylindricity is shown in Figure 9.29.

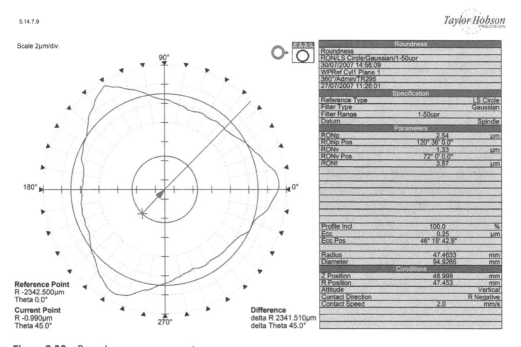

Figure 9.28 Roundness measurement.

Table 9.4 Results of roundness measurements.

The results for roundness measurement of test workpieces.

Workpiece	OORs measured by the probe using the calibrated constant (µm)	OORs measured by Taylor Hobson (µm)	Error between the OORs measured by the probe and Taylor Hobson (µm)
#1	1.68	1.43	0.25
#2	1.77	1.73	0.04
#3	2.45	2.88	−0.43

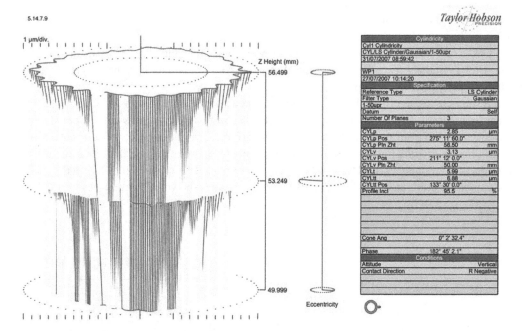

Figure 9.29 Cylindricity measurement

In-Process Measurement of Component Roundness

a. Principle of Measurement   The principle of roundness measurement is centered on the light intensity that varies with the displacement from the reflective object. With the aim to measure the OOR of the part, the configuration of the current in-process probe for radius measurement can be simply reconfigured by blocking one beam for adapting itself to measure the displacement (Fig. 9.30). A least-square circle (LSC) is applied to evaluate the OOR.

Principle of Measurement   The principle of the probe is based on the laser differential Doppler technique. The radius equation is given by

$$r = \frac{15\lambda f_d}{\pi N \sin \theta} \tag{9.9}$$

where f_d is the Doppler frequency, θ is the angle between a beam and the optical axis, N is the rotational speed, and λ is the laser wavelength.

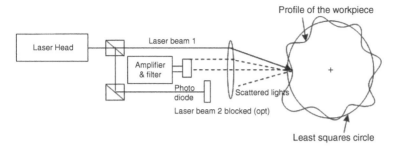

Figure 9.30 Configuration for roundness measurement

b. Experimental Results and Discussions The output signal of the reconfigured probe is voltage. The reconfigured probe was calibrated with a reference workpiece to define a calibration constant for converting the output voltage to displacement. The calibration constant for the probe is 1.061. Any errors that deviate from this constant are considered OOR as defined in ISO 12181. The OOR results of the probe were compared with those of the CMM. The errors are 0.12 μm, 0.2 μm, and 0.16 μm for a diameter of 80 mm, 90 mm, and 100 mm, respectively.

There are 4 circumstances in roundness measurement. First, the workpiece and the rotation system are perfect. Second, the workpiece is perfect, but the rotation system is not perfect. Third, the workpiece is not perfect, but the rotation system is perfect. And finally, the workpiece and the rotation system are not perfect.

The first circumstance provides an error of zero, while the second circumstance provides a constant error caused by the system, such as runout. The third circumstance will only provide the error of the workpiece; in this case it is OOR error. The last circumstance will include all errors of the workpiece and the rotary system. The second and fourth circumstances will be considered, while the first and the third will not be considered, since it is unlikely that there will be a perfect rotation system in a real situation. In the second case, the reference component will be a perfect circle. The measured output voltage is the radial stroke of the rotation system. In this case, the reference component directly provides the calibration constant of the probe. When measuring any workpiece, any errors outside this constant will be the OORs of the workpieces. However, perfect circularity is expensive. The realistic circumstance is the last one. The reference part is the part from which the OORs are known. The output of the probe contains the OORs of the reference workpiece and the errors from the rotary system. With these known OORs, the calibration constant of the probe can be determined using the optimization method. The error caused by the rotation system can be estimated from the calibration constant.

To measure the errors in circular shape, 24 points are taken with an equal angle around the workpiece. These points are analyzed to find the maximum variation between the peak and the valley. The value will include all errors from the workpiece and the rotation system, which can be expressed as:

$$V = OOR + \text{Radial throw of the rotation system}, \tag{9.10}$$

where V is the voltage.

At each roundness measurement, 20 individual measurements were taken in order to reduce the statistical uncertainty by averaging. The maximum deviation obtained from the roundness measurement is 0.020 volts, which can be converted into the deviation of 0.25 μm in terms of the displacement.

Table 9.5 The measured results and the uncertainty of measurements

Radius [mm]	Radius measured by the CMM (mm)	Radius measured by the probe [mm]	Uncertainty (μm)
15	15.0516	15.0505	1.8
25	25.0746	25.0742	2.3
35	35.0772	35.0776	3.7
45	45.0641	45.0653	5.1

Experimental Results and Discussions The radius measurement results, shown in Table 9.5, were compared to those obtained from the CMM. The errors are respectively $-1.1\,\mu m$, $-0.4\,\mu m$, $0.4\,\mu m$, and $1.47\,\mu m$ for a radius at 15 mm, 25 mm, 35 mm, and 45 mm, respectively. The average uncertainty of the radius measurement over a diameter range of 90 mm is $3.23\,\mu m$.

At the diameter of 90 mm, the workpiece was rotated at different rotation speeds, which are around 200, 500, and 800 rpm. The deviations in those uncertainties of measurements and in the errors are $0.21\,\mu m$ and $0.65\,\mu m$, respectively. With this slight change in the deviations, the performance of the probe is consistent.

According to the formula in Equation 9.9, the uncertainty of the radius measurement originates from laser wavelength instability, rotation speed variation, beam crossing angle, and Doppler frequency. The laser wavelength instability is less than one-third micrometer; hence, it can be neglected. The beam crossing angle can be eliminated through the calibration process. As the rotation speed and the Doppler frequency are correlated, the effect due to the variation of the rotation speed can be minimized through the correlated factor. The main parameter that affects the uncertainty is the technique used to analyze the Doppler signal. Different techniques provide different results. With this application on the solid surface and considering only the frequency domain analysis in the Gaussian curve-fitting techniques widely used in fluid velocity measurement, the Gaussian smooth technique developed in-house gives deviations of 5200 Hz and 75 Hz at a diameter of 90 mm ($f_d \sim 520\,kHz$). Therefore, the Gaussian smooth technique performs better and is more suitable for this application on the solid surface.

9.5 Angular Measurements

Angle measurement accuracy is critical in assessing parts features such as gears, jigs, fixtures, tapers of bores, flank angle and included angle of a gear, and the angle made by a seating surface of a jig with respect to a reference surface.

The main goal of angle estimation is not only to calculate angles. This may sound odd, but it is true in the measurement of system element alignment. The measurement of machine parts' straightness, parallelism, and flatness necessitates the use of extremely sensitive instruments such as autocollimators. The angle reading from such an instrument is an indicator of alignment error. There are several instruments available, ranging from basic scaled instruments to advanced styles that use laser interferometry techniques. The basic forms are simple improvised protractors with improved discrimination (lower count), such as a Vernier protractor. These devices have a mechanical support or a basic mechanism to correctly align them against the given workpiece and lock the reading.

In metrology applications, instruments that use the same basic concept as a spirit level but have a higher resolution, such as traditional or electronic clinometers, are common.

Collimators and angle dekkors, which are part of the family of instruments known as optical tooling, are by far the most accurate instruments.

9.5.1 Line Standard Angular Measuring Devices

A line standard is a length that is measured as a distance between two lines. As an extension, measurement can be carried out if the two lines form an angle, or the angle can be measured using the printed scale on the rotating element with respect to the other. Several examples will be shown next.

Protractor

This instrument is composed of two faces one of which is moving to measure angles by reading the value on the circular printed scale (Fig. 9.31) showing a Vernier index and types of measured angles (Fig. 9.32). The standard protractor can measure within 360° in various configurations of the angles and has a precision that can reach up to 5′. There are various types in the market, but they have a similar principle of measurement.

9.5.2 Face Standard Angular Measuring Devices

These are instruments measuring angles with two faces of the measuring instruments. These include usually the sine bar and angle gauges.

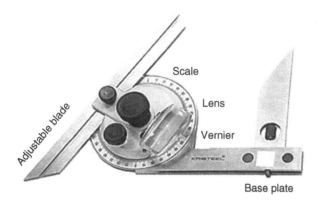

Figure 9.31 Universal bevel protractor (optionally optical).

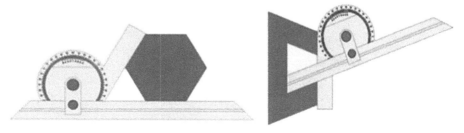

Figure 9.32 Samples of angle measurements.

Sine Bar

The sine bar is a tool used to measure angles. It consists of a hardened steel rod with a flat surface and is mounted on two precision cylinders or rollers in grooves especially designed for this purpose. The other is also parallel to the upper flat surface (Fig. 9.33 and 9.34), but the sine bar is our indirect measurement tool. The oblique triangle is here, and the size of the two sides determines the size of the third side and the two sharp corners. Using this tool can achieve high accuracy, and for the maximum angle of 45°, the angle measurement error is less than 2 seconds. Typically, a right-angled triangle is obtained using a precisely horizontal flat plate, on which measurement blocks are arranged in a direction perpendicular to the flat plate plane. It is an important part of angle measurement. The actual measurement is to compare the plane of the top feature of the part with the plane of the reference plate. Usually mechanical or electronic height gauges are used.

Sources of error in sine bar:

- Error in distance between cylinders or rollers;
- Error in the size and form of the cylinders or rollers;
- Error in the parallelism of axes between cylinders and the upper surface of the flat bar;
- Error in the flatness of the upper flat surface,
- Error in the slip gauges combination used for angle setting.

Angle Gauges

These are preconceived gauges with known angle values used for inspection and measuring these known standard angles values (Fig. 9.35). A combination of these gauges can be used for more values to measure.

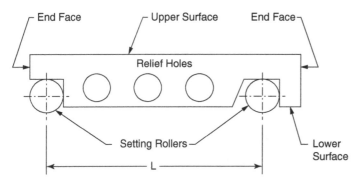

Figure 9.33 Sine bar.

Figure 9.34 Sine center.

Figure 9.35 Samples of angle measurements.

9.5.3 Measurement of Inclines

Spirit Level

This is a tool to indicate how parallel or perpendicular is a surface to earth. It has a mineral spirit solution with fluorescent color inside the level, as shown in Figure 9.36. This instrument can also have an electronic leveling showing the inclination in degrees.

Inclinometer

It consists of a precision pendulum using a flexural mount or magnetic bearing. When the inclinometer is tilted with an angle α, a position sensor generates an electrical signal, which is amplified. A galvanometer (Fig. 9.37) receives this signal and produces a torque that attempts to keep the pendulum mass in its original position with respect to the position sensor. The current applied is then proportional to the sine of the angle α; therefore, it gives an indication of the position after transforming it to voltage.

	Inclinometer
Resolution	0.1 arcsec → 0.01° arcsec
Measuring range	$\pm 1° \rightarrow \pm 90°$
Frequency domain	0.5 Hz on 1° and 40 Hz on 90° range

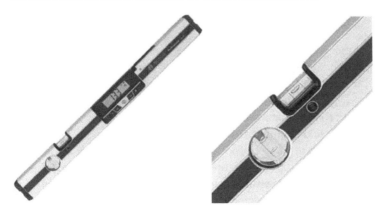

Figure 9.36 Professional digital inclinometer with a resolution of 0.1 degree in range of measurement from 0 to 360 deg.

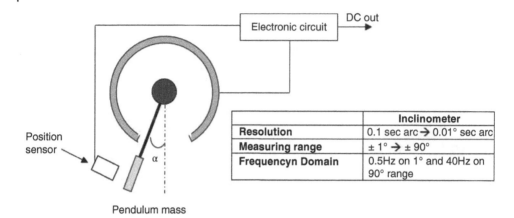

	Inclinometer
Resolution	0.1 sec arc → 0.01° sec arc
Measuring range	± 1° → ± 90°
Frequencyn Domain	0.5Hz on 1° and 40Hz on 90° range

Figure 9.37 Inclinometer sensor.

On the other side, a clinometer is a tool to measure the angle of elevation, or with respect to ground, as shown in Figure 9.38. This can be a Vernier scale associated to a right-angle clinometer, or an electronic angular scale mostly used with good precision. Also, recently mobile apps with digital clinometers have been developed (Fig. 9.39).

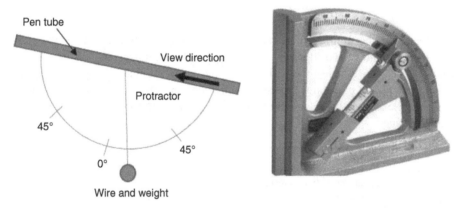

Figure 9.38 Vernier clinometer

Figure 9.39 Clinometer apps on google play.

9.5.4 Optical Instruments for Angular Measurement

Among noncontact instruments to measure angles for better precision are the optical instruments and laser interferometry instruments, allowing to measure with a resolution of arcsecs.

Optical Angle Comparators
a) **Optical Comparators**

These are the type of optical measuring system that is stand-alone as a desktop. It uses the principle of optical microscopy. The object to be measured is put on the table with light shining on it. The shadow is generated and projected on the screen. To allow precision measurement, a telecentric optical system is operated (Fig. 9.40).

Figure 9.40 Optical comparators

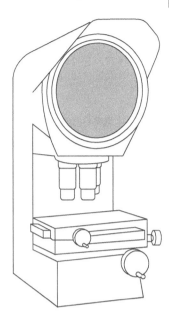

b) **Autocollimators**

This is a very reliable and accurate instrument to measure small angles. This is a combination between and collimator and an infinity telescope.

The autocollimator shown in Figure 9.41 combines in general the function of a collimator and a telescope. They share the same optical path using a beam splitter. Figure 9.41 shows the schematic setup of an autocollimator with straight viewing, a physical beam splitter, and infinity adjustment.

The image of the collimator reticle is projected to infinity by the autocollimation telescope.

A target tilting mirror returns the projected image to the autocollimator and makes a new image of the collimator reticle via the beam splitter in the eyepiece reticle shown.

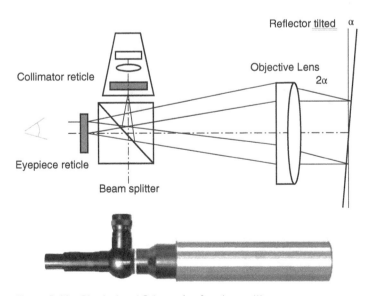

Figure 9.41 Physical and Schematic of an Autocollimator.

9.6 Metrology for Complex Geometric Features

Since there is no specific instrument that can be used to measure accurately complex features, the only possible techniques are the following.

9.6.1 Edge Detection Techniques Using a CCD Camera

This is based on taking images and applying several existing algorithms to detect edges of features.

With complex geometries, especially for small parts (Fig. 9.42), there is a need to measure mesoscale features within tolerances in the micrometer or nanometer range. Advanced image processing can improve inspection performance in terms of accuracy and processing time. Although scanning probe microscopes could achieve nanometer-scale uncertainties, they are limited in range to a few microns.

In addition, contact inspection methods such as CMM have the disadvantage of potentially damaging the surface in the micrometer range despite their good accuracy. It is believed that a non-contact technique such as machine vision inspection is the alternative for such a scale. Various techniques have been proposed for many applications, mainly in image processing and computer vision. Accurate edge detection in machine vision inspection for dimensional measurement is extremely important [1]. Many attempts have been made with CCD cameras but with uncertainties in the submillimeter range. Most of the problems in vision inspection are related to the low accuracy compared to other methods, which is caused by the low resolution of a camera and the

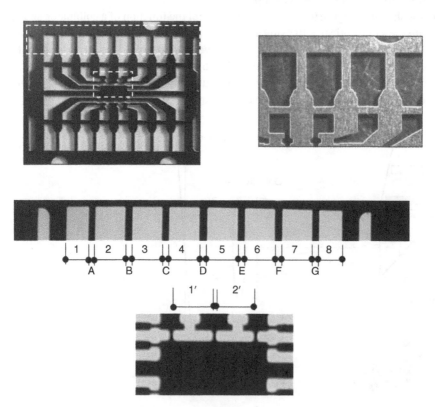

Figure 9.42 Selected region for the inspection.

poor performance of image processing techniques. For vision inspection systems in manufacturing processes, a camera with a high frame rate is often preferred, but it usually has a low resolution. Inspection with a high-resolution still CCD camera could be an alternative approach to improve the accuracy of vision inspection [2].

This is a low-cost inspection vision system based on a subpixel edge detection technique. It is used for feature extraction and measurement, for instance, inspection of mesoscale features in metrology.

For better performance in accuracy and processing time, the combination of the pixel accuracy edge detector and the subpixel accuracy edge detector was implemented along with camera calibration. This procedure revealed smooth edge boundaries closer to true contour of the specimen compared to a Canny edge detector. An example of measurement is shown. Figure 9.42 shows a sample of an electronic circuit with a couple of features to be measured only by this method of edge detection after taking a high-resolution image. The Canny edge technique has been employed in an algorithm of edge detection, and the results were compared to the Mitutoyo machine available in the National Physical Laboratory (NPL), UK. The results can be seen in Table 9.6 [3].

9.6.2 Full Laser Scanning for Reverse Engineering

This is another technique that has been partially discussed in previous chapters. It consists of either a touch probe or laser head scanner mounted on a CMM or robot arm. Using a software to collect a cloud of points scanned out of the whole surface of the part with complex features, a full 3D geometry can be constructed, and CAD files can be generated with fairly good precision depending on the process of scanning and data processing from the cloud of points. Once the 3D CAD is generated, it is possible to read, dimension, or even modify the features.

Figure 9.43 shows a reverse engineered part through 3D scanning while Figure 9.44 shows the many available possibilities to edit and measure any feature of this part. Total inspection can be carried out as detailed in Chapter (Fig. 9.45).

Table 9.6 Comparison of inspection results.

Measured point	Results of proposed method [μm]	NPL [μm]	Difference [μm]	Measured point	Results of proposed method [μm]	NPL [μm]	Difference [μm]
1	1446	1397.1	48.9	A	390	448.6	−58.6
2	1862	1813.9	48.1	B	426	479.8	−53.8
3	1873	1821.6	51.4	C	415	466.9	−51.9
4	1890	1833.1	56.9	D	445	493.4	−48.4
5	1895	1840.6	54.4	E	410	460.7	−50.7
6	1871	1821.0	50.0	F	427	481.2	−54.2
7	1853	1806.3	46.7	G	404	462.3	−58.3
8	1428	1390.2	37.8	1'	1186	1180.0	6.0
				2'	1141	1131.5	9.5

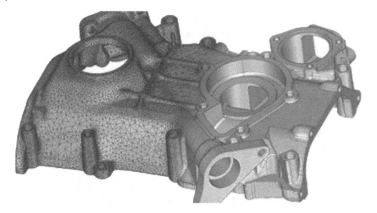

Figure 9.43 Reverse engineered scanning to obtain 3D CAD file (Courtesy of Creaform) *Source:* Courtesy of Creaform.

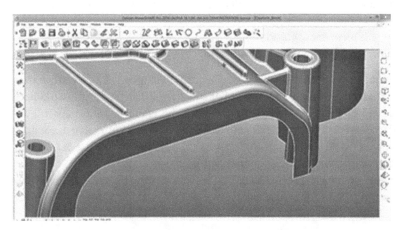

Figure 9.44 Features measurement and modification if needed. (Courtesy of Creaform) *Source:* Courtesy of Creaform.

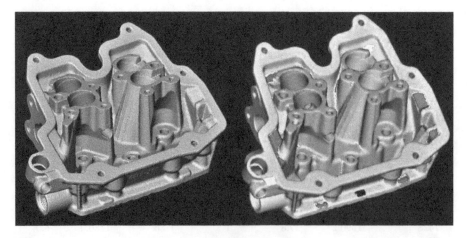

Figure 9.45 Full digital inspection of a reversed part.

9.7 Measurement Surface Texture

9.7.1 Geometry of Surface

The complex texture of any surface results from the combination of irregularities of various types. The three main causes are:

a) Those resulting from machine tool inaccuracies, deformation of work pieces due to cutting forces;
b) Those resulting from the inherent action of the particular production process; and
c) Those resulting from the vibration of the machine and workpiece.

A cross section of a typical surface is given in Figure 9.46. In general the surface consists of roughness, waviness, and errors of form.

Roughness: defines the irregularities on the surface obtained by each federate, particle, or spark if manufactured. There are two types: Roughness related to the unit event and micro-roughness caused by disturbances within each unit event: tearing, depositing, or tool wear.
Waviness: is related to the texture on which the roughness is superimposed, so it is less related to the event itself but more to the function of the machine tool on which the workpiece is produced.
Form: is related to the machine tool (e.g., large deviations). The main causes are thermal deflection and flexure machine/artifact.

Form, waviness, and roughness are never separated in practice. They are superimposed on top of each other. The surface finish of relevance is the roughness as it is a measure of the process capabilities, although often the waviness is as important [4].

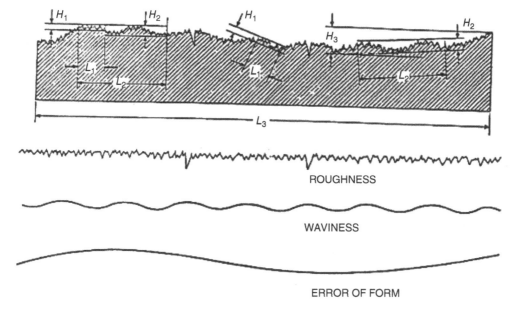

Figure 9.46 A surface texture representing the combined effects of several causes.

9.7.2 Surface Integrity

The manufacturing process when producing a surface could influence the surface functional performance:

a) Rough surface;
b) Tensile residual stress;
c) Features needed on the surface for specific applications.

The link between manufacturing process, surface aspects, and functional performance is important and is assessed as surface integrity.

Abusive machining produces poor surface integrity, while turning usually produces acceptable surface integrity provided the tools are sharp and machine conditions employed promote high tool life.

9.7.3 Specification of Surfaces

If we consider that an average roughness value is required for the surfaces, and this is specified in a similar way to straightness and flatness, then as can be seen in Figure 9.46 top, the value of H will depend on the corresponding value of L. Figure 9.46 bottom also shows surface profiles having the same amplitude but different wavelengths of the irregularities. The functional effect of the surfaces having profiles shown in Figure 9.47 will be completely different, and therefore, it is necessary to consider not only amplitude but also wavelength.

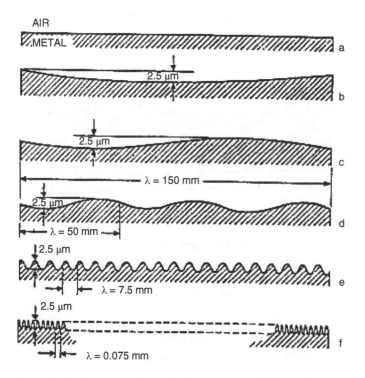

Figure 9.47 Various profiles having the same amplitude at different wavelengths (Not to scale).

The surface texture parameters are separated into three basic types:

Amplitude parameters are measures of the vertical characteristics of the surface deviations.
Spacing parameters are measures of horizontal characteristics of the surface deviations.
Hybrid parameters are combinations of spacing and amplitude parameters.

The following definitions are ISO 11562.

- **The mean line** is a least-squares line of nominal type fitted across the primary profile, with the regions of the profile above and below this line being equal and separated to a minimum. The mean tines for the roughness and waviness profiles are described by profile fitters as specified in ISO 11562.
- **Cut-off** - A cut-off is a filter that removes or reduces unnecessary data in order to examine wavelengths in the area of interest, using either electronic or mathematical methods.
- **Bandwidth** is the ratio between upper cut-off (Ls) to higher cut-off (Lc)
- **Sample duration** - The profile is separated into sample lengths I, which are long enough to contain statistically significant amounts of data. For roughness and waviness analysis, the sample length is equal to the selected cut-off (Lc) wavelength. The sample length is also known as the cut-off length.

Evaluation Length - The length in the direction of the X-axis used for assessing the profile under evaluation. The evaluation length may contain one or more sample lengths. For the primary profiles the evaluation length is equal to the sample length.

NOTE: Almost all parameters are defined over one sample length; however, in practice more than one sample length is assessed (usually five) and the mean calculated. This provides a better statistical estimate of the parameter's measured value.

All parameters using roughness, waviness, or primary profiles conform to the following assumptions:

T = Type of profile, R (roughness), W (waviness), or P (primary);
n = Parameter suffix, e.g., q, t, p, v;
N = Number of measured sampling lengths.

When a parameter is displayed as Tn (e.g., Rp), then it is assumed that the value has been measured over 5 sampling lengths. If the number of measured sampling lengths is other than 5 sampling lengths, then the parameter shall display this number thus TnN, for instance, Rp2.

Max Rule - If a parameter also displays a max, for example, Rz1 max, then the measured value shall not be greater than the specified tolerance value. If max is not displayed (e.g., Rp) then 16% of the measured values are allowed to be greater than the specified tolerance value. See ISO 4288-1996 for more details of the max and 16% rules. ISO 3274-1996, ISO 4287-1997, ISO 4288-1996, ISO 11 562, and other international standards are followed where appropriate by Taylor Hobson equipment.

9.7.4   Sampling Length

This is defined as the length of profile selected for the purpose of making an individual measurement of surface texture. The recommended sampling lengths (ISO 4288, 1996) are as follows (Table 9.7) h]:

Table 9.7 Relationship between surface parameters and sampling length (ISO 4288-1996).

Periodic profiles	Recommended cut-off ISO 4288-1996		Cut-offs	Sampling length/ evaluation length
	Non-periodic profiles			
Spacing Distance Sm (mm)	R_z (µm)	R_a (µm)	λc (mm)	λc/L (mm)
> 0.013 to 0.04	(0.025) to 0.1	(0.006) to 0.02	0.08	0.08/0.4
> 0.04 to 0.13	> 0.1 to 0.5	> 0.02 to 0.1	0.25	0.25/1.25
> 0.13 to 0.4	> 0.5 to 10	> 0.1 to 2	0.8	0.8/4
> 0.4 to 1.3	> 10 to 50	> 2 to 10	2.5	2.5/12.5
> 1.3 to 4	> 50 to 200	> 10 to 80	8	8/40

For fine machined surfaces, the sampling length of 0.8 mm is assumed. Larger sampling lengths in general would indicate a rougher surface, but the sampling length should be specified if different from 0.8 mm.

Filters A filter is a mean of separating roughness from waviness (Fig. 9.48). The common filters used are:

a) 2CR (2 Cap. and 2 Resist);
b) Phase corrected (ISO 11562);
c) The valley suppression (VS) (ISO 13565-1).

Control parameters The parameters generally used to specify a surface profile and used to control the process are summarized in Figure 9.49.

- Rq is the RMS value of the waveform.
- Ra is the most commonly use.

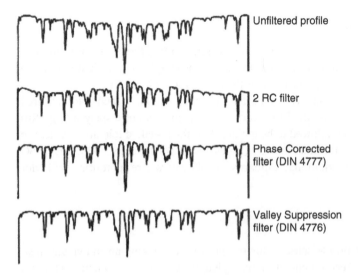

Unfiltered profile

2 RC filter

Phase Corrected filter (DIN 4777)

Valley Suppression filter (DIN 4776)

Figure 9.48 Comparison of the above filters.

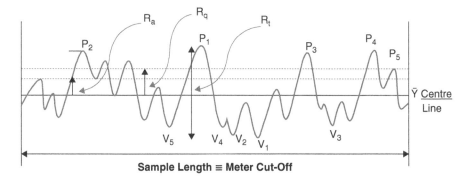

Figure 9.49 Surface profile parameters

The R_z, R_b, and R_p relate to the height of the peaks of the profile.

Parameter	Definition	Ease	Reliability
R_q	$\left(\dfrac{1}{N} \sum \left[Y_i - \overline{Y}\right]^2\right)^{\frac{1}{2}}$	5	1
R_a	$\dfrac{1}{N} \sum \left|\left[Y_i - \overline{Y}\right]\right|$	4	2
R_{3Z}	$P_3 - V_3$	3	3
R_z	$\dfrac{P1 + P2 + P3 + P4 + P5 - V1 - V2 - V3 - V4 - V5}{5}$	2	3
R_t	$P_1 - V_1$	2	4
R_p	P_1	1	5

Some of the parameters describing the surface roughness are given by order of reliability:

- R_q (μm) is the root mean square roughness;
- R_a (μm) is the average roughness;
- R_z (μm) is the average maximum height of the profile;
- R_t (μm) is the maximum height of the profile;
- R_p (μm) is the maximum profile peak height;
- R_v (μm) is the maximum profile valley depth;
- R_{sk} is the skewness;
- R_{ku} is the kurtosis;
- R_{max} (μm) is the maximum roughness depth;
- R_{pm} (μm) is the average maximum profile peak height;
- R_{vm} (μm) is the average maximum profile valley depth.

Surface Finish Characterization A profile can be analyzed by a variety of means. A large range of parameters is used and the user has to select from hundreds of parameters proposed over the years and/or follow the international standards (e.g., ISO) for standards definitions and good practice in measurements.

9.7.5 Instruments and Measurement of Roughness

ISO 3274 defines the various elements of a typical stylus instrument. Figure 7 shows the interrelationship between the elements. The stylus instrument is the most common acceptable instrument quantifying roughness. A sharp diamond stylus (2.5 μm radius) having a 90° included angle is traversed across the irregularities of the surface and generally follows the surface profile (peaks and valleys, etc.). The displacement of the stylus in a normal direction to the surface is magnified up to 10S times and recorded on a strip chart or meter. A general diagram of a stylus instrument in shown in Figure 9.50.

The magnitude of the stylus deflection is representative of the deviation of the surface profile from a reference datum. Where precise profile measurements are essential, the stylus uses a datum reference, which is an integral part of the instrument. This datum must of course be aligned to the general direction of the profile being tested. On workshop instruments, acceptable profiles can be achieved using the surface itself to act as a datum.

This is achieved using a skid (Fig. 9.51), which has a large radius and is supported by the surface peaks. The practical advantage of this method is that the skid protects the stylus from accidental damage.

The sampling length as outlined in roughness specifications is achieved in practice on analog-type stylus instruments by the use of high pass electrical filters. The filter characteristics are such that the low frequency components of the surface profile are not transmitted through the filer. The term meter cut-off is equivalent to the break frequency point on the filter characteristic at about 80% transmission level. When carrying out practical measurements on a surface, the measurement trace should always be in a direction normal to the surface lay; otherwise, vastly different results can be obtained.

Computer Aided Measurements The Ra (center line average) can be obtained directly from a meter reading on basic analog-type instruments. The statistical parameters must be obtained by numerical analysis on the surface profile. This is achieved by converting the basic analog signal from the

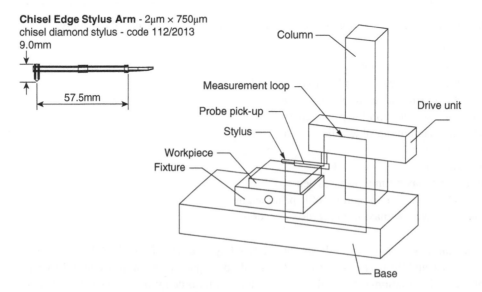

Figure 9.50 Stylus instrument (ISO 3274).

Figure 9.51 Handy surface finish equipment (courtesy Taylor-Hobson) *Source:* Courtesy of Taylor-Hobson

stylus to digital values via an A.D.C. and then to read the digital values by a microcomputer. Most of the specified values can then be calculated from one digitized sample.

Multiple Choice Questions of this Chapter

Multiple Choice Questions are given for each chapter with solutions in an online extension of this book. Please use link: www.wiley.com\go\mekid\metrologyandinstrumentation\

References

1 Mutneja, V. "Methods of image edge detection: A review." *Journal of Electrical & Electronic Systems* 4. 2015. doi:10.4172/2332-0796.1000150.

2 Muntarina, K., Shorif, S., and Uddin, M. (2021). "Notes on edge detection approaches." *Evolving Systems* 1-14. 2021. doi:10.1007/s12530-021-09371-8.

3 Mekid, S., and Ryu, H.S. "Rapid vision-based dimensional precision inspection of mesoscale artefacts." *Proceedings of the Institution of Mechanical Engineers, Part B: Journal of Engineering Manufacture*, 221(4), 2007, pp. 659–672.

4 Griffith, B. *Manufacturing Surface Technology*. Penton Press 2001.

5 Leach, R. "Measurement good practice guide, the measurement of surface texture using stylus instruments," NPL 2001.

10

Mechanical Measurements and Calibration

"It is important to not go over the pipetteman volume limit because that will decalibrate the machine."

—wiktionary.org

10.1   Importance of Mechanical Measurements

Measurement constitutes a fundamental basis for all daily transactions, research, and development. Hence, it becomes necessary to measure different parameters and be able to compare them to theoretical values the system was designed for. The next interest is the very usefulness of parameters measurement in applications, for instance, control systems, power, and production plants where temperature, pressure, humidity, current, and so forth, are important. It goes without saying that health monitoring can only be done through measurements.

As it has been introduced in previous chapters, measurements are never true until we consider the associated errors and uncertainties, which depend on many factors we have discussed. Since the instrument is the means of measurement, it will need regular adjustment and comparison to known references, which is called calibration.

Calibration is the comparison between a known measurement referred to a standard and the measurement carried out using your instrument. The accuracy of the standard should be typically ten times the accuracy of the measuring device being tested.

10.2   Mechanical Measurements and Calibration

Mechanical calibration is meant to be carried out using measurements executed with relatively simple tools to record changes in the dimensions of an object because of breakdown or wear during use. Whenever it is a matter of comparing results among measurements, the measurements performed with these instruments must be converted to the metric system. The advantage of using these instruments is their relative sturdiness and the fact that their accuracy is not affected by the onset of wear, for example, or any other effect.

There is no doubt that mechanical instruments need calibration services. This covers several measurands such as mass, force, dimension, angle, volume, flatness, torque, and vibration, which are calibrated in a controlled temperature facility to maintain thermal stability of both the reference standards and instruments under test.

The most frequently tested instruments and sensors are:

- Force measurement: Load cells and force gauges;
- Length measurement: Micrometers, Verniers, height gauges;
- Mass measurement: Scales/balances;
- Torque measurement: Torque wrenches;
- Weight and mass sets;
- Hydrometers;
- Volume and density;
- Vibration measurements: accelerometers;
- Acoustic measurements;
- Radiation.

The performance and accuracy of the physical test equipment is critical to the organization success; hence, high quality, reliable reference standards are needed for calibration tasks. It is required that the mechanical calibration services should be conducted by professionals to ensure accuracy, reliability, and traceability of the measurement instruments.

The frequent handling of instruments exposes them to mechanical shock and to various atmospheric conditions, so it is inevitable that mechanical instruments will be subject to drift. Hence, a regular calibration schedule is required to assess the measurement shift between subsequent calibrations and to determine tolerance limits for the measurement uncertainty budget. As examples, the force gauge calibration in tension and compression up to x kN ($1 \leq x \leq 10$) based on the lab's equipment for the calibration of load cells, force gauges and push/pull gauges.

10.3 Description of Mechanical Instruments

10.3.1 Mass Measurements

The mass is used to measure the weight of objects, such as fruits or a workpiece. There are several possible units, but the metric SI is the most used and hence the mass is either in kilogram [kg] or simply in gram [g]. The new kilogram standard adopted recently was introduced in the first chapter of this book.

There are several types of instruments measuring mass with configuration system as shown in Figure 10.1.

a) **Gravitational-Based Instruments**
   These are the standard balances using gravitation by a comparison technique to other known weights. They are known as analytical or digital balances.
b) **Transducer-Based Measurements**
   As opposed to mass comparison for measurement, these are transducer-based instruments that can be configured in specific format and calibrated capable of measuring a mass. Examples of transducers are piezoelectric, strain gauges, or smaller such as MEMS sensors. The transducer is part of an electronic circuit to deliver a signal to be converted to a mass.
c) **Gravitational Interaction–Based Instruments**
   This is based on the fundamental interaction between two bodies. In measuring stars and galaxies, researchers determine the distance between the bodies, their trajectory time related, and sometimes speed of rotation to extract the mass.

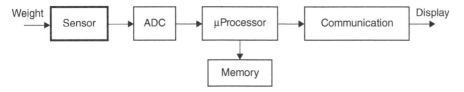

Figure 10.1 Open loop transducer weight system.

d) **Measurement Device Based on Space Linear Acceleration Mass**

In space, the body mass of the astronaut is a major factor in health monitoring. For this type of measurements, space agencies use a linear acceleration method (LAM) whose principle is for the device to generate a constant traction force and the astronaut is accelerated on a parallelogram motion guide that rotates at a large radius to achieve a nearly linear trajectory. The computation of the acceleration is carried out by regression analysis of the path-time trajectory, and the body mass is calculated by the formula m = F/a. However, in actual flight the device is so unstable that the deviation between runs can be 6–7 kg [1].

e) **Vibrating Tube Mass Instrument**

One of the most challenging activities is the measurement of the mass at microscopic level, for instance, micrograms. The process is elaborated as follows:

1) Determination of the buoyant mass of the sample through the density of the hoisting fluid.
2) Determination of the mass by measuring the previous buoyancy mass in different fluid density.

10.3.2 Force Measurements

The force is the measure of the interaction between two bodies. The force is measurable over a large range starting from atomic forces toward electromagnetic and gravitational forces. The attributed SI unit is the newton [N]. Physically, the force can have various forms as acting or reacting force based on Newton's laws and others. The weight is sometimes considered as mass and as force where it is expressed in newtons (Table 10.1). The load is also expressed as force.

In terms of measurement, the principle of the force measurement is mostly based on transducers, for example, load cells based on piezoelectric transducers, or strain gauges (Figs. 10.2–10.3).

Figure 10.3 shows an example of bolt tension tightening and loosening using a strain gauge configuration of Figure 10.2. The Wheatstone bridge helps balancing the voltage and hence reflects the difference as the resistivity of the strain gauge changes.

10.3.3 Vibration Measurements

Vibration measurements are data measured from any body in dynamic motion or oscillatory motion about an equilibrium position. Several parameters can be measured including displacement [m], speed [m/s], acceleration [m/s^2] of the object, and frequency, which is the number of times a complete motion cycle takes place within one period of one second. This frequency is given in hertz [Hz].

The accelerometer mounted on the mass m will deliver the displacement y of the mass due to displacement of the surface support x (Figure 10.4). An equation of motion can be written as shown in Equation 10.1. The output signal is materialized by y while the input can be in x.

$$m\frac{d^2y}{dt^2} + c\frac{dy}{dt} + ky = c\frac{dx}{dt} + kx. \tag{10.1}$$

Table 10.1 Force transducer types and ranges [2].

Device type	Typical range of rated capacities	Typical uncertainty % of reading	Typical temperature sensitivity and operating range % of reading per °C
Strain gauge load cells: Semiconductor gauges	0.01 N to 10 kN	0.2 to 1	0.02 (−40 °C to +80 °C)
Thin film gauges	0.1 N to 1 MN	0.02 to 1	0.02 (−40 °C to +80 °C)
Foil gauges	5 N to 50 MN	0.02 to 1	0.01 (−40 °C to +80 °C)
Piezoelectric crystal	1.5 mN to 120 MN	0.3 to 1	0.02 (−190 °C to +200 °C)
Hydraulic	500 N to 5 MN	0.25 to 5	0.05 (+5 °C to +40 °C)
Pneumatic	10 N to 500 kN	0.1 to 2	0.05 (+5 °C to +40 °C)
LVDT, capacitive, tuning-fork, vibrating wire	10 mN to 1 MN	0.02 to 2	0.02 (−40 °C to +80 °C)
Magnetostrictive	2 kN to 50 MN	0.5 to 2	0.04 (−40 °C to +80 °C)
Gyroscopic**	50 N to 250 N	0.001	0.0001 (−10 °C to +40 °C)
Force balance**	0.25 N to 20 N	0.0001	0.0001 (−10 °C to +40 °C)

Source: Courtesy of Tmssoftware

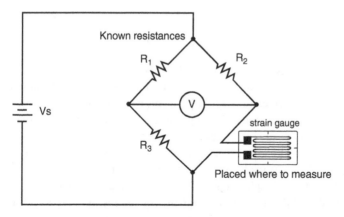

Figure 10.2 Strain gauge in Wheatstone bridge.

Equation 10.1 is a second-order ordinary differential equation, which can be solved in terms of the system parameters m, c, and k. The waveform of x is known; it can be defined as $x = x_0 \sin(\omega t + \varphi)$ leading to a solution in y.

The required accelerometer is an electromechanical device measuring acceleration forces ($m \cdot a$), where m is the mass and a the acceleration.

10.3.4 Volume and Density

Various volumes can be measured depending on the material. Liquids and solids are part of them but air is also included and sometimes a combination of all. Standard geometric forms, such as cubes, can be processed by just using known equations. But if we are dealing with random shapes, then the solids are immersed in liquid, and the volume is measured indirectly.

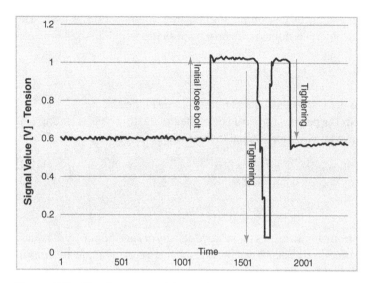

Figure 10.3 Signal outcome of the strain gauge in Wheatstone bridge.

Figure 10.4 Dynamic mass-spring system.

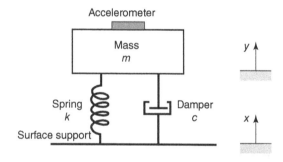

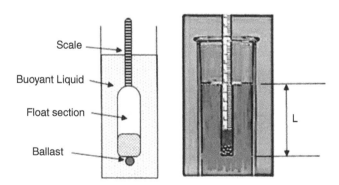

Figure 10.5 Hydrometer.

Graduated cylinders, pipets, and burets are typically used to measure volumes of liquids. In general, a graduated cylinder for direct determination of liquid volumes can be used. The measurements of volumes of solid objects is not as straightforward as it is with liquids.

- Cube: $V = L^3$ [m^3]
- $1\ m^3 = 1000\ l$

On the other side the density is defined as the mass per unit volume of the object.

Archimedes' principle is used as the general physical law related to floating objects stating that a body immersed in a fluid is buoyed up by a force equal to the weight of the fluid displaced.

$$F_b = \delta_{density} \times M_{immersed} \times A_{gravity}. \tag{10.2}$$

10.3.5 Hydrometers

As suggested by the name, the hydrometer measures the relative density or specific gravity of liquids using the concept of buoyancy or flotation as explained previously. It is a sealed hollow tube made from glass. The main components and functions of common hydrometers are:

a) **Gravity bulb:** To lower the center of gravity so that hydrometer floats vertically in a liquid.
b) **Floating bulb:** To increase up thrust on hydrometer.
c) **Narrow stem:** To increase the sensitivity of hydrometer.

Example: Test tube floated inside water jar with water density ρ and giving a height L_w. The tube has a cross section a (Figure 10.5).

In water:

The water density is ρ_w
 The volume displaced is $V = a \cdot L_w$ and the corresponding weight of water is V. $\rho_w \cdot g$.

In another liquid:

If the liquid parameters are density ρ_0, height L_0, then $V_0 = L_0 \cdot a$ and weight $= V_0 \cdot \rho_0 \cdot g$.
 Since the weight of this displaced water is equal to the weight of the loaded tube, we have:

$$L_w \cdot \rho_w = L_0 \cdot \rho_0 \tag{10.3}$$

$$\rho_0 / \rho_w = L_w / L_0. \tag{10.4}$$

This is the density of the second liquid with respect to water. The error of the final results are based on the manipulation and reading of values.

10.3.6 Acoustic Measurements

The microphone is the common sensor to measure acoustics. This is part of the sound and vibration domain that affects most of the environments. Vibration measurement have been discussed previously; sound can be very detrimental to human and measuring systems. There are limits not to exceed for safety; hence the importance of measurement.

The speed of elastic waves depends on elasticity and density, the two properties of the medium. The speed [m/s] is the speed of propagation, c, where the sound wave travels through a media.

The wavelength of sound in the medium denoted by λ [m] is the distance between successive waves of the same sense (Eq. 10.4).

$$\lambda = cT \tag{10.5}$$

The frequency f [Hz], is defined as the number of cycles per unit time, T (Eq. 10.5).

$$f = 1/T \tag{10.6}$$

The sound pressure propagation is given by Equation 10.6:

$$p(t) = p_0 sin(\omega t + \varphi) \tag{10.7}$$

or with respect to barometric pressure $p_b(t)$ with $p_{ins}(t)$ varying around $p_b(t)$:

$$p(t) = p_{ins}(t) - p_b(t), \tag{10.8}$$

where p_0 is amplitude of sound pressure fluctuation in P_a, ω is frequency of sound pressure fluctuation in rad/sec, φ is phase of the sound signal, and $p_{ins}(t)$ is instantaneous pressure.

In real practical applications, the quantity of the sound pressure is determined by Equation 10.8.

$$p = p_{rms} = \sqrt{\frac{1}{T} \int_0^T p^2(t)dt} \tag{10.9}$$

With sound intensity in free sound field given as:

$$I = \frac{p^2}{\rho c}$$

The next concern is how to measure these parameters. Figure 10.6 shows a digital sound level meter that is easy to use and can measure sound level in decibels (dB). A sample of acquired sound data from a compressor are shown in Figure 10.7. Data was collected using a data logger with microphones. It is important to understand that noise can also be modeled and simulated and later validated through measurement as shown in the noise modeling example generated in pressurized air flow in a pipe using Helmholtz resonators (Fig. 10.8).

Figure 10.6 Noise measurement instrument.

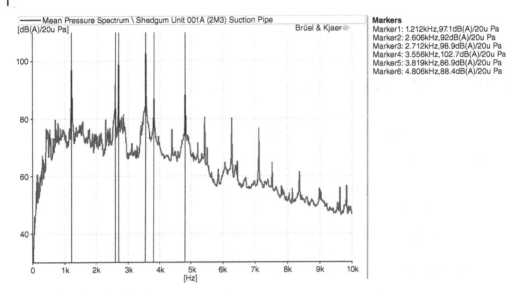

Figure 10.7 Sample of noise measurement in a compressor.

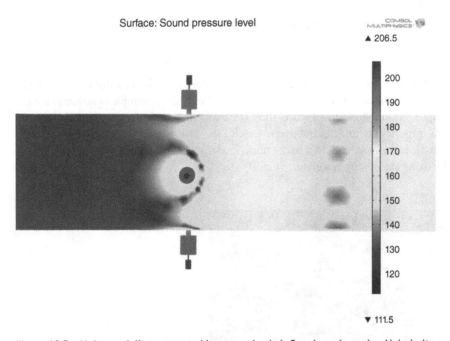

Figure 10.8 Noise modeling generated in pressurized air flow in a pipe using Helmholtz resonators.

10.4 Calibration of Mechanical Instruments

10.4.1 When Is Equipment Calibration Needed?

Calibration of instruments is a necessity, as it is in many respects enshrined in law. Whether they are used in factories, laboratories, or at home, the devices must be calibrated regularly to ensure that they give accurate results. Nonetheless, it is important to mention that measurement results are usually of critical value with legal requirements and liability implications in factories and laboratories.

They must provide consistent control of products. Any deviations from accuracy can not only affect productivity performance but also pose a risk to the lives of employees involved in the process. It becomes vital that instrument calibration is performed carefully and at regular intervals.

10.4.2 When Is There No Need for Calibration?

Usually, it is not needed to calibrate the instrument if it is not in use, when used as null detector or transfer device where measurement or output value is not explicitly used.

It is also not needed if the equipment is an absolute or fundamental standard or is fail safe, or when the instrument makes measurements or provides known outputs that are monitored by a calibrated device during use.

10.4.3 Process of Equipment Calibration

Instrument calibration has an exact process that varies depending on the type of instrument, how critical its role is in the operation, and the standards used for calibration. The following is a typical process to follow for instrument calibration:

- **Device design**: When performing calibration, special attention should be paid to the design of the device being calibrated.
- **Calibration instructions**: The instructions given for performing the instrument calibration should be followed exactly. Deviation from the instructions or use of an incorrect calibration value may affect accuracy.
- **Given accuracy**: Maintaining the accuracy ratio is also critical to a calibration process. This describes the accuracy of the test standard compared to the accuracy of the instrument being calibrated. It is important to maintain an accuracy ratio of at least 4:1. This means that the accuracy of the standard should be at least four times greater than that of the device being calibrated.
- **Tolerance specs**: The tolerance value of the instrument should be observed. It should be noted that each calibrator has a specific tolerance value due to normal variations in instrumentation and the quality control process. The tolerance value varies depending on various factors, including the industry and even the country in which the calibration process is to be performed.
- **International standards**: Adherence to internationally recognized standards is essential. Therefore, all standard procedures established under nationally or internationally recognized standards must be followed when calibrating instruments.
- **Uncertainty analysis**: At the end of the calibration process, an uncertainty analysis must be performed. This will help in evaluating any factors that may have influenced the results of the calibration.

10.5 Equipment Validation for Measurement

Under the aspects of measurement, the equipment validation should be established in an official document stating the accurate functioning of the equipment. This archived official document shows a validation process exhibiting evidence that the components critically contribute to the accurate functioning of the equipment and consistently meet the predefined specifications and operational attributes.

The process involves the identification and quantification of each constituent that has a bearing upon the result produced during analytical measurements. The validation is a commercial-grade assurance that internal malfunctioning of the equipment will not adversely impact the quality of the result.

10.5.1 Is There a Need of Equipment Validation?

Equipment validation is the hallmark of assurance that certifies the accurate functioning of an instrument under the prescribed range of operating environment and conditions, while steadfastly adhering to the correct operating specifications.

Validation facilitates meeting industry and regulatory standards set down to govern the accuracy of instrumentation. The validation protocol's format and content can be tailored to comply with the user's requirements. This further helps in assuring the accuracy level to a high degree.

The non-compliance to GMP or other regulatory body's requirements may render the instrumentation of a company ineligible for industry use. This can spell significant losses for the company, further underlining the necessity of validation.

Validation generates confidence in the analytical results generated by the equipment. This is critical as even a minor deviation from the standard operating course can bring about failures having the potential of inviting an industrial catastrophe or mishap.

10.5.2 Features and Benefits of Validation

The intended validation of a device is guided by a validation protocol. It is a written plan that expresses the method for performing the validation. The components are test parameters, product characteristics, production equipment, and critical stages that define the scope of acceptable test results.

Validation of analytical methods aims to establish, through laboratory testing, that the performance characteristics of the instrument are consistent with the analytical requirements of the desired applications.

Ongoing validation is intended to confirm the accuracy of method control activities to prove that validation results are genuine and authentic.

In the device design life cycle, validation is performed prior to shipment of a new or refurbished product or a device manufactured to revised production standards, assuming that the revisions may affect the new characteristics of the device.

Integrity testing is intended to ensure that devices meet industry and regulatory specifications. This validation activity is a cost-effective way to dramatically reduce the time delays in introducing a new product to become part of the life cycle.

Testing can be performed on-site during manufacturing or by simulating the operating environment and conditions with standard samples. Validation is a demonstration of the integrity of the device from the consumer's point of view.

10.5.3 Process of Validation of Equipment

The process of validation can significantly delay the introduction of a product to the market, as this process is very time consuming. Lack of proper understanding of regulatory requirements and instrument accuracy can compound this problem. The general steps established for this validation process may be as follows:

- Identify the design, user, and functional requirements;
- Establish the detailed validation protocol and develop the project schedule;
- Verify the suitability of the test environment and operating conditions, and simulate ideal functional conditions, if necessary;
- Assess the risks so that the necessary safeguards can be put in place;
- Collect standard results from certified traceable equipment to check against the test samples;
- Execute the test cases and produce a summary report;
- Establish a framework for managing deviations and initiate change control procedures to address offending components;
- Regularly re-validate.

The appropriate validation protocols are specified below:

- **Installation Qualification** (IQ) includes identification and verification of all components against the manufacturer's list. Documentation of the working environment conditions is performed to verify their suitability for the smooth operation of the system.
- **Operation Qualification** (OQ) is concerned with checking all functions of the equipment to ensure that they do not deviate from the manufacturer's specifications. The process requires extensive use of certified traceable simulators and standards to verify that input parameters are processed reliably and accurately.
- **Performance Qualification** (PQ) verifies equipment performance through its routine analytical use to measure conformance to specifications. The measured values are checked against a certified simulator or control standards. Standards are used that have similar values to the test samples. This establishes the suitability of the device for its intended use.

10.6 Difference between Calibration and Validation of Equipment

Equipment calibration deals with assessing the accuracy of equipment's results by measuring the variation against a defined standard to decide upon the relevant correction factors. The equipment is accordingly adjusted to fine-tune its performance to tally with accepted standards or specifications.

Equipment validation is a documented assurance that each constituent of the equipment is complying with the manufacturer's specification under a defined operating environment and standard. This is pulled off by checking the performance against traceable control standards.

10.7 Difference between Calibration and Verification

Calibration and verification are often used interchangeably across the scientific disciplines, but that is incorrect. They are not the same thing. Calibration is where instruments are tested for accuracy against a specified standard, and thereby the lack of trueness is compensated while correction is applied. Verification is the maximum accepted error and is not possible without the preceding calibration. Both terms are very closely related and hence used interchangeably accidentally.

Qualification means documented assurance that equipment (or facilities, utilities, etc.) are suitable for intended use. Validation is objective evidence that a PROCESS or PROCEDURE (including analytical methods) result in a consistent output meeting quality specifications. Qualification and validation are therefore related but are not the same at all. Please see "FDA Process Validation Guideline" (Jan. 2011) and/or ASTM E 2500-07 for details.

10.8 Calibration of Each Instrument

Usually, the mechanical, dimensional, and physical calibrations are conducted under the guidelines of ISO/IEC 17025 throughout the laboratory process, including certificate reporting, measurements, and traceability. A range of onsite calibration services are available for the support of automotive and aviation repair workshops. If you require, your calibrations can be added to our recall database for email reminders 30, 60, and 90 days prior to calibration due date.

10.8.1 Mass Calibration

Mass calibration is carried out to compare weight from the scale readout against a known standard of mass value. The process determines the maintenance level of the accuracy of measurements. Hence, it is possible to adjust the instrument to balance within accepted tolerances. The environment should be controlled to offer a wide range of testing capabilities. Metrology laboratories should be ISO/IEC 17025 accredited.

Balance and scale calibration ranges from 1 mg to 500 kg for the calibration of laboratory balances, benchtop balances, benchtop scales, floor scales, aircraft scales, hanging scales, personal weigh scales, and load cell platforms.

10.8.2 Force Calibration

The process of force calibration is the comparison of applied forces indicated by a measurement system under test to a calibration standard. The next concern is the traceability of the standard that defines the resulting range and accuracy of the system being tested.

For example, the ISO-13679 standard ("Procedures for testing Casing and Tubing Connections") addresses measurement range and accuracy by specifying that the calibration standard must be traceable to the National Institute for Science and Technology (NIST), accurate to within 1% (of reading) and that the calibrated range must encompass the loads to be used in the test program [2].

Torque Wrenches and Screwdrivers Torque wrench calibration covers usually a range of 1 to 1000 Nm for the calibration of torque wrenches and torque screwdrivers. Calibration can be undertaken to a selection of standards upon request, from BS 7882:2008, DIN 51309:2005, or BS ENISO 6789-1.

10.8.3 Pressure Calibration

This is one of the most frequently performed types of equipment calibration. Under pressure calibration service gas and hydraulic pressure are typically measured across a variety of sectors. Various types of pressure balances and calibrators along with a number of pressure gauges are used for carrying out pressure calibration. For the purpose of pressure calibration, it is vital that ISO 17025 UKAS accreditation (UK) and national standards be adhered to when performing pressure calibration. Pressure instruments that are frequently calibrated include:

- Analogue pressure gauges;
- Barometers;

- Digital indicators;
- Digital pressure gauges;
- Test gauges;
- Transmitters.

The pressure calibration laboratory is usually equipped with a range of precision pressure balances, dead weight testers, digital calibrators, and digital barometers enabling the company to offer pneumatic, gas, hydraulic, and oxygen clean pressure gauge calibration.

A typical accredited pressure range covers both absolute and gauge measurement from −0.9 to 35 bar for pneumatic gauge pressure, up to 1100 bar hydraulic gauge pressure and 0.01 to 36 barA for absolute and barometric sensor pressure.

All types of pressure devices can be calibrated, covering absolute, gauge, and differential pressures, such as manometers, micromanometers, inclined manometers, bourdon tube gauges, vacuum sensors, transducers, transmitters, barometers, piezoelectric, piezoresistive, quartz, relief valves, switches, magnehelic gauges, oxygen gauges, pressure calibrators, caisson gauges, and blood pressure meters with output signals of voltage, mA, frequency, charge or RS232. All calibrations have complete line of traceability to national standards and certificates and come with all the requirements stipulated in the ISO/IEC 17025 standard.

The range of high accuracy balances and portable electronic calibrators allow the laboratory to offer cost-effective and timely calibrations to suit all customer requirements.

- Gauge pressure is the deviation of pressure away from atmospheric pressure and can be expressed as positive or negative pressure.
- Absolute pressure is the pressure relative to a complete vacuum and will always be expressed as a positive reading. A barometer measures absolute pressure.
- Differential pressure is the difference between two pressures and can be positive or negative. A manometer measures differential pressure.
- Pressure units can be shown in different forms such as Torr, mm Hg, mbar, Pa, bar, psi. We can calibrate to all pressure units.

Example of calibration ranges

- Pneumatic 0–35 bar g
- Pneumatic 0.1–36 barA
- Hydraulic 0–1100 bar g
- Barometric 5–1300 mbarA
- Piezoresistive calibration
- Piezoelectric calibration
- HART communication configuration
- On-site calibration facility, UKAS accredited

Oil Free, Oxygen Clean Calibration

- Ultrapure nitrogen 0–20 bar
- Oxygen use fluid 0–400 bar
- Full oxygen cleaning service
- Particle assessment
- ASTM G93

Pressure Calibration Sensors

- Digital pressure gauges
- Bourdon tube gauges
- Piezoelectric
- Piezoresistive
- Barometers
- Pressure indicators
- Manometers
- Micromanometers
- Magnehelic
- Transmitters
- Altitude sensors
- Test gauges
- Caisson gauges
- Transducers
- Transmitters

10.8.4 Vibration Measurements

A) Calibration of Accelerometer The acceleration is measured using Vernier accelerometers along the line marked by the arrow on the label. The unit is usually meters per second per second (m/s^2).

Note that some accelerometers sense the effect of gravity; hence, there is an easy way to calibrate them. They can also be used as inclinometers to measure angles. Its reading will change as its orientation is changed from horizontal to vertical. Angles can be measured to the nearest degree.

By experience, it is not needed to calibrate an ab accelerometer. They are calibrated prior to being shipped, hence using the default calibration, but then use the software's zeroing option and zero the sensor along the axes.

The sensor can be calibrated for measuring accelerations in the horizontal direction following this procedure:

- Position the accelerometer with the arrow pointing down for the first calibration point.
- Define this as $-9.8\,m/s^2$ or -1 g.
  Rotate the accelerometer so the arrow points up and use the reading for the second calibration point.
- Define this as $+9.8\,m/s^2$ or $+1$ g. The accelerometer will then read 0 with no acceleration when held horizontally.

Accelerometer calibration can be provided for test accelerometers from 1 Hz to 10 kHz or more covering, for example, Dytran, PCB, Environmental Equipment, Bruel and Kjaer, Analog, and Omega models. Vibration testing can be conducted for products up to 10 kg of mass.

The standards used in this calibration are:

a) ISO 9000 company quality management system;
b) ISO 17025 calibration laboratory quality system;
c) ISO 16063 "Methods for the Calibration of Vibration and Shock Transducers" including the following parts only:
   - Part 11. Primary vibration calibration by laser interferometry;
   - Part 13. Primary shock calibration by laser interferometry;
   - Part 21. Vibration calibration by comparison method;
   - Part 22. Shock calibration by comparison method.

Calibration data is meaningless without knowing the measurement uncertainty, and the measurement uncertainty is often not well stated as it may not include components from reference standard or may not even be compliant with relevant standards in case the shaker transverse or drift are neglected, for example.

The systematic uncertainty includes reference and signal conditioner uncertainty, transverse motion of exciter (shaker), reference sensitivity or drift, and measurement channel inaccuracies.

The random uncertainty includes the mounting and cabling, operator technique, electrical noise, lab conditions, such as temperature, and the relative motion between the sensor under test (SUT) and reference.

Table 10.2 extracted from ISO 17025 A2LA shows an uncertainty budget depending on the range of frequencies

Absolute Calibration Method The test is subjected to a known, accurate, and reliable standard of nature: drop test, gravimetry, or laser interferometry. A signal conditioning with analyzer, voltmeter, and so forth, are needed (Fig. 10.9).

10.8.5 Volume and Density

Under a constant temperature and atmospheric pressure, the density of distilled water is constant. The volume of water can be determined by weighting dispensed water. The calibration of a pipette is carried out by a gravimetric method. When determining the volume of water, the accuracy of measurements is affected by ambient temperature, atmospheric pressure, and relative humidity. These factors are usually combined to give the Z factor, used in the calculation of volume of water. Then the calculated volume of water is compared with the theoretical volume to determine the accuracy and precision of the pipette.

Table 10.2 Excerpt from The Modal Shop's published ISO 17025 A2LA certified uncertainty budget.

Table 7: *Uncertainty Budget*

Uncertainty Component	Relative Uncertainty Contribution						
	10–99 Hz	100 Hz	101– 920 Hz	921– 5000 Hz	5001– 10,000 Hz	10 kHz to 15 kHz	15 kHz to 20 kHz
Primary standard calibration	0.25	0.10	0.25	0.50	0.75	1.50	2.00
Primary standard stability	0.06	0.06	0.06	0.06	0.06	0.06	0.06
Ratio error	0.31	0.29	0.29	0.29	0.29	0.29	0.29
Random (Type A)	0.30	0.09	0.14	0.10	0.18	0.24	0.63
Combined standard uncertainty (%) u_c	0.51	0.33	0.42	0.59	0.83	1.55	2.04
Expanded uncertainty (%) U (k=2)	1.02	0.66	0.84	1.18	1.66	3.10	4.08

Table 8: *Published Best Expanded Uncertainty, k = 2.*

Published uncertainty (%) U (k=2)	**1.15**	**0.75**	**1.00**	**1.35**	**1.85**	**3.30**	**4.30**

Source: Guide to the measurement of force, NPL, 2013.

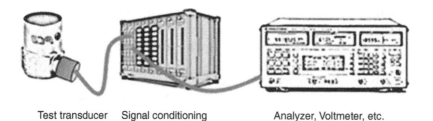

Test transducer   Signal conditioning   Analyzer, Voltmeter, etc.

Figure 10.9   Absolute configuration

10.8.6 Hydrometers

Hydrometers are specifically calibrated to work at a given temperature. Temperature correction is necessary because the density of water varies with temperature. Most hydrometers are calibrated to a reference temperature of 60 °F (15.6 °C). When you purchase the hydrometer, the manual should tell you the temperature at which the hydrometer can be used since some of them are calibrated at different temperatures.

10.8.7 Acoustic Measurements

Sound measurements progressing in air are executed over a wide range of frequencies ranging from infrasound to ultrasound, that is, from a tenth of a Hertz to about 200 kHz. Sound is also measured over a wide dynamic range that starts below the threshold of human hearing, lower than 20 μPa or 0 dB and goes up to more than 20 kPa or 180 dB. These wide ranges and the different types of sound field that occur cause a need for many different models of microphones and several different calibration and test methods. There are a few calibration techniques used as follows [3].

- Principle of reciprocity calibration;
- Primary pressure reciprocity calibration;
- Primary free-field reciprocity calibration;
- Free-field corrections for LS microphones.

Acoustic calibration services are traceable according to IEC and ANSI international standards. Supported standards are IEC 61672, IEC 60651/ 60804, BS 7580, ANSI S1 4, ANSI S1.43, IEC 61260, IEC 60225, IEC 60942, IEC 61094-4, and IEC 61094-6.

Multiple Choice Questions of this Chapter

Multiple Choice Questions are given for each chapter with solutions in an online extension of this book. Please use link: www.wiley.com\go\mekid\metrologyandinstrumentation\

References

1 Yan, H., Li, L., Hu, C. et al., "Astronaut mass measurement using linear acceleration method and the effect of body non-rigidity," *Sci. China Phys. Mech. Astron.* 54, 2011, pp. 777–782.
2 "Guide to the measurement of force," NPL, 2013.
3 Frederiksen, E., "Acoustic metrology—an overview of calibration methods and their uncertainties," *International Journal of Metrology and Quality Engineering* 4, 2013, pp. 97–107. doi:10.1051/ijmqe/2013045.

11

Thermodynamic Measurements

"Is temperature measurement really reliable to detect coronavirus?"

one asks himself.

11.1 Background

Thermodynamic properties of any material or solution are valuable not only for estimating the usefulness of the material or the feasibility of reactions in solution, but they also provide one of the best methods for investigating theoretical aspects related to the material or solution structure. Thermal properties of materials can be measured directly or indirectly. This includes temperature, developed pressure, calorimetry, and thermal conductivity. The measurements are made with a thermodynamic instrument, which is any device that allows for quantitative measurement. The reservoir and the meter are both thermodynamic instruments:

a) Any instrument capable of measuring any parameter of a thermodynamic system is referred to as a thermodynamic meter.

b) A thermodynamic reservoir is a system that is so large that it does not change its state parameters significantly when it comes into contact with the test system.

Pressure, temperature, and volume characterize the thermodynamic state of matter, whereas work does not. Understanding experiments and instruments with measurement rules is critical in thermodynamic metrology. For example, temperature can be measured through:

i) LIG thermometers (liquid in glass);
ii) RTDs (resistive temperature detectors);
iii) SPRTs (standard platinum resistance thermometers);
iv) Thermocouples;
v) Thermal imaging cameras.

Except for LIG and thermal cameras, most of these instruments use an external instrument to read signals from the sensing element while it is in operation.

11.2 Scale of Temperature

The units have previously been introduced, but it is worth refreshing memories here. The kelvin (K) is the International System of Units' base unit of temperature (SI). The Celsius scale (centigrade °C), Fahrenheit scale (°F), and Kelvin scale (K) are the most commonly used scales (Figure 11.1).

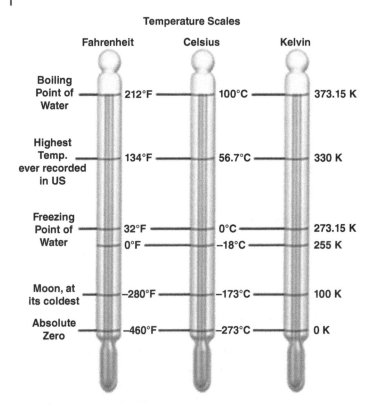

Figure 11.1 Temperature scale.

11.2.1 Ideal Gas Law

The equation of state of an assumed ideal gas is the ideal gas law. Although it has several limitations, it accurately describes the behavior of many gases under a wide range of conditions. Émile Clapeyron first stated it in 1834 as a combination of empirical Boyle's law, Charles's law, Avogadro's law, and Gay-Lussac's law. The law is as follows:

$$PV = nRT \tag{11.1}$$

Where P, V, and T are the pressure, volume, and absolute temperature, respectively; n and R are the number of moles of gas and the ideal gas constant, respectively. It is the same for all gases.

11.2.2 Vacuum

A vacuum is a space that is devoid of matter, in other words, a space with much lower pressure than the atmosphere. Torr (Torr) is the most commonly used pressure unit, defined as 1/760 of a standard atmosphere (101 325 Pa). Torr is not part of SI units. Perfect vacuum is measured at 0 K, which is −273.16 °C (Table 11.1).

11.2.3 Gas Constants

The heat capacity is the amount of heat needed to raise a given amount by one degree. It is defined for a gas as a molar heat capacity C_i: the amount of heat required to raise the temperature of 1 mole of the gas by 1 K.

$$Q = nC_i\,\varDelta T \tag{11.2}$$

Table 11.1 Pressure units' conversion.

Units	Pascal	Bar	Technical atmosphere	Standard atmosphere	Torr	Pounds per square inch
Symbol	(Pa)	(bar)	(at)	(atm)	(Torr)	(lbf/in²)
1 Pa	$\equiv 1$ N/m²	10^{-5}	1.0197×10^{-5}	9.8692×10^{-6}	7.5006×10^{-3}	0.000 145 037 737 730
1 bar	10^5	$\equiv 100$ kPa	1.0197	0.98692	750.06	14.503 773 773 022
1 at	98066.5	0.980665	$\equiv 1$ kgf/cm²	0.967 841 105 354 1	735.559 240 1	14.223 343 307 120 3
1 atm	101325	1.01325	1.0332	1	760	14.695 948 775 514 2
1 Torr	133.322 368 421	0.001 333 224	0.001 359 51	$1/760 \approx 0.001$ 315 789	1 Torr $\approx$ 1 mmHg	0.019 336 775
1 lbf/in²	6894.757 293 168	0.068 947 573	0.070 306 958	0.068 045 964	51.714 932 572	1

The value of the heat capacity Ci depends on whether the heat is added at constant volume, C_V, or constant pressure; C_P.

a) Heat Capacity Assuming Constant Volume

$$Q = n \cdot C_V . \Delta T, \tag{11.3}$$

where $C_V = (3/2)R$ for a monatomic ideal gas.

b) Heat Capacity Assuming Constant Pressure

At constant pressure, the ideal gas takes more heat to achieve the same temperature change than it does at constant volume.

$$Q = n \cdot C_P . \Delta T, \tag{11.4}$$

At constant volume, all heat added is used to raise the temperature. At constant pressure, some of the heat is used to perform work.

$C_P = (5/2)R$ for a monatomic ideal gas

The ratio C_P / C_V of the specific heats is an important number. It is represented by the symbol γ. For a monatomic ideal gas:

$$\gamma = C_P/C_V = (5R/2)/(3R/2) = 5/3. \tag{11.5}$$

The heat capacity at constant pressure C_p is the product of the mass and C_p.

c) Density and Specific Gravity

These are used to specify the mass of substances, but they are not similar; hence, their definitions are given below:

1) The density is an absolute quantity defined as mass per unit volume. Its SI unit is kg m^{-3} or kg/m^{-3};

2) The specific gravity is a relative quantity with no units and is defined as the ratio of a material's density with that of water at 4 °C (where it is most dense and is taken to have the value of 999.974 kg·m^{-3}).

11.3 Power

Power is defined as the rate of doing work (transferring heat) per unit of time. The SI unit of power is joule per second (J/s), also known as watt. It can also be in SI-based unit as kg·m^2·s^{-3}.

$$P = dW/dt, \tag{11.6}$$

where P is the power, W is work, and t is time. Work is also the product of force applied over a distance x.

$$W = F \cdot x. \tag{11.7}$$

Other forms of power are as follows:

i) $P = E/t$
ii) $P = F \cdot v$
iii) $P = T \cdot \omega$

For example, electrical power is defined as the instant power in watts or joules per second, as seen in Equation 11.8.

$$P(t) = I(t) \cdot V(t), \tag{11.8}$$

where I(t) is the current in amperes, and V(t) is the potential difference in volts.

11.4 Enthalpy

Enthalpy is a property of a thermodynamic system. It has the dimensions of energy (joules or ergs), and its value is determined entirely by the temperature, pressure, and composition of the system. It is the sum of the internal energy and the product of the pressure and the volume of the system.

$$H = E + PV. \tag{11.9}$$

The variation of enthalpy for any process under constant pressure is the change in the system's internal energy added to the pressure-volume work done by the system. Hence, it is positive for an expanding system and negative for a contracting system. In other words, it is either heat absorbed or released by the system under this process. The specific heat of a substance at constant pressure is defined as the rate of change of the enthalpy with respect to temperature.

Some basic information:

i) Gas is chosen as the standard thermometric substance;
ii) A real gas behaves as an ideal gas when pressure approaches zero;
iii) The standard fixed point of thermometry is the triple point of water.

11.5 Humidity Measurements

Humidity is the amount of water vapor present in the air. This is usually given as percentage of concentration, and the humidity measurement is very important for other metrology measurements, for instance, laser interferometry. It depends on the temperature and pressure of the system.

Warm air holds more moisture than cold air, so with the same amount of absolute/specific humidity, cooler air has a higher relative humidity compared to warmer air. The dew point is then introduced as the temperature at which air must be cooled to become saturated with water vapor.

Definitions and units of humidity:

a) *Vapor pressure:* This is defined as the partial pressure of water vapor in the air and expressed in [hPa] unit.
b) *Saturation vapor pressure:* This is the vapor pressure that is in thermodynamic equilibrium with the surface of water or ice, expressed in [hPa] unit.
c) *Dewpoint temperature:* As discussed previously, this is the air temperature at which the moist air saturates with respect to water at a given pressure. It is usually equal to or lower than the actual air temperature. The temperature at which moist air saturates with respect to ice, on the other hand, is known as the frost point temperature. The usual unit is [°C].
d) *Relative humidity (H):* This is the ratio of the vapor pressure (e) of the moist air to its saturation vapor pressure (e_s) at its temperature. This is expressed as a percentage, as shown below:

$$-H = (e/e_s) \times 100\% \tag{11.10}$$

$$-H_w = (e/e_{sw}) \times 100\% \tag{11.11}$$

$$-H_i = (e/e_{si}) \times 100\%, \tag{11.12}$$

where H_w and e_{sw} are the saturation vapor pressure with respect to water, and H_i and e_{si} are, the saturation vapor pressure with respect to ice.

11.6 Methods of Measuring Temperature

In contrast to basic standards such as length and mass, temperature is measured by indirect comparison. To determine temperature, a standardized calibrated device is required. Various primary effects causing temperature changes can be used to measure temperature, such as variations in physical or chemical conditions, electrical properties, radiation ability, or physical dimensions. The temperature signal has a response affected by:

a) Level of thermal conductivity and heat capacity of the element and the fluid surrounding the element;
b) Coefficient of heat transfer of the film;
c) Surface area per unit mass;
d) Mass velocity of the hosting fluid.

Fundamentally, thermometry can be based on:

a) Thermal expansion where the materials exhibit a change in size with change in temperatures;
b) The changes of the electrical resistance of a conductor;
c) Thermoelectric technique mostly known as thermocouples;
d) Radiative temperature measurement.

These fundamental measurement techniques result in two types of sensors that can measure temperature: in contact and noncontact.

In the case of contact sensing, the inference is drawn on the assessment of temperature. It is assumed that the object and the sensor are in thermal equilibrium. Examples are:

a) Thermocouples with all types;
b) Pressure thermometers;
c) Liquid-in-glass thermometers;
d) Bimetallic strip thermometers;
e) Resistance temperature detectors (RTDs);
f) Thermistors.

For noncontact-type sensors, we measure the radiant power of the infrared or optical radiation received by the object. Instruments such as radiation or optical pyrometers determine the temperature; the noncontact-type sensors include:

a) Fiber-optic thermometers;
b) Radiation pyrometers;
c) Optical pyrometers.

11.7 Temperature Measured through Thermal Expansion Materials

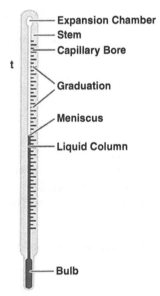

Figure 11.2 Liquid-in-glass thermometer.

11.7.1 Liquid-in-Glass Thermometer

Temperature can be measured basically by the thermal expansion of a liquid. Example is the liquid-in-glass (LiG) thermometer. A LiG thermometer possesses a glass capillary tube inside a stem with a liquid-filled bulb at one end.

As the liquid's temperature located in the reservoir increases, it expands and rises into the capillary tube. The level of the liquid in the column after calibration corresponds to a specific temperature, which is marked on the glass. The liquid contained inside the thermometer can vary, but commonly used liquids include mercury, toluene (or a similar organic substance), and low-hazard biodegradable liquids (Figure 11.2).

In calibration, this thermometer is subject to one out of three measuring environments:

a) Complete and entire immersion thermometer;
b) Total immersion thermometer but up to the liquid level;
c) Partial immersion thermistor to a predetermined level.

11.7.2 Bimetallic Thermometer

Metals expand or contract in response to temperature changes. As a result, the bimetallic thermometer employs a bimetallic strip to convert temperature into mechanical displacement. The elongation of the bimetallic strip depends on the thermal expansion property of the metal. Every metal has a different temperature coefficient. The temperature coefficient shows the relation between the change in the physical dimension of metal and the temperature change. The expansion or contraction of metal depends on the temperature coefficient, that is, at the same

Figure 11.3 Bimetallic thermometer.

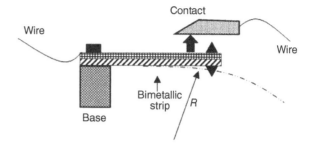

temperature the metals have different changes in the physical dimension. This allows for the bending of the bimetallic beam and, if properly calibrated, measurement (Fig. 11.3).

$$R = \frac{d}{(c_{th1} - c_{th2})(T_2 - T_1)},$$ (11.13)

where:

R   is the curvature radius;
C_{th} is the thermal expansion coefficient;
T   is the temperature;
d   is the plate thickness.

The working principle of bimetallic thermometers depends on the two fundamental properties of metals:

1) Metals have the property of thermal expansion, that is, they expand and contract with temperature change.
2) The temperature coefficient is different for different metals. Metals expand and contract differently at the same temperature.

The advantages of bimetallic thermometers include:

a) Simple and robust design;
b) Less expensive than other thermometers;
c) They are fully mechanical and do not require any power source to operate;
d) Easy installation and maintenance;
e) Nearly linear response to temperature change;
f) Their accuracy is between ±2 and 5% of the scale;
g) Suitable for wide temperature ranges.

Some disadvantages of bimetallic thermometers are:

a) They are not recommended for use at extremely high temperatures;
b) They may require frequent calibration;
c) May not give an accurate reading for low temperature;
d) They are limited and not recommended for temperatures higher than 400 °C;
e) Calibration is disturbed if roughly handled.

11.7.3   Electrical Resistance Thermometry

The electrical resistance of a conductor varies with temperature. This is simple in principle, but it results in two classes of resistance thermometers:

1) Resistance temperature detectors RTD (conductors);
2) Thermistors (semiconductors).

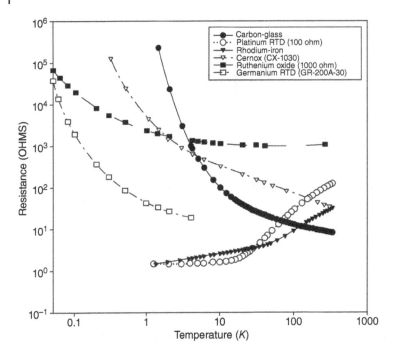

Figure 11.4 Temperature based on resistance for selected materials [1].

High-accuracy resistance thermometers use extensively platinum as it is capable of resisting high temperatures while sustaining excellent stability and exhibiting good linearity. Figure 11.4 shows the relationship between resistance and temperature limits depending on the conductor and semi-conductor materials. Germanium looks better in low temperatures up to 10 K, whereas ruthenium covers the entire temperature range from very low to high.

11.7.4 Resistance Temperature Detectors

C.H. Meyers built the first classical RTD out of platinum in 1932. A platinum helical coil was wound on a crossed mica web, and the entire assembly was housed inside a glass tube, allowing the strain on the wire to be minimized and its resistance to be maximized (Fig. 11.5). Because of poor thermal contact between the material and the measured point, slow thermal response time and fragility of the structure limited its application.

Because of technological advancement, most RTDs were developed later to be rugged. In 1968, the International Practical Temperature Scale was developed, and henceforth pure platinum RTDs have been used as the standard instruments for interpolating between fixed points of the scale. The triple point of water (0.01 °C), the boiling point of water (100 °C), the triple point of hydrogen (13.81 K), and the freezing point of zinc (419.505 °C), are examples of fixed points.

In practice, the RTD element or resistor must be close to the area where the temperature is to be measured, and it must transmit an electrical current. The resistance of the RTD element is then measured using an instrument. This will be correlated to the temperature value. The scale is nearly linear over a wide temperature range of 200–650 °C. Because of their toughness, they are extremely useful.

Several materials are used to build resistance thermometers, such as platinum (Pt), nickel (Ni), and copper (Cu), which are inserted in a bulb. Platinum is the most popular. They are usually called

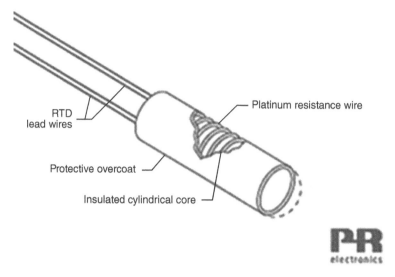

Figure 11.5 Wire-wound RTD.

platinum resistance thermometers. Its coefficient of resistivity at 20 °C is 0.003927 °C^{-1} compared to aluminum (Al), which is 0.00429 °C^{-1}.

The following factors contribute to platinum's popularity:

a) Temperature and resistance with linear relationship;
b) Large temperature coefficient of resistance, resulting in readily measurable values of resistance changes due to variations in temperature;
c) High stability as the temperature resistance remains constant over a long period of time.

The selection of a suitable material for RTD elements depends on the following criteria:

a) To allow formation into small wires, ductility is recommended for the material;
b) Expected to have a linear temperature with respect to resistance;
c) Corrosion resistant;
d) Greater stability and sensitivity;
e) Good reproducibility;
f) Inexpensive.

Three configurations exist for RTDs:

1) A partially supported wound element is a small coil of wire inserted into a hole in a ceramic insulator and attached along one side of the hole;
2) A wire-wound RTD element is made by twisting a platinum or metal wire on a glass or ceramic bobbin and sealing it with a coating on molten glass (Fig. 11.6).
3) Thin film RTD, which elements are created by depositing or screening a platinum or metal glass slurry film on a ceramic substrate that is small and flat.

Temperature sensing can use thin film RTDs as shown in Figure 11.6 that are made by depositing a thin layer of resistive material such as platinum or ceramic substrate (around 10–100 Å). The platinum layer is coated with epoxy or glass to protect the deposited platinum film and acts as a strain reliever for external lead wires.

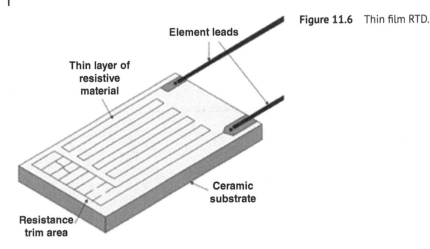

Figure 11.6 Thin film RTD.

Thin film sensors were initially considered unreliable due to their instability and susceptibility to mechanical failure. Breakage of lead wires was observed. Thin film RTD is preferred for its high accuracy over time and improved reliability because the thin film is sensitive to temperature due to its low thermal mass and simple manufacturing of small sensors. A thin film RTD is shown in Figure 11.6. RTDs have the following advantages when compared to other types of temperature sensors:

a) Resistance elements is used for differential temperature measurement;
b) Temperature-sensitive resistance elements is easily replaced;
c) It has higher resistance versus temperature linearity characteristics of RTDs;
d) Characterized by greater accuracy (as high as ±0.1 °C). Standard platinum resistance thermometers have ultra-high accuracy of around ±0.0001 °C;
e) Very good stability;
f) High flexibility is exhibited by RTDs with reference to the choice of measuring equipment, interchangeability of elements, and assembly of components;
g) An instrument can host multiple resistance elements;
h) No loss of accuracy of RTDs over small to wide working range;
i) Well suited for remote indication and stand-alone applications.

RTDs disadvantages are defined as follows:

a) The RTDs are expensive due to the use of platinum compared to other temperature sensors;
b) The nominal resistance is low for a given size, and the change in resistance is much smaller than other temperature sensors;
c) The thermistors keep the highest temperature sensitivity compared to RTDs.

11.7.5 Examples for Discussion

The Wheatstone bridge shown in Figure 11.7 is composed of equal resistances having one of them as RTD with the following data:

a) The fixed resistances, R_2 and R_3, are equal to 25 Ω;
b) The RTD has a resistance of 25 Ω at a temperature of 0 °C and is used to measure a temperature that is steady in time;

Figure 11.7 RTD Wheatstone bridge arrangement.

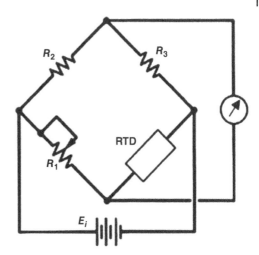

The RTD resistance over a small temperature range may be expressed, as in Equation 11.14:

$$R_{RTD} = R_o[1 - \alpha(T - T_o)]. \tag{11.14}$$

We are assuming the coefficient of this RTD to be 0.003925 °C^{-1}. With the placement of the RTD we balance the bridge by adjusting R_1. The value of R_1 is then 37.4 Ω. What is the temperature captured at the RTD? When designing the bridge to measure RTD, the temperature uncertainty must be specified. It is important to know that the bridge can be very precise depending on the precision of its components and the way it is controlled. If there is a need for the uncertainty in the measured temperature to be less than 0.5 °C, would a 1% total uncertainty in each of the resistors that make up the bridge be acceptable? As a result, we can choose the quality of the components based on this uncertainty while ignoring the effects of lead wire resistances.

The problem is to determine the RTD temperature where the uncertainty in the measured temperature is 1% (total uncertainty) in each of the resistors composing the bridge circuit with a 95% confidence level.

The RTD resistance is measured by balancing the bridge as shown below:

$$R_{RTD} = R_1 \frac{R_3}{R_2}.$$

The RTD resistance is found to be 37.4 Ω. With the previous information, Equation 11.14 becomes:

$$37.4 \ \Omega = 25(1 + \alpha T) \ \Omega.$$

The RTD temperature is found to be 126 °C.

On the next investigation, let us perform a design-stage uncertainty analysis.

Assumptions:
1) The resistances have a total uncertainty of 1%,
2) Initial values of the resistances in the bridge is equal to 25 Ω,

The uncertainties of the design-stage are:

$$u_{R_1} = u_{R_2} = u_{R_3} = (0.01)(25) = 0.25 \ \Omega.$$

We apply the propagation of uncertainty in each resistor to the uncertainty when determining the RTD resistance by:

$$u_{RTD} = \sqrt{\left[\frac{\partial R}{\partial R_1} u_{R_1}\right]^2 + \left[\frac{\partial R}{\partial R_2} u_{R_2}\right]^2 \left[\frac{\partial R}{\partial R_3} u_{R_3}\right]^2},$$

where:

$$R = R_{RTD} = \frac{R_1 R_3}{R_2}.$$

Thus, the RTD will have the uncertainty of the resistance:

$$u_{RTD} = \sqrt{\left[\frac{R_3}{R_2} u_{R_1}\right]^2 + \left[\frac{-R_1 R_3}{R_2^2} u_{R_2}\right]^2 + \left[\frac{R_1}{R_2} u_{R_3}\right]^2}$$

$$u_{RTD} = \sqrt{(1 \times 0.25)^2 + (1 \times -0.25)^2 + (1 \times 0.25)^2} = 0.433 \ \Omega,$$

while the uncertainty in temperature can be determined as:

$$R = R_{RTD} = R_0[1 - \alpha(T - T_0)],$$

and:

$$u_T = \sqrt{\left(\frac{\partial T}{\partial R} u_{RTD}\right)^2}.$$

As initially defined, $T_0 = 0\,°C$ with $R_0 = 25\ \Omega$, and neglecting uncertainties in T_0, α, and R_0, we have:

$$\frac{\partial T}{\partial R} = \frac{1}{\alpha R_0}$$

$$\frac{1}{\alpha R_0} = \frac{1}{(0.003925°C^{-1})(25\,\Omega)}.$$

Then the design-stage uncertainty in temperature becomes:

$$u_T = u_{RTD} \left(\frac{\partial T}{\partial R}\right) = \frac{0.433\ \Omega}{0.098\ \Omega/°C} = 4.4°C.$$

The desired uncertainty in temperature is not achieved with the specified levels of uncertainty in the pertinent variables.

Following the application on the uncertainty chapter, it is clear here that the uncertainty analysis can prevent the carrying out of measurements that may yield meaningless results.

11.7.6 Thermistors

The thermistor, like an RTD, is based on a material whose resistance varies with temperature (Fig. 11.8). These are also called semiconductors. When a thermistor is used to measure temperature, its resistance decreases as the temperature rises. Materials used in thermistors for temperature measurements have very high temperature coefficients (8–10 times higher than platinum and copper) and high resistivity. Hence, they are very sensitive to small variations in temperature and respond very quickly.

The possible relationship between temperature and resistance for a thermistor is assumed to be:

$$R = R_0 e^{\beta(1/T - 1/T_o)},$$

where β varies between 3500 and 4600 K depending on the material, temperature, and the way the sensor is made.

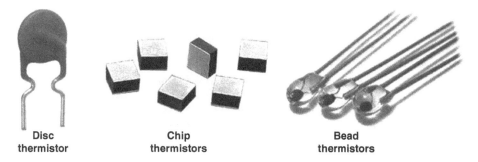

| Disc | Chip | Bead |
| thermistor | thermistors | thermistors |

Figure 11.8   Thermistor types.

11.8   Thermoelectric Temperature Measurement or Thermocouples

11.8.1   Basic Thermocouples

Measuring temperature with thermocouples is very popular. It is a common technique for measuring and controlling temperature that employs an electrical circuit called thermocouple. They are active sensors that are used for this type of measurement and consist of two electrical conductors made of dissimilar metals with at least one electrical connection. The thermoelectric effect is caused by the direct conversion of temperature differences to an electric voltage.

When two dissimilar metals are joined to form two junctions, for example, a hot junction at a higher temperature than the other junction, the cold junction serves as a reference. As a result, a net emf is generated, which establishes the flow of current and can be measured with a voltmeter.

The magnitude of the thermoelectric emf voltage produced is a function of the junction temperature and depends on the materials used to form the two junction layers. The thermoelectric voltage results from the combination of the Peltier effect and the Thomson effect. Peltier discovered that when two dissimilar metals are connected to an external circuit in such a way that a current is drawn, the thermoelectric voltage can be slightly altered, a phenomenon now called the Peltier effect. In other words, there is always a potential difference between two dissimilar metals in contact with each other. Thomson found that the emf at a junction undergoes an additional change when there is a temperature gradient along one or both metals. The Thomson effect states that a potential gradient also exists in a single metal if a temperature gradient is present. These two effects form the basis for a thermocouple used in temperature measurement. When two dissimilar metals are connected to form a closed circuit, that is, a thermocouple, current will spontaneously flow through the circuit provided one junction is kept at a different temperature than the other. This effect is called the Seebeck effect. In Figure 11.9, if the temperatures at the hot junction (T_1) and the cold junction (T_2) are equal and opposite (hot vs cold) at the same time, there is no current flow. However, if they are unequal, then the emfs do not equalize, and hence current flows. It is worth mentioning that the voltage signal is a function of the junction temperature at the measurement end, and the voltage increases as the temperature increases. The variations in thermoelectric voltage are calibrated with respect to temperature; the devices used to record these observations are called thermocouple pyrometers and can operate on the following principles:

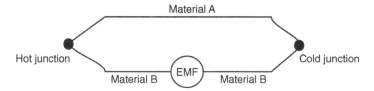

Figure 11.9   Basic thermocouple circuit.

a) Peltier Effect

This is the case when two dissimilar metals are joined together to form two junctions; the emf is generated within the circuit due to the different temperatures of the two junctions of the circuit. The Peltier effect plays a prominent role in the working principle of the thermocouple.

b) Thomson Effect

In this case, when two unlike metals are joined together forming two junctions, the potential exists within the circuit due to temperature gradient along the entire length of the conductors within the circuit. Typically, the emf suggested by the Thomson effect is very small and can be eliminated with proper material selection.

c) Seebeck Effect

When two different or unlike metals are joined together at two junctions, an electromotive force (emf) is generated at the two junctions. The amount of emf generated is different for different combinations of the metals.

11.8.2   Fundamental Thermocouple Laws

In addition to the three basic effects discussed previously that form the foundation of thermo-electric emf generation, three laws of thermocouples that govern this phenomenon must be studied to understand their theory and applicability in bringing useful information on temperature measurement.

a) Law of Homogeneous Materials

It is stated here that a thermoelectric current cannot be sustained in a circuit of a single homogeneous material, regardless of the variation of its cross section and by the application of heat alone. This law shows that two dissimilar materials are required to build any thermocouple circuit.

b) Law of Intermediate Metals

If an intermediate metal is inserted at any point in a thermocouple circuit, the net voltage is not affected provided the two junctions introduced by the third metal are at the same temperature. This law permits the thermoelectric voltage to be measured by inserting a component into the circuit at any point without affecting the net emf, provided that the additional junctions inserted are all at the same temperature.

Figure 11.10 shows that when a third metal, material C, is introduced into the system, two additional junctions, R and S, are formed. If these two additional junctions are kept at the same temperature, say T_3, the net voltage of the thermocouple circuit remains unchanged.

c) Law of Intermediate Temperatures

When a thermocouple circuit produces a thermoelectric voltage v1 when its two junctions have temperatures T_1 and T_2, and v2 when its two junctions have temperatures T_2 and T_3, then the thermocouple produces a thermoelectric emf voltage of v1 + v2 when its junction temperatures are held at T_1 and T_3 (Fig. 11.11).

This law relates to the calibration of the thermocouple and is important for reference junction compensation. This law allows us to make corrections to the thermocouple readings when the reference junction temperature differs from the temperature at which the thermocouple was calibrated. Normally, when creating a thermocouple calibration chart, the reference junction or cold junction

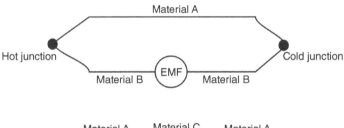

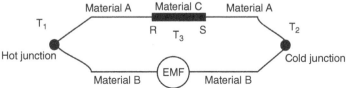

Figure 11.10 intermediate metals.

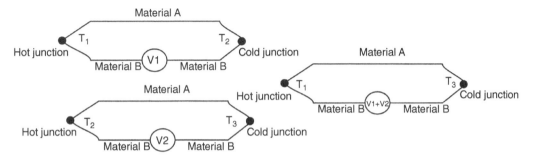

Figure 11.11 Intermediate temperatures.

temperature is set to 0 °C. However, in practice, the cold junction is rarely held at 0 °C but is usually held at ambient conditions. Therefore, the third law can be used to determine the actual temperature using the calibration chart.

11.9 Thermocouple Materials

In theory, any two materials can be joined together to form a thermocouple. However, only a few are suitable for temperature measurement. Combinations of different thermocouple materials and their temperature ranges are listed in Table 11.2.

Thermocouples made of base metals such as copper-constantan have a high resistance to condensate moisture corrosion. The iron-constantan type is essentially a low-cost thermocouple that can withstand oxidizing and reducing atmospheres. The chromel-alumel thermocouple can withstand oxidizing atmospheres.

Thermocouple materials are divided into base metals and rare, noble, or precious metals. Thermocouples made of platinum (platinum-rhodium) are called precious metal thermocouples; all other thermocouples belong to the group of base metals. These thermocouples are used with caution in a reducing atmosphere. Material combinations such as tungsten-tungsten-rhenium, iridium-tungsten, and iridium-iridium-rhodium are called special types of thermocouples and are used for a high temperature range of 1500–2300 °C.

Table 11.2 Basic thermocouple materials.

Type	Positive Material	Negative material	µV/°C	Usable range (°C)	Comments
E	chromel 90% nickel, 10% chromium	constantan 57% copper 43% nickel	68.00	0–800	Highest output thermocouple
T	copper	constantan	46.00	−187–300	Often used for boiler flues
K	chromel	alumel	42.00	0–1100	General purpose.
J	Iion	constantan	46.00	20–700	Used with reducing atmospheres. Tends to rust
R	platinum with 13% rhodium	platinum	8.00	0–1600	High temperatures
S	platinum with 10% rhodium	platinum	8.00	0–1600	As Type R. Used outside the UK
V	copper	copper/nickel	—		Compensating for type K to 80 °C
U	copper	copper/nickel	—		Compensating for types R and S to 50 °C

For high temperature measurements, the thermocouple wire should be thicker. However, as the wire thickness increases, the response time of the thermocouple to temperature variations decreases. Thermocouples are assigned a letter and grouped based on the temperature range that they can measure. Base metals, which can measure up to 1000 °C, are designated Type K, Type E, Type T, and Type J. Precious metals, which can measure up to 2000 °C, are classified as Type R, Type S, or Type B. Refractory metals are designated as Type C, Type D, or Type G. The following characteristics should be met:

1) Ability to produce a reasonable linear temperature-EMF relationship;
2) Generation of sufficient thermoelectric voltage per degree of temperature change to facilitate detection and measurement;
3) Ability to withstand sustained high temperatures, rapid temperature changes, and the effects of corrosive environments;
4) Good sensitivity to detect even small temperature variations;
5) Very good repeatability, allowing easy replacement of the thermocouple with a similar one without the need for recalibration;
6) Good calibration stability;
7) Cost effectiveness.

11.9.1 Advantages and Disadvantages of Thermocouple Materials

The following are some distinct advantages that merit the use of thermocouples:

a) Temperature can be measured over a wide range;
b) Thermocouples are self-sustaining and do not require an auxiliary power source;
c) Fast and good response can be obtained;
d) The readings obtained are consistent and therefore always repeatable;

e) Thermocouples are robust and can be used in harsh and corrosive conditions;
f) They are inexpensive;
g) They can be easily installed.

However, thermocouples also have certain disadvantages, which are listed below:

a) They have low sensitivity compared to other temperature measurement devices such as thermistors and RTDs;
b) Calibration is required because there is some nonlinearity;
c) Temperature measurement may be inaccurate due to changes in reference junction temperature; therefore, thermocouples cannot be used for precise measurements;
d) To increase the life span of thermocouples, they should be protected from contamination and must be chemically inert.

11.9.2 Thermocouple Voltage Measurement

The Seebeck voltage for a thermocouple circuit is measured without current flow. From our discussion of the Thomson and Peltier effects, the thermoelectric voltage deviates from the open circuit value when there is current flow in the thermocouple circuit. Therefore, the best method for measuring thermoelectric voltages is to use a device that minimizes current flow. For many years, the potentiometer served as the laboratory standard for measuring voltage in thermocouple circuits. When a potentiometer is balanced, it has almost no load error. Modern voltage measurement devices, such as digital voltmeters or data acquisition cards, have large enough input impedance values to be used with minimal load error. These devices can also be used in static or dynamic measurement situations if the load error generated by the meter is acceptable for the particular application. For such requirements, high impedance voltmeters have been incorporated into commercially available temperature indicators, temperature controllers, and digital data acquisition systems (DAS).

Case Study

The thermocouple circuit shown in Figure 11.12 is used to measure temperature T1 in that location. The reference junction of the thermocouple has a temperature of 0 °C, which is maintained inside an ice bath. The output voltage is measured with a potentiometer showing 9.669 mV. Determine T1.

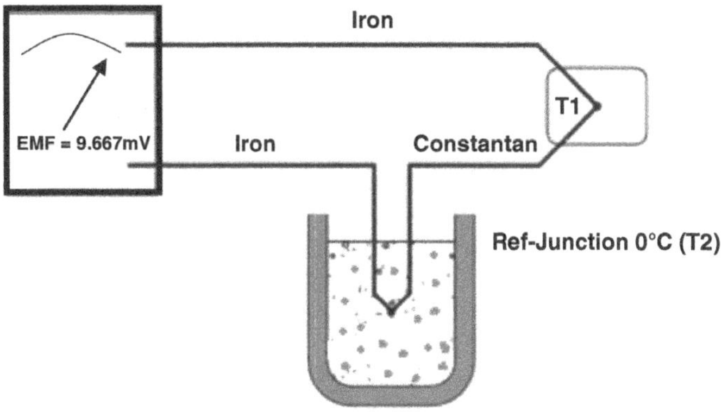

Figure 11.12 Type J thermocouple circuit.

(Continued)

(Continued)

In this case, the reference junction temperature is 0 °C. Therefore, the temperature corresponding to the output voltage can be easily determined to be 180.0 °C from any standard thermocouple reference table for type-J that are referenced to 0 °C.

Using the law of intermediate metals, the junctions formed at the potentiometer have no effect on the voltage measured for the thermocouple circuit as stated previously, and the voltage output accurately reflects the temperature difference between junctions 1 and 2.

In the next part, we suppose that the thermocouple circuit from the previous example has maintained reference junction 2 at a temperature of 30 °C and produces an output voltage of 8.132 mV. What temperature is tapped at the reference junction?

According to the law of intermediate temperatures, the output emf for a thermocouple circuit with two junctions, one at 0 °C and the other at T1, would be the sum of the emfs for a thermocouple circuit between 0°C and 30 °C and between 30 °C and T1. Thus:

$$\text{emf}_{0-30} + \text{emf}_{30-T1} = \text{emf}_{0-T1}.$$

Using this relationship, the voltage reading can be converted from the nonstandard reference temperature to a 0 °C reference temperature by adding $\text{emf}_{0-30} = 1.537$ to the existing reading. This gives an equivalent output voltage referenced to 0 °C as:

$$1.537 + 8.132 = 9.669 \text{ mV}$$

This thermocouple clearly measures the same temperature as the previous example, 180.0 °C.

It should be noted that an increase in the reference junction temperature results in a decrease in the output voltage of the thermocouple circuit. Negative voltage values, compared to the polarity, shown in Table 11.2, indicate that the measured temperature is lower than the reference junction temperature.

11.10   Multi-Junction Thermocouple Circuits

A thermocouple circuit consisting of two terminals made of different metals produces an open circuit emf related to the temperature differences between the two terminals. More than two junctions may be used in a thermocouple circuit, and the thermocouple circuit may be designed to measure temperature differences or average temperature or possibly to amplify the output voltage of a thermocouple circuit (Fig. 11.13).

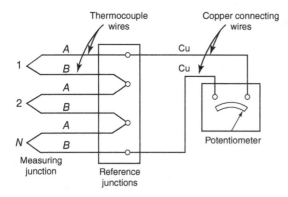

Figure 11.13   Thermopile arrangement [2]. *Source:* Benedict, R. P. Fundamentals of Temperature, Pressure and Flow Measurements, 3rd ed., John Wiley & Sons, 1984.

11.11 Thermopiles

The term "thermopile" refers to a thermocouple extension. A thermopile is made up of a series of thermocouples with hot junctions arranged side by side or in a star pattern. The number of individual emfs determines the total production in such situations. Combining thermocouples to create a thermopile has the advantage of producing a much more sensitive element. Consider the following illustration:

For instance, a sensitivity of 0.002 °C at 1 mV/°C can be accomplished with a chromel–constantan thermopile comprising 14 thermocouples. In the event that n indistinguishable thermocouples are joined to form a thermopile, the total emf will be n times the yield of the single thermocouple. Thermopiles are created using a series of semiconductors for specific applications such as estimating the temperature of sheet glass. Thermocouples can be connected in parallel for normal temperature estimation. During the construction of a thermopile, it is necessary to ensure that the hot intersections of the individual thermocouples are adequately protected from each other. Figures 11.14a and 11.14b represent, separately, a thermopile having an arrangement association and one having a star association.

Thermocouples in Parallel

When a spatially averaged temperature is desired, multiple thermocouple junctions can be arranged, as shown in Figure 11.14. In such an arrangement of N junctions, a mean emf is produced, given by:

$$\overline{emf} = \frac{1}{N} \sum_{i-1}^{N} (emf)_i.$$ (11.15)

The mean emf is indicative of a mean temperature:

$$\overline{T} = \frac{1}{N} \sum_{i-1}^{N} T_i.$$ (11.16)

11.12 Radiative Temperature Measurement

This noncontact method of measuring temperature is dependent on detecting the object's thermal radiation to determine its temperature. The emission of electromagnetic waves from an object's surface is referred to as radiation.

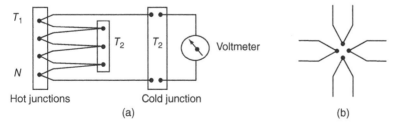

Figure 11.14 Thermopiles in series and in star connection. *Source:* Richard S. Figliola, Donald E. Beasley, Theory and Design for Mechanical Measurements, John Wiley & Sons, 2014.

Thermal Imaging Cameras

A thermal camera is based on noncontact measurement detecting infrared energy (heat) and converting it into a visual image. This is done after calibration with a known reference.

Thermal imagers, as opposed to regular cameras and human eyes, make pictures using heat rather than visible light (Fig. 11.15). Heat, also known as infrared or thermal energy, and light are both components of the electromagnetic spectrum; however, a camera capable of detecting visible light will not detect thermal energy, and vice versa. Thermal cameras capture the infrared energy and use the data to create images through digital or analog video outputs as shown in the example taken for a thermo cooler Peltier in Figure 11.16.

The thermal cameras' specifications include a range of measurements from low to high, similar to furnace temperatures. Then, depending on the camera lens, the field of view (FOV) shows the extent of a scene that the camera will see at any given time. The resolution is the other spec related to the number of pixels the camera can have on the scene (Fig. 11.17). The thermal sensitivity showing the smallest temperature difference that can be measured, for example, for a resolution area 160×120 pixels with a thermal sensitivity less than 120 mK, can show temperature differences around $0.12\,°C$.

The focus may be fixed or automatic for the camera depending on the user need. The spectral range is the range of the wavelengths that the sensor in the camera detects and is measured in micrometer, for instance, $8–14\,\mu m$, for most applications.

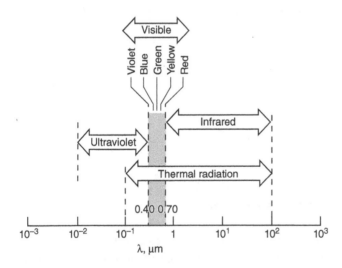

Figure 11.15 Thermal radiation spectrum.

Figure 11.16 Capture of the temperature distribution of a Peltier cooler. Range between $-21\,°C$ and $+29\,°C$.

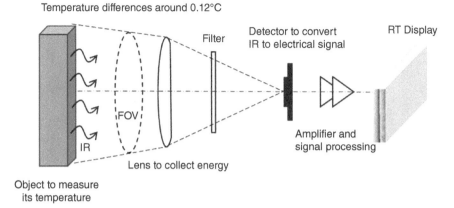

Figure 11.17 Principle of thermal camera.

Multiple Choice Questions of this Chapter

Multiple Choice Questions are given for each chapter with solutions in an online extension of this book. Please use link: www.wiley.com\go\mekid\metrologyandinstrumentation\

References

1 Figliola, R. S., and Beasley, D. E., *Theory and Design for Mechanical Measurements*, 6th ed., 2015, Wiley.

2 Benedict, R. P., *Fundamentals of Temperature, Pressure and Flow Measurements*, 3rd ed., John Wiley and Sons, New York.

12

Quality Systems and Standards

12.1 Introduction to Quality Management

Quality management and metrology are important components in manufacturing enterprises. They become essential in computer integrated manufacturing (CIM) production where automation takes place; hence, the intelligent production system will consider various types of quality embedded in the system leading to "zero defect" or "zero error". This is only done with self-learning and adjusting.

The limits of measurements uncertainty are of concern to both engineers and to end users. They inspect and evaluate this uncertainty, support expected valid data, and reduce the use of poor data for any interpretation. The experience of metrologists has demonstrated that data reliability is best achieved by a well-designed and operational quality assurance program. Hence is born the certification of capability of accepted measurements requiring the existence of a quality assurance program as one of the criteria of certification.

According to ISO 8402, quality is defined as the totality of features and characteristics of a product or service that bears on its ability to satisfy stated or implied needs.

Alternatively, quality is known as a perception and is to some extent a subjective attribute to a product or service. The meaning of quality has developed over time because of business interests. The understanding and interpretation is varying between concerned people.

Focusing on the key processes that provide customers with dependable and high-quality products and services will benefit a company the most. Producers can assess the conformance quality or degree to which a product or service was manufactured in accordance with the required specifications. On the other hand, customers may check and compare the quality specifications of a product or service.

In a modern global marketplace, quality is a key competency from which companies derive competitive advantage. Achieving quality is fundamental to competition in business in propelling business into new heights. Quality has had several interpretations; some of them are shown below along their source:

a) "Degree to which a set of inherent characteristics fulfills requirements" (ISO 9000:2008, International Organization for Standardization)
b) "Number of defects per million opportunities" (Six Sigma; Motorola University).
c) "Quality combines people power and process power" (Subir Chowdhury, 2005).
d) "Conformance to requirements" (Philip B. Crosby, 1979).
e) "Fitness for use" (Joseph M. Juran, American Society for Quality).

f) "Uniformity around a target value and the loss a product imposes on society after it is shipped" (Genichi Taguchi, 1992).

g) "Quality in a product or service is not what the supplier puts in. It is what the customer gets out and is willing to pay for" (Peter Drucker, 1985).

h) "The efficient production of the quality that the market expects, and costs go down and productivity goes up as improvement of quality is accomplished by better management of design, engineering, testing and by improvement of processes" (Edwards Deming, 1986, and Walton, Mary, and Edwards Deming, 1988).

Because of the increased emphasis on quality, organizations must structure, streamline, and standardize quality management systems that can assist them in managing all quality processes while mitigating operational risks. However, not every quality management solution is suitable for achieving total quality management and business objectives.

12.2 Quality Management

Quality management oversees all activities and tasks that must be accomplished to maintain a desired level of excellence. This includes the determination of a quality policy, creating and implementing quality planning and assurance, and quality control and quality improvement. In general, quality management focuses on long-term goals through the implementation of short-term initiatives.

Both total quality management (TQM) and quality management systems (QMS) represent popular approaches to preserving quality in the efforts of organizations across the world. It is, however, important to distinguish between them [1].

12.2.1 Total Quality Management (TQM)

TQM is a management strategy that focuses on long-term success through customer satisfaction. TQM originated in manufacturing, but due to its popularity, it has spread to nearly every industry. TQM was primarily concerned with improving processes in order to increase customer satisfaction, frequently by following a cycle of "Plan, Do, Check, and Act" (PDCA). TQM was often managed by individuals, rather than being implemented across entire organizations.

The principles of TQM are listed below with some similarities to those of QMS and some differences.

- It adopts a process-driven approach;
- It adopts a strategic and systematic approach;
- It focuses on customers;
- It involves people;
- It integrates organization systems;
- It ensures communication with stakeholders;
- It adopts a factual approach to decision making;
- It commits to continual improvement.

TQM needs three stages for inspection and testing, according to ISO 9001

12.2.2 Quality Management System (QMS)

QMS is described as a set of business processes geared toward delivering products and services to a consistently high standard. To be effective, QMS aligns with a plan for continuous measurement and improvement, as well as a commitment to quality to meet customers' expectations.

A QMS is frequently implemented throughout an organization, bringing all operations in line with the same requirements and standards in order to deliver consistency and quality at all levels. After implementing a QMS, most organizations are audited by a certification body, which ensures that their QMS meets standards and awards them certification such as ISO 9001. The key components are:

- It focuses on customers;
- It involves people;
- It provides strong leadership;
- It adopts a systematic approach to management;
- It commits to continual improvement;
- It adopts a process-driven approach;
- It adopts a factual approach to decision making;
- It recognizes that supplier relations are mutually beneficial.

12.2.3 TQM Is Essential to Complete TQS

The definition of quality standards relates mainly to any documents that provide requirements, specifications, guidelines, or characteristics used consistently to ensure that materials, products, processes, and services are fit for their purpose.

The standards provide organizations with the shared vision, understanding, procedures, and vocabulary required to meet their stakeholders' expectations. The standards provide a neutral and authoritative foundation for organizations and consumers all over the world to communicate and conduct business (Fig. 12.1).

12.2.4 ISO-Based QMS Certification

The ISO-based QMS is a management standard that presents a series of steps designed to increase business efficiency and customer satisfaction, thereby assisting organizations in meeting the needs of customers and other stakeholders, as well as the regulatory requirements associated with their products/services.

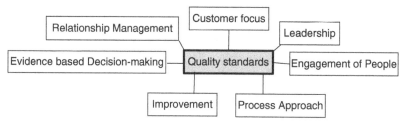

Figure 12.1 Principles of quality standards.

Table 12.1 Standards.

Topic	Standard
Quality management	ISO 9000
	ISO 9001
Auditing	ISO 19011
Environmental management	ISO 14000
	ISO 14001
Risk management	ISO 31011
Social responsibility	ISO 26000
Sampling by attributes	Z1.4
Sampling by variables	Z1.9
Food safety	ISO 22000

It is a quality management system (QMS) designed to monitor the functions and data of a specific organization, assisting in the maintenance and preservation of the proper balance within it. The ISO-based QMS is an internationally recognized standard that can be implemented by any organization, large or small; this certification can benefit both marketing and promotional aspects, as well as organizational improvement.

Several manufacturing companies were early adopters of the ISO-based QMS. This strategy has recently gained favor with a number of service-based industry associations. ISO 9001 certification became popular for a variety of reasons, including the fact that an ISO-based QMS, when properly implemented, can help to improve efficiency and thus business competitiveness.

In short, the benefits of properly implementing an ISO-based QMS may include:

a) Cost savings: helps optimize the organization's operations;
b) Customer satisfaction: helps improve quality, enhance customer satisfaction, and increase sales;
c) New markets: assists in overcoming trade barriers and opening up and access to global markets;
d) Market share: helps increase productivity and competitive advantage;
e) Environmental benefits: aids in reducing negative impacts on the environment.
   A couple of standards are needed for updates when dealing with quality. Table 12.1 gives an overview.

12.3 Components of Quality Management

12.3.1 Quality System (QS)

A quality system comes on top of an organization and is defined as the organizational structure, responsibilities, processes, procedures, and resources for implementing quality management. Quality management includes all aspects of the overall management function governing and implementing the company quality policy and quality objectives. Both quality control and quality assurance are parts of quality management.

12.3.2 Quality Assurance (QA)

Part of quality management is quality assurance (QA), focusing on providing confidence that quality requirements will be fulfilled.

The confidence provided by quality assurance has two aspects: internally to management and externally to customers, government agencies, regulators, certifiers, and third parties. An alternate definition is "all the planned and systematic activities implemented within the quality system that can be demonstrated to provide confidence that a product or service will fulfill requirements for quality."

12.3.3 Quality Control (QC)

Quality control is concerned with the inspection aspects and identification of defects of *quality* management. This will fulfill quality requirements. QC has to comply with QS as described in ISO 9001.

- Practices management: a well-managed laboratory is essential for reliable measurements. Data quality is influenced by the level of management staff and the operational policies. Maintaining a good level of the staff, develop policies and provide resources to do this.
- Competence of staff is essential for quality measurements. There is minimum educational background for each staff supplemented by continuous training, including skills, qualifications, and knowledge
- Quality operations should be conducted in a reliable and consistent manner. These will need good lab practices, good measurements practices, and standard operations procedures.

Quality control is focused on fulfilling quality requirements; it encompasses the operational techniques and activities undertaken within the quality assurance system to verify that the requirements for quality of the trial-related activities have been fulfilled. The general relationship between QS, QA, and QC is shown on Figure 12.2

12.3.4 Quality Assessment

Quality assessment describes the activities and procedures utilized to monitor the effectiveness of quality control and to evaluate rigorously the quality of the data output.

Figure 12.2 Quality system, quality assurance and quality control association.

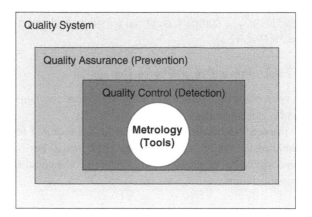

To achieve this, two internal and external approaches are needed with some lab audits:

- **Internal approach:** are all repetitive measurements that are key to evaluating the precision.
- **External approaches:** after the assessment of the precision that is internally carried out, the lab should seek external evaluation of its measurement accuracy.
- **Lab audits:** are valuable techniques for quality assurance and may include evaluation of the internal and external approaches.

Documentation is of high merit to be considered for quality assurance. Metrologists should keep records in hard or soft copies in a safe storage of the following:

- What is measured?
- Who measures?
- When are measurements made?
- How are measurements made?
  - Equipment
  - Calibration
  - Methodology
- Data obtained;
- Calculations;
- Quality assurance support;
- Reports.

12.4 System Components

12.4.1 Quality Audits

The quality audit is the practice of systematic inspection and examination of a quality system. It has to be carried out by either an internal or an external auditor. This quality audit is a program designed to evaluate the existing quality systems of your laboratory and suppliers.

When an audit is applied on the supplier and their facility, it will determine if they can produce your products according to certain directives and requirements.

Since it is one of the key steps in the sourcing process, the auditing will be important to ensure that new or existing manufacturers/suppliers can deliver products with the expected quality, undertake continuous improvements, and operate efficiently in compliance with social and environmental standards.

When examining a factory's structure, its organization, quality process, and experience, the audit enables you to compare potential suppliers and select the best possible one for your production needs (Fig. 12.3).

12.4.2 Preventive and Corrective Action

As part of the maintenance procedures, several methods are used; among them there are preventive and corrective actions.

Preventive action is a proactive action required to solve a potential issue ahead of time and avoiding the system to stop and generate further issues. Preventive action is usually programmed on a regular basis but also include deciding when quality performance data indicates trends toward possible failure.

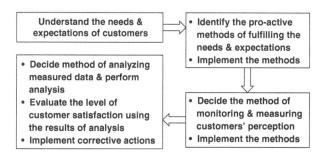

Figure 12.3 Evaluation of needs.

Example: Bulbs can be changed regularly based on their known lifetime even if they are still lighting.

The corrective action and a reactive action are required to correct an issue in the system and eliminate its cause. This is considered as a too late action.

Example: the previous bulbs can also be changed only at the time they are damaged (not lighting anymore).

A request for action (RFA) is then needed to initiate and record the identified non-conformity (NC) or opportunity to improve (OFI). These can be categorized as follows:

a) **Pre-Analytical Phase**
  - Incorrect test request or test selection;
  - Incomplete laboratory request forms;
  - Incorrect specimen collection, labelling and transportation 2 Laboratory Quality Standards and their Implementation.
b) **Analytical Phase**
  - Use of faulty equipment, improper use of equipment;
  - Use of substandard or expired reagents;
  - Incorrect reagent preparation and storage;
  - Incorrect technical procedures; non-adherence to standard operating procedures (SOPs) or internal quality control (IQC).
c) **Post-Analytical Phase**
  - Inaccurate reporting and recording;
  - Inaccurate calculations, computation, or transcription;
  - Return of results to the clinician too late to influence patient management;
  - Incorrect interpretation of results.

12.4.3 Occupational Safety Requirements

This is usually known as the Occupational Safety and Health Act (OSHA). This was set to prevent accidents at work and therefore, it becomes mandatory to be provided by the employers. OSHA provides assistance and help to employers and employees. Occupational safety (OS) is defined as the science of the prevision, recognition, evaluation, and control of hazards arising from the workplace that can put at risk the safety and well-being of workers. This includes the promotion and maintenance of the highest degree of physical, mental, and social well-being of workers in all occupations. It may consider the possible impact on the surrounding communities and the general environment.

The central focus of OS is to assess and manage occupational risks through the application of preventive and protective actions discussed here. This covers many different areas of activity and related standards that frequently need to be adjusted in line with changes in technology and research regarding potential workplace risks for human health.

Hence, it is necessary to establish a safety committee inside the department to monitor the applications of safety rules and maintenance. It is worth mentioning that the safety in the instructional laboratory is shared between all parties.

12.4.4   Housekeeping Practices

It is clear to see that housekeeping practices have an impact when meeting OSHA's general requirements for walking and working surfaces (1910.22). The rule states:

a) Floors are to be clean and dry.
b) Housekeeping is to be clean, orderly, and sanitary.
c) Aisles and passageways are to have sufficient clearance. They are to be kept clear, without obstructions that could create a hazard.
d) Permanent aisles are to be marked.

Some examples of violations of these rules are blocked aisles, material lying across an aisle or on the floor, wet or oily floors, or material overhanging high shelves.

12.5   Quality Standards and Guides

The quality system has an integral part consisting of the quality standards that are designed to support laboratories to meet regulatory requirements [2]. The latter requirements are important and cover the local health regulations and monitoring laboratory functions to secure laboratory safety and consistency of performance [3].

Quality standards help to improve the quality of the service provided to customers; they are used to:

a) **Identify Gaps and Areas for Improvement**
    • Conduct an initial assessment of the provided services against quality standards to plan quality improvement projects;
    • Understand the priority area to focus on;
    • Identify and prioritize improvement needs for the coming business cycle;
    • Identify potential areas for local audit.
b) **Measure the Quality of Service**
    • Measure levels of service delivered;
    • Develop tools and metrics;
    • Measure quality improvement after implementing changes.
c) **Using Quality Measures**
    This usually include three components:
    • Structure - the environment or setting;
    • Process - the activity carried out;
    • Outcome - the result or impact.

d) **Data Sources**

We indicate where national quality assured indicators currently exist within each quality standard and that can measure the quality statement. National indicators can include those developed by a national bureau of metrology or trademark through their indicators for quality improvement program.

If the national quality indicators are not available, the quality measures should form the basis for local audit and service review. Audit standards or levels of expected achievement should be decided locally by any office, unless otherwise stated.

e) **Understand How to Improve the Service**

This is simply to develop an action and commissioning plan. In the expected action plan, there is a need to detail the steps needed to improve services so they align with the quality standard.

The plan should contain a list of actions to improve the service in-line with a quality statement. To be more active, an action plan with assignment is as follows:

- A person is responsible;
- There is a deadline;
- There is a way to measure progress.

f) **Demonstrate That the Laboratory Can Provide Quality Service**

You can provide assurance to your board, commissioners, regulators, and patients by demonstrating that you meet the levels of care specified in a quality standard.

An initial assessment of services against a quality standard can highlight areas where you are providing quality care as well as areas where you can improve.

Using quality standards as part of an assurance process (that meets the needs of your board and commissioners) saves time and eliminates ad hoc data requests. Producing an annual report is a good way to show how you use evidence-based guidance such as quality standards and nice guidance. You could demonstrate:

- How your service meets the levels of care set out in quality standards, informed by readily available evidence;
- The progress with quality improvement projects aligned to quality standards.

g) **Commission Quality Services**

You must understand how services compare to quality standards in order to conduct an evidence-based assessment of their quality. Compare commissioned services to standards, for example, to identify service gaps and inform joint strategic needs assessments at the local level. Standards are also used:

- To contribute to the advancements outlined in local health and well-being strategies and national outcomes frameworks;
- As a source of evidence-based quality indicators for incentive schemes for local quality improvement;
- To establish quality care levels in contracts;
- To develop key performance indicators to assess service performance;
- To fulfill statutory requirements for commissioning services.

Multiple Choice Questions of this Chapter

Multiple Choice Questions are given for each chapter with solutions in an online extension of this book. Please use link: www.wiley.com\go\mekid\metrologyandinstrumentation\

References

1 Kim-Soon, Ng., 2012, *Quality Management System and Practices*, 10.5772/36671.
2 https://www.nice.org.uk/standards-and-indicators/how-to-use-quality-standards.
3 https://asq.org/quality-resources/iso-9000.

13

Digital Metrology Setups and Industry Revolution I4.0

13.1   Introduction

Among the new requirements needed in industry are the digitalization, automation, and fast measurements that are becoming extremely important in this era of digital manufacturing.

Digital measuring instruments are those having self-contained devices that collect measurements, process it, and show it on a digital display, such as digital calipers and digital voltmeters, but most importantly, transfer data to cloud for various objectives such as storage, analysis including many post-processing, and decisions making.

The measurement system is a combination of a real-time control system and a system for data transmission. Digital computing is the tool for data processing. The technology readiness for most measurement instrumentation exists, and moreover we witness currently virtual instruments very common with sensors, data acquisition, processing, and decision making. The whole system can be transferred inti a microchip to be installed in the physical equipment.

In large manufacturing systems, there is a need to obviously produce parts and systems but with an optimized input of resources, creating better product quality, and if all is well-established, the production can be automated and inspection can be in-process or post-process with data management.

13.1.1   What Is a Digital Measurement?

Digital measurement, or even the concept of digital, is not defined in the *International Vocabulary of Basic and General Terms in Metrology* (VIM). The term "digital" is mentioned indirectly: An analog (continuous or discontinuous line) or digital display can be used. More specifically, the terms analog measuring instrument and digital measuring instrument are defined as follows: An analog measuring instrument is a measuring instrument in which the output is a continuous function of the measurand or of the input signal.

A digital measuring instrument, on the other hand, is a measuring instrument that has a digitized output or display. This definition describes the display but not the operating principle.

13.1.2   Metrology and Digitalization

Innovative activities with corresponding confidence in any efficient quality industrial or institutional infrastructure are the pillars for a stable and successful economy. This will allow obtaining valid data based on high precision measurements. In the current digital transformation, comprehensive operations are in progress to be transformed under this platform. Hence, data analytics and management can be executed in a comprehensive way. Real-time monitoring and measurement with high-quality computers using the Internet of Thing (IoT) and multiple sensors have really opened opportunities for large interaction and communication between entities considered impossible years ago.

In this digitalization of the economy, metrology plays a fundamental role not only in the verification and approval but also in a constant access to data and measurement any time locally or remotely to guarantee success. Analytic tools are always associated to this role, for instance, data acquisition and mathematical and statistical procedures in a secured platform.

The prerequisite to a successful digital transformation are the metrology standards, accreditation, and legal metrology to become highly stable and effective to be linked to the economy, industry, and society.

There is a strong need complementary to previous ones by validating the measurement data and tracing them to the International System of Units. This is mainly the role of the many international bureaus or national metrology institutes, such as NPL, NIST, PTB, to act as a legal body in front on the industry and society.

This digital age of metrology is being pushed through research programs by these metrology institutes and collaboration with universities to include cloud computing, big data, and cybersecurity with a reliable and fast wireless communication.

Digital measurements will ensure the following:

a) Eliminating manual error or inconsistency;
b) Increasing productivity;
c) Minimizing waste;
d) Enabling greater factory automation.

The current interest and worldwide trend is on the following:

A) **The Digital Transformation of Metrological Services**
   This activity will develop a wireless reference for measurements; it will also approve statistical procedures for predictive maintenance and platforms for digital calibration, building a digital metrology cloud for reference data.

B) **Metrology in the Analysis of Large Quantities of Data**
   This is to develop an analytical method for big data and related analysis techniques for various metrology applications in industry, especially for large data information.

C) **Metrology of the Communication Systems for Digitalization**
   The communication of the metrological data is required to be secured and reliable under any condition. The consideration covers the following: traceability of complex, high-frequency measurands for 5G networks, derived measurands in digital communication systems, nonlinear and statistical measurands in high frequency, and complex antenna systems.

D) **Metrology for Simulations and Virtual Measuring Instruments**
   With the previous discussion on items developed, this is a great opportunity for interconnected measuring systems leading to the simulation of complex measuring systems with post-processing of data as a result of experiments, procedures, and standards in automated processes.

13.1.3 Implementation Strategy

The national German institute of metrology, PTB, has suggested three pillars to the digital transformation in the economy and society:

a) **Metrology Cloud**
Establishing a trustworthy core platform for a digital quality infrastructure by coupling existing data infrastructures and databases and providing all partners with customized access for digitally upgrading legal metrology.

b) **Digital Calibration Certificate**
Developing a secure and standardized digital information structure for universal use in calibration, accreditation, and metrology as well as digitally upgrading the whole calibration hierarchy in the quality infrastructure.

c) **Virtual Experiments and Mathematics-Aided Metrology**
Developing an interdisciplinary, virtual competence group to metrologically support the paradigm change for the use of simulations and data analysis as essential components of measurement procedures.

Moreover, the metrological research for modern high-frequency networks (5G), the expansion of the quality infrastructure to online surveillance, and the metrological support to digitalized precision production are some of the tasks that will have to be furthered in the long term in the respective departments.

13.2 Data Acquisition

Data acquisition is commonly known as DAQ and represents the process of collecting data measurements of the real, physical world with the process of sampling the signals and converting them to digital form that can be handled by software.

The basic components of a DAQ as presented previously are sensors, signal conditioning, analog-to-digital converter, and a PC with DAQ software for logging and post-treatment (Table 13.1).

Several measurands can be recorded with a DAQ system including distance and displacement, temperature, voltage, current strain, and pressure. Data can be digitalized to be stored properly (Fig. 13.1).

Table 13.1 Sensors with related signal conditioning needed.

Signal conditioning:	Amplification	Filtering	Excitation	Linearization	Bridge
LVDT/RVDT	v	v	v	v	
Thermocouple	v	v	v	v	
Strain gauge	v	v	v	v	v
RTD	v	v	v	v	
Accelerometer	v	v	v	v	
	v	v	v	v	
	v	v	v	v	

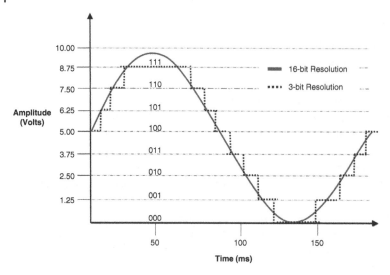

Figure 13.1 Measurement Process.

13.3 Setup Fundamentals for Measurement and Data Acquisition

13.3.1 Length Measurement in Open Loop

Several types of equipment and sensors can be used to measure length in an open- or closed-loop system. The distinction is that in the first example, data is collected and recorded for later processing, whereas in the second example, data is inspected while it is being measured. LVDTs, lasers, and any other go-no-go configuration can be used as sensors.

Measurement and quantification are the basic concepts of metrology. This considers explicit and internationally accepted definitions, principles and standards. The fundamentals of metrology are valid for all science and engineering fields constantly looking for the best design technologies that minimize direct involvement of humans in measurement processes. The purpose of any measurement system is to provide the user with a numerical value corresponding to the variable being measured by the system as simple as the one in Figure 13.2.

Metrologists consider measurement as a complex structured process in which a measurement task is carried out by applying theoretical and methodological principles and by executing agreed procedures and techniques (methods) for data generation.

The measurement task must be defined by specifying the objects under measurement, the property of interest, and more importantly, the measurands. The collected data can be either displayed for real-time analysis or stored (Fig. 13.2).

Figure 13.3 exhibits the characteristics of typical sensors and ADCs mean that the data acquisition part of a typical modern instrumentation system can be split into the three functional blocks,

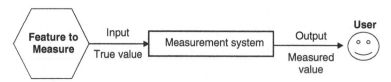

Figure 13.2 Measurement Process.

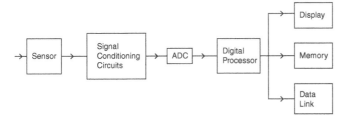

Figure 13.3 A block diagram of a typical instrumentation system with several different output devices.

a sensor, signal conditioning circuits, and an ADC. The digital output from the ADC can then be processed in a programmable digital processor to extract information that can be displayed to an operator, stored in a memory, transmitted via a data link, or used in feedback control. The costs of all the components are continually falling. It is therefore becoming economically viable to collect an increasing amount of data, and hence hopefully information, from an ever expanding range of host systems.

This technology trend and its impact on every conceivable system means that all engineers should be familiar with instrumentation systems.

If the data acquired is to be stored in a cloud or to be sent through a network, the configuration is shown in Figure 13.4.

13.3.2 Thermal Measurement and Data-Acquisition Considerations

In Chapter 11, an overview of temperature measurement techniques was presented. The usage of thermocouples with data-gathering systems is becoming increasingly popular. However, the thermocouple's properties, such as the need for a reference (cold junction) and the low signal voltages produced, make it difficult to utilize.

Nonetheless, with a little care and a reasonable estimate of achievable precision, the systems are perfectly enough for most monitoring and moderate accuracy measurements.

Following the selection of the proper thermocouple type, the cold junction compensation mechanism must be considered. The thermocouple's two connection points to the data acquisition system (DAS) board create two new thermocouple connections. External cold junction methods between the thermocouple and the board remove this problem; however, the thermocouple is more typically connected directly to the board and uses built-in electronic cold junction compensation.

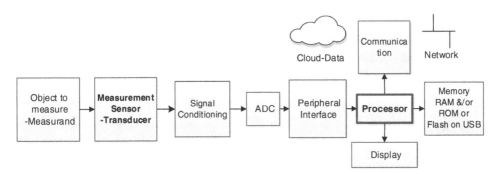

Figure 13.4 Basic sensor configuration in measurement with data transmission.

A separate thermistor sensor is employed to identify the cold junction error and provide an appropriate bias voltage adjustment, either directly or through software, by measuring the temperature at the system connection point. The internal correction approach has a usual inaccuracy of the order of 0.5–1.5 °C, which is an essential concern. This error is directly passed on to the measurement as an offset (systematic error).

Internal polynomial interpolation may also be used to translate observed voltage to temperature on these boards. If not, this can be programmed into the data-reduction strategy using information from Table 11.2, for example. Nonetheless, a "linearization" error is introduced, which is a function of thermocouple material and temperature range and is normally defined with the DAS board.

Thermocouples are frequently utilized in severe industrial environments with a lot of EMI and radiofrequency noise. To reduce noise, thermocouple wire pairs should be twisted. A differential-ended connection between the thermocouple and the DAS board is also preferred.

The thermocouple, however, becomes an isolated voltage source in this configuration, which means there is no direct ground path to keep the input within the common mode range. As a result, a common criticism is that the output level of the measured signal may drift or unexpectedly jump.

Typically, a 10- to 100-k resistor is placed between the low terminal of the input and low-level ground to eliminate this interference behavior.

The signal must be conditioned using an amplifier because most DAS boards employ analog-to-digital (A/D) converters with a 5 V full scale.

A gain of 100 to 500 points is usually sufficient. Accurate measurements necessitate high gain, low noise amplifiers. Consider a 12-bit A/D converter with a signal conditioning gain of 100 for a J-type thermocouple. This results in a full-scale input range of 100 mV, which is suitable for the majority of measurements.

Then, the A/D conversion resolution is

$$\frac{E_{FSR}}{(G)(2^M)} = \frac{10V}{(100)(2^{12})} = 24.4 \ \mu V/bit \tag{13.1}$$

A J-type thermocouple has a sensitivity of ~55 μV/°C. Thus, the measurement resolution becomes

$$(24.4\mu V/bit)/(55\mu V/°C) = 0.44°C/bit. \tag{13.2}$$

In comparison to the average sample rate capabilities of general-purpose DAS boards, thermocouples have large time constants. High-frequency sampling noise is a possible source of temperature changes that are larger than expected. Simple fixes include lowering the sample rate or employing a smoothing filter. The averaging period should be on the same scale as the time constant of the thermocouple.

Study Example

Suppose there is a need to develop a commercially available temperature measuring system for a PC-based control application. Figure 13.5 shows a schematic representation of the proposed temperature-measuring system. The following components make up the temperature-measurement system:

- A PC-based DAS that measures analog input voltage signals using a data acquisition board, a computer, and relevant software.
- A reference junction compensator and a J-type thermocouple. The reference junction compensator is used to give the thermocouple an output emf that is equal to the difference in temperature between the measuring junction and a reference temperature of 0 °C.
- An uncalibrated J-type thermocouple that meets NIST standard limits of error.

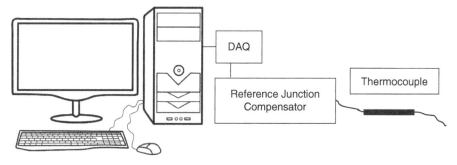

Figure 13.5 PC based temperature and data acquisition system.

Table 13.2

Component	Characteristics	Accuracy Specifications
Data-acquisition board	Analog voltage input range: 0 to 0.1 V	12-bit A/D converter accuracy: ± 0.01 % of reading
Reference junction compensator	J-type compensation range from 0° to 50°C	± 0.5 °C over the range 20° to 36°C
Thermocouple (J-type)	Stainless steel sheathed ungrounded junction	Accuracy: ± 1.0°C based on NIST standard limits of error

The system is intended to monitor and control a process temperature that varies slowly over time in comparison to the DAS sampling rate. The procedure runs at a temperature of 185 °C. The measurement system components must meet the following standards (Table 13.2): The temperature measurement must have a total uncertainty of less than 1.5 °C for the purpose of the measurement system.

Does this measuring system meet the overall accuracy requirement based on a design stage uncertainty analysis?

Explanation The design stage uncertainty for this measuring system is calculated by converting the uncertainty of each system component to a temperature equivalent and then aggregating these design stage uncertainties. The full-scale voltage range is divided into 2^{12}, or 4096, equal-sized intervals by the 12-bit A/D converter. As a result, the A/D resolution in measuring voltage is as follows:

$$\frac{0.1V}{4096 \text{ intervals}} = 0.0244 \text{mV}.$$

The DAS's uncertainty is defined as 0.01 percent of the reading. The thermocouple voltage must have a known or defined nominal value. The nominal process temperature in this example is 185 °C, which translates to a thermocouple voltage of about 10 mV. As a result, the DAS's calibration uncertainty is 0.001 mV.

The temperature measurement system from the DAS will contribute to the total uncertainty as follows. We combine the resolution and calibration uncertainties as

$$u_{DAS} = \sqrt{(0.0244)^2 + (0.001)^2} = 0.0244 \text{mV}.$$

The relationship between uncertainty in voltage and temperature is provided by the static sensitivity, which is obtained from standard materials tables at 185 °C as 0.055 mV/°C. Thus, an uncertainty of 0.0244 mV corresponds to an uncertainty in temperature of

$$\frac{0.0244\text{mV}}{0.055\text{mV/°C}} = 0.444°\text{C}.$$

This uncertainty can now be combined directly with the ice point uncertainty and the uncertainty interval associated with the standard limits of error for the thermocouple, as

$$u_T = \sqrt{(0.44)^2 + (1.0)^2 + (0.5)^2} = \pm1.2°\text{C} \quad (95\%).$$

Remark

In some circumstances, calibrating the thermocouple against a laboratory standard, as a RID, which has a calibration traceable to NIST standards, may be useful. In comparison to the uncalibrated value of 1 °C, a thermocouple uncertainty level of 0.1 °C can be attained with acceptable expense and care.

If the thermocouple is calibrated properly, the overall temperature measurement uncertainty is decreased to 0.68 °C, implying that by reducing the thermocouple's uncertainty contribution by a factor of ten, the whole system uncertainty is lowered by a factor of two.

Study Example

Providing a fixed temperature point at which the sensor outputs may be compared is an effective means of analyzing data acquisition and reduction errors associated with the usage of several temperature sensors within a test facility. Assume that the outputs of M thermocouple sensors (all of which are T-type) are to be measured and recorded on an M-channel data acquisition device.

Each sensor is calibrated to the same reference junction temperature (for example, the ice point) and operates normally. The sensors are kept at a constant and consistent temperature. For each of the M thermocouples, N values are recorded (about 30 for example). What kind of information can you get from the data?

Given	M ($j = 1, 2, \dots, M$) thermocouples
	N ($i = 1, 2, \dots, N$) readings measured for each thermocouple

Approach: The mean value for all readings of the ith thermocouple is given as

$$\overline{T}_j = \frac{1}{N}\sum_{i=1}^{N} T_{ij}.$$

The pooled mean for all the thermocouples is given as

$$(\overline{T}) = \frac{1}{M}\sum_{j-1}^{M} \overline{T}_j.$$

The difference between the pooled mean temperature and the known temperature can be used to evaluate the systematic uncertainty that can be expected during data collecting from any channel. The disparities between each $\overline{T}_j$ and $(\overline{T})$, on the other hand, must indicate random uncertainty among the M channels.

The data acquisition and reduction instrumentation system have a standard random uncertainty of

$$s_{\overline{T}} = \frac{s_T}{\sqrt{M}} \quad \text{where} \quad (s_T) = \sqrt{\frac{\sum_{j=1}^{M} \sum_{i=1}^{N} (T_{ij} - \overline{T}_j)^2}{M(N-1)}}$$

with $v = M(N-1)$ degrees of freedom.

Discussion The errors components considered in the estimates are:

- Random errors of reference junction;
- Random errors in the known temperature;
- Random errors of the data acquisition system;
- Systematic errors in the extension cable and connecting plug;
- Systematic errors of the thermocouple emf-T correlation; and
- Instrument calibration and probe errors.

When parts are being inspected, digital measurements are taken in an open loop. As demonstrated in Figure 13.6, data can be saved for archiving and reporting official results. Interface devices to host sensors, measure, and wirelessly transport data are required for closed-loop digital measurement. These are primarily IoT devices, which will be covered later.

13.3.3 Data Transfer to Cloud

As an example of transferring of data and monitoring data, the strategy of data handling can be done in three steps: data collection, transfer to cloud for either storage or post-processing of data with prognostic, and decision making. The third step is the client server for post-processing if not planned in the cloud and decision-making assignment. This is demonstrated in Figure 13.7.

Using Nb-IoT (narrowband IoT) technology and the RS485, industrial communication can be built on IoT and multiband wireless communication. Nb-IoT is distinguished by excellent indoor and outdoor coverage, low latency, low connectivity costs, low battery consumption, and a network design that is optimized [1]. It removes restrictions on consumption, interference, and congestion.

Nb-IoT is built on 4G and benefits from current infrastructure because of its wide radio coverage. Furthermore, the frequency spectrum employed allows for better penetration inside structures and underground. Following that, the fundamentals of an IoT system are introduced.

13.3.4 Internet of Things (IoT) Metrology

The principle of building IoT devices is usually to be used as a support to any other instrument or equipment to enhance a particular aspect by directly connecting and transferring data wirelessly. There is a need for any application in metrology to evaluate the quality of data transferred or to follow a dedicated standard [2].

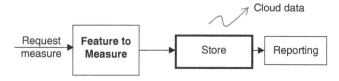

Figure 13.6 Open loop inspection and reporting.

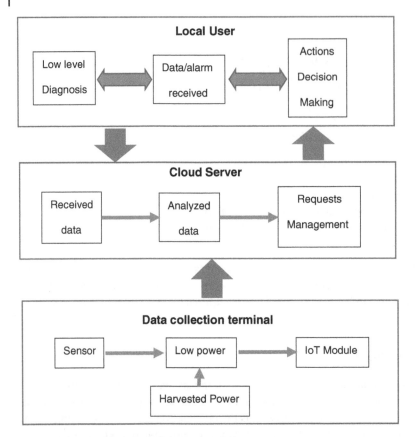

Figure 13.7 Various stages of data transmission.

National metrology institutes (NMIs) have a long history in ensuring accuracy of measurement data, and they are now working to digitalize these practices. Figure 13.8 shows the initial block of sensors set up with local decision-making and alarms to be sent while data is sent to a cloud. On the other side, power supply is harvested in this particular case through radio frequency (RF) to charge a battery and keep the IoT alive.

13.3.5 Closed-Loop Data Analysis- (In-Process Inspection)

While inspection of dimensional metrology can be open loop, for instance, quality control as shown previously, the next objective is to inspect while processing the materials, usually known as "in-process inspection" to avoid any loss of material and achieve zero-defect production.

The next generation of intelligent machine tools will require embedded sensors to measure displacement, temperature, and vibration for all moving axes in the machine, as well as in-process inspection of workpieces to compensate for machining errors (Fig. 13.9) and thus ensure the production of high-quality products at a low cost and in a short period of time.

Higher performance will necessitate fast automatic error compensation to obtain true dimensional tolerances, which will be easily facilitated by in-process accurate sensors. Manual inspection methods and statistical sampling procedures have traditionally been used to control the quality of manufactured parts. Its disadvantages include the release of some defective parts and the use of an inspection area.

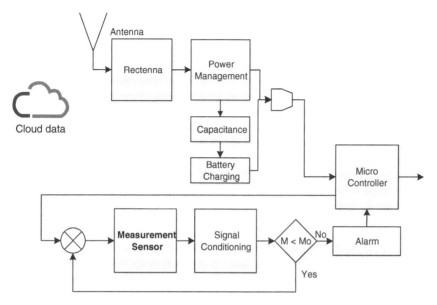

Figure 13.8 Advanced sensor configuration in measurement.

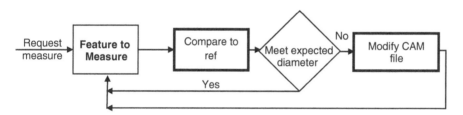

Figure 13.9 Advanced sensor configuration for in-process inspection.

To overcome these drawbacks, over the last two decades, in-process measurement techniques have been proposed to control the quality of a workpiece while it is being machined. The use of a measuring sensor and a control system governing measurement and adjustment of machining parameters is the trend principle in workpiece accuracy inspection. In mass production, the diameter is usually one of the significant parameters to be checked.

The set of hardware, software, procedures, and activities are integrated in the manufacturing system in order to provide measurements of dimensional characteristics of manufactured products and tools during the manufacturing process.

Example: To achieve dynamic error compensation in CNC machine tools, a noncontact laser probe capable of dimensional measurement of a workpiece (Fig. 13.10) while it is being machined has been developed. The measurements are automatically fed back to the machine controller for intelligent error compensations. Based on a well resolved laser Doppler technique and real-time data acquisition, the probe delivers a very promising dimensional accuracy at a few microns over a range of 100 mm. The developed optical measuring apparatus employs a differential laser Doppler arrangement allowing acquisition of information from the workpiece surface. In addition, the measurements are traceable to standards of frequency allowing higher precision.

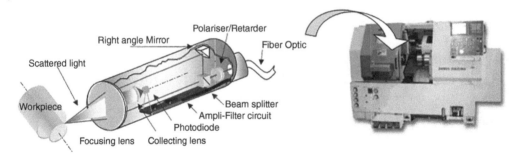

Figure 13.10 Differential Doppler technique probe and its location on Takisawa lathe CNC machine.

The resolution of this laser-based sensor depends on the spacing of the interference fringes. It is almost proportional to Doppler signal pulses number counted from the spindle rotation and estimated to be 0.5 μm. However, the accuracy of this system is about ±15 μm over a range of 80–100 mm and a corresponding repeatability of 5 μm. The main limitation is mainly due to weak signal filling technique inducing extra pulses and other error sources (e.g., vibration of the tool holder of the machine tool).

13.4   Digital Twin Metrology Inspection

Similarly to the principle of digital twin manufacturing, digital twin metrology inspection is the combination of a software model of the product as 3D CAD with semantic tolerances. It is proposed that inspection is integrated within the manufacturing system, known as in-process inspection [3, 4].

Coordinate metrology is key in manufacturing and is largely integrated in recent CNC machine tools, for instance, CNN CMMs having CMM integrated in CNC machine tool. This is very effective currently, but still the software configuration does not have all the context of the ongoing manufacturing process. There is a need to integrate process parameters with the hosting machine into the CMM governing software. Hence, a digital twin principle can be applied effectively. This needs data transfer and processing of 3D models with related DG&T for tolerance checks.

If the in situ CMM is not used, then in-process inspection can operate for specific functional features. Many in-process instruments are integrated, such as cameras and laser heads for dimensional measurement.

In Figure 13.11, the author proposes a new integrated concept for the next generation of machine tool centers. The performance of control systems will be extended toward self-controlled manufacturing based on the knowledge gained and the features extracted, with the goals of cost-effective, high-quality, fault-tolerant, and more flexible systems with better process capability. New intelligent control systems must be created and integrated with open architecture controllers such as OpenCNC or OSACA-based CNCs.

In order to allow an automated error-free production with near zero down time, open interfaces with API developments, learning capabilities, and self-tuning and self-adjusting mechanisms as well as sophisticated model-based prediction instruments have to be implemented at these layers. Quality inspection could operate in situ with environmental conditions taken into account. For the first time, the concept of self-healing with e-maintenance could be operational.

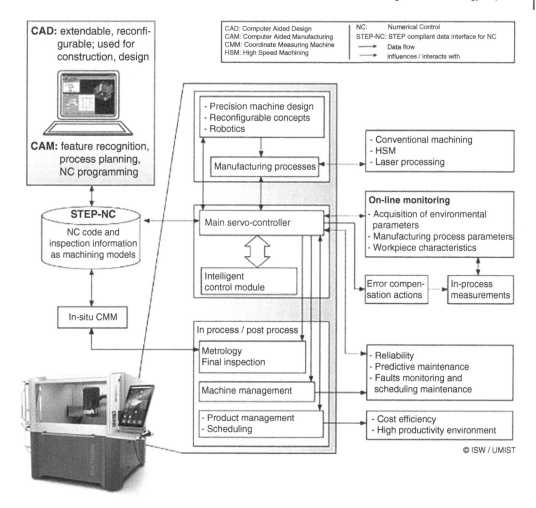

Figure 13.11 Integrated concept for the next generation of machine tool centers.

Cyber physical manufacturing systems (CPMS) are built on the IoT and cloud technologies to integrate and interconnect cyber-physical systems (CPS). They are a high-tech methodology for the construction of a new generation of factories that are increasingly intelligent, flexible, and self-adaptive [5, 6, 7, 8, 9, 10, 11, 12].

Through horizontal integration (value added networks) and a functional hierarchy of resources, CPMS generate a large amount of data.

During the generation of a CPM3 framework we have faced a number of issues referring to the big data analysis, that is, extraction of useful information from data sets, and finding the relevant structure from unstructured data sets. The first problem represented the extraction of GFs' parameters from a neutral CAD format (Fig. 13.12).

We encountered a number of difficulties related to large data analysis throughout the development of the CPM3 framework, including the extraction of meaningful information from data sets and the discovery of important structure from unstructured data sets. The extraction of GFs' parameters from a neutral CAD format was the initial issue. The issue is solved by creating a set of rules that included analytical procedures for each GF type and its parameters.

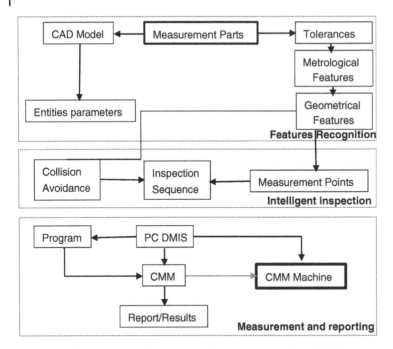

Figure 13.12 Concept of Cyber-Physical Manufacturing Model (CPM³).

The structuring of measurement points generated from GFs and the construction of an ideal measurement path was a more difficult task. Hemmersley sequences were used to choose measurement sites on regular surfaces; however, extraction of optimal measurement sites from free form surfaces required a more sophisticated methodology.

To determine the number of control sections and measurement points on free form surfaces, regression analysis with ANOVA was used.

Nonetheless, the obtained measurement points set in both cases was unstructured and difficult for measurement path construction. We have to use ant colony optimization to discover the spatial organization of points in the set in order to build the best measurement path.

There is a plan to develop CPM3 modules that will allow us to create virtual models of parts based on measurement findings and connect them to CPMS' downstream and upstream processes.

Multiple Choice Questions of this Chapter

Multiple Choice Questions are given for each chapter with solutions in an online extension of this book. Please use link: www.wiley.com\go\mekid\metrologyandinstrumentation\

References

1 Jahid, A., and Hossain, M.S., "Dimensioning of zero grid electricity cellular networking with solar-powered off-grid bS," *Proceedings of the IEEE International Conference on Electrical & Electronic Engineering (ICEEE)*, 2017, Rajshahi, Bangladesh.

2 Bhaskaran, P.E., Maheswari, C., Thangavel, S., Ponnibala, M., Kalavathidevi, T., and Sivakumar, N.S., "IoT based monitoring and control of fluid transportation using machine learning," *Computers & Electrical Engineering* 89, 2021.

3 Mekid, S., and Vacharanukul, K., "Differential laser Doppler based non-contact sensor for dimensional inspection with error propagation evaluation," *Sensors* 6(6), 2006, pp. 546–556.

4 Vacharanukul, K., and Mekid, S., "In-process dimensional inspection sensors," *Measurement: Journal of the International Measurement Confederation* 38(3), 2005, pp. 204–218.

5 Monostori, L., Kádár, B., Bauernhansl, T., Kondoh, S., Kumara, S., Reinhart, G., Sauer, O., Schuh, G., Sihn, W., and Ueda, K., "Cyber-physical systems in manufacturing," *CIRP Annals - Manufacturing Technology*, 2016, 65, pp. 621–641.

6 Mekid, S., "Integrated nanomanipulator with in-process lithography inspection," *IEEE Access* 8, art. no. 9097243, 2020, pp. 95378–95389.

7 Mekid, S., and Hussain, R., "Battery performance and low power challenges for standalone mechatronic devices," *14th International Multi-Conference on Systems, Signals and Devices, SSD*, 2017, pp. 499–504.

8 Baroudi, U., Al-Roubaiey, A., Mekid, S., and Bouhraoua, A., "The impact of sensor node distribution on routing protocols performance: A comparative study," *Proc. of the 11th IEEE Int. Conference on Trust, Security and Privacy in Computing and Communications, TrustCom-2012 - 11th IEEE Int. Conference on Ubiquitous Computing and Communications, IUCC-2012*, 2012 art. no. 6296190, pp. 1714–1720.

9 Ryu, H.S., and Mekid, S., "Low cost and rapid precision measurement at mesoscale using sub-pixel edge detection technique," *EUSPEN 2005 - Proceedings of the 5th International Conference of the European Society for Precision Engineering and Nanotechnology*, 2005, pp. 273–276.

10 Mekid, S., and Ryu, H.S., "Rapid vision-based dimensional precision inspection of mesoscale artefacts," *Proceedings of the Institution of Mechanical Engineers, Part B: Journal of Engineering Manufacture* 221(4), 2007, pp. 659–672.

11 Mekid, S., "Further structural intelligence for sensors cluster technology in manufacturing," *Sensors* 6(6), 2006, pp. 557–577.

12 Mekid, S., Schlegel, T., Aspragathos, N., and Teti, R., "Foresight formulation in innovative production, automation and control systems," *Foresight* 9(5), 2007, pp. 35–47.

Index

Page numbers followed by *f* and *t* refer to figures and tables, respectively.